Electrical Energy Storage for Buildings in Smart Grids

Series Editor
Bernard Multon

Electrical Energy Storage for Buildings in Smart Grids

Benoît Robyns
Arnaud Davigny
Hervé Barry
Sabine Kazmierczak
Christophe Saudemont
Dhaker Abbes
Bruno François

iSTE

WILEY

First published 2019 in Great Britain and the United States by ISTE Ltd and John Wiley & Sons, Inc.

ISTE Ltd
27-37 St George's Road
London SW19 4EU
UK

www.iste.co.uk

John Wiley & Sons, Inc.
111 River Street
Hoboken, NJ 07030
USA

www.wiley.com

Library of Congress Control Number: 2019937199

British Library Cataloguing-in-Publication Data
A CIP record for this book is available from the British Library
ISBN 978-1-84821-612-9

Contents

Foreword

In this third volume, the final work in a definitive survey of electrical energy storage, Professor Robyns and his colleagues discuss the environmentally responsible energy solutions which are currently available for use in the building sector, for residential or tertiary usages. As in the previous volumes, the authors have applied a rigorous methodology for designing supervisors using fuzzy logic, a means of managing energy flows in an optimal manner, taking account of a large and varied range of constraints.

The task that the authors set for themselves was not an easy one, as their ambitions for the work grew over time, building on their accumulated experiences. Their aim in this book is to offer innovative solutions for systems which are extremely complex, a result of the dense network of interconnection and of the number of actors involved. One example is that of eco-neighborhoods, which, in addition to the capacity to be self-sufficient in energy, are designed to enable newcomers to slot in easily using a "plug-and-play" model. For instance, the smooth integration of charging facilities for the increasing number of electric and hybrid vehicles on our roads – a number set to increase substantially over the coming years – is essential. For this reason, as the authors rightly note in the introduction, it is also crucial that we take account of the public acceptability of new energy solutions: these will affect the whole population, not just in the public sphere but also in the home. The current debate concerning the large-scale rollout of smart meters to measure energy consumption is a striking illustration of this.

Once the authors have risen to the challenge that they set themselves, producing yet another exceptional book, featuring a clear and accessible presentation of the issues alongside a selection of relevant examples, rigorously examined using a comprehensive methodology. Anyone concerned with the ongoing shift in the energy paradigm, a crucial concern for our society, is sure to draw inspiration from this work to support their own work and reflection.

Eric MONMASSON
University of Cergy Pontoise
SATIE Laboratory
Paris, September 29, 2018

Introduction

In France, in 2016, residential and tertiary sector buildings represented 45% of total final energy use. The proportion of electrical energy continues to increase, currently representing approximately 37% [MIN 17]. There is thus much to be gained by increasing energy efficiency in this area, equipping buildings to produce and store energy and establishing intelligent energy management systems, interacting with the distribution grid.

Current developments in the sphere of renewable energy and the trend toward self-production and self-consumption of electrical energy produced onsite have led to increased interest in the means of storing electrical energy, a key element of sustainable development. Self-consumption provides a stimulus for better mastery of energy consumption and leads to a reduction in electric bills (reducing costs associated with connection to the main distribution grid, subscribed power and, potentially, taxes). Collective self-consumption can result in additional optimizations, grouping together buildings with different consumption profiles in terms of time. Considerable gains may also be made through load management, modulating consumption by adjusting loads or through local production and self-consumption, with or without a storage system. Finally, in addition to these financial aspects, collectives may benefit from using renewable forms of self-consumption (one of the main aims in such cases), as there are several potential sources of production (notably solar panels on roofs). The consumption of locally produced energy also prevents or limits losses associated with the transportation of energy over long distances.

The increase in popularity of electricity as an energy carrier for buildings can be attributed to the flexibility which it offers, as well as to the potential to avoid pollution at the usage site. In the coming years, an increasing proportion of these buildings will be equipped with storage systems, providing emergency backup, compensating for natural variations in renewable energy supplies, and will also be able to provide services for the wider electric system. Storage systems are expensive, and shared usage offers a means of spreading the cost, while contributing to the management of system aging. At the time of writing, studies are being carried out with regard to using the storage capacity of electric vehicles to provide services to the electric distribution grid or to the buildings where they recharge: these solutions are known as Vehicle to Grid (V2G) and Vehicle to Home (V2H). Similar solutions would be possible for integrated storage in commercial and tertiary (with offices) buildings, or, indeed, whole residential neighborhoods.

The aim of this book is to increase awareness of the potential offered by these developing technologies, in the context of buildings, groups of buildings and/or neighborhoods, integrated into large "smart grids" or forming smaller "micro grids", particularly with regard to their management and valorization.

Storage will form an essential element of future smart grids, but these networks will be unable to attain their full "smart" potential without collecting large amounts of data, via connected meters, among other things. The installation of these meters raises ethical questions with regard to the protection of the data which they generate, which should give a precise indication of the energy usage habits of consumers, but is also affected by questions of cybersecurity.

The development of self-consumption of locally produced energy raises other ethical questions of a fundamental nature: energy, particularly electricity, has become essential to maintaining the lifestyles of industrialized societies, for comfort, sanitation, security, education and more. Self-consumption challenges the current electrical supply model, which is highly centralized in terms of both production and management. We are effectively facing an energy revolution. In extreme cases of self-consumption, in which public network management entities are left out of the picture altogether, this could be compared to the "uberization" (an exchange of services between private individuals to the exclusion of larger companies, enabled through the use of Internet applications) recently seen in the contexts of urban automobile transport and short-term lets. However, access to electricity is essential to the operation of our societies, which are highly dependent on this energy supply. Self-consumption could also undermine the French principles of energy solidarity and equal access to energy (in terms of cost). These last points raise further ethical questions, particularly with regard to an increased risk of energy poverty and even energy-based communitarianism. There is a danger that self-consumption may simply benefit those consumers who are already in a strong position – for example wealthier households with the financial capacity to install solar panels on the roofs of their houses.

Furthermore, self-consumption is largely based on the use of "new" renewable energy sources (essentially solar, as well as wind power), which are, by their very nature, variable and weather-dependent, fluctuating significantly with the seasons and from day to night. This being so, climate change is a source of additional uncertainty with regard to the future behavior of these new technological solutions.

For these reasons, we would do well to adopt an ethical rule set out in [GIO 18]: "Do not leave your children to solve problems which you yourself voluntarily created, which are of vital importance for your descendants, and for which you are not sure that a realistic solution exists or will be found in the future. Furthermore, any advances resulting from the scientific discoveries and/or technological developments in question should support the common good and promote the restoration of original ecosystems, if these systems created balance and harmony, wherever possible".

This does not mean that we should limit research into the development of smart grids and self-consumption; instead, these projects should be subject to regular ethical review in connection with the questions set out above (even though the risks seem smaller and of a different nature to those associated with the development of nuclear power). An interdisciplinary approach to these questions is necessary, connecting science and sociology, economics, ethics and even, where applicable, legal considerations. Law-makers have a key part to play in providing an "ethical buttress" [GIO 18] for new methods of energy production and consumption.

In Europe, Germany leads the way in terms of electrical self-consumption, with 500,000 installations in 2018, compared to 20,000 in France, where a regulatory framework has yet to be fully defined. Debate centers on the notion of locality as it relates to self-consumption, a notion that may be defined in various ways. It may be limited to part of the distribution grid (e.g. downstream of a medium-voltage to low-voltage transformer substation [CRE 18] serving part of a residential neighborhood) or to a distance, for example a one-kilometer radius around a production facility [MIN 18] enabling energy exchanges between large-scale service buildings in addition to homes. There are also questions regarding taxation: for example, in France, a tax is levied to support the development of renewable energy, and self-supply installations of under 9 kW [CRE 18] or 1 MW [MIN 18] may be exonerated. Finally, the charges for use of the public distribution grid by collective self-consumption, which only use a small portion of this network, need to be determined; these entities must remain connected to the grid to ensure that supply is maintained even though their renewable systems are not producing electricity and there is no power stored on-site.

The aims of this book are:

– to highlight the importance of storing electrical energy in the context of sustainable development, smart buildings, smart grids and smart cities;

– to demonstrate the variety of services which electrical energy storage may provide;

– to consider the socio-economic questions associated with changes stemming from the emergence of smart buildings and smart grids, providing elements of response;

– to present methodological tools for the design of a management system for stored energy, following a generic and pedagogical approach. These tools are based on causal approach, artificial intelligence and explicit optimization techniques. They will be presented throughout the book, in the context of real-world case studies;

– to illustrate these methodological approaches through the use of various real-world examples, used as a basis for clearly explaining the integration of renewable energy and electric vehicles into our environment (buildings, energy sharing between residential and tertiary buildings, urban neighborhoods and rail energy hubs).

In Chapter 1, we will describe the issues surrounding electrical energy storage in buildings, blocks and neighborhoods, whether integrated into a large smart grid or forming their own micro grid. We will highlight the storage requirements for these applications, alongside the services which they may provide. The socio-economic aspects of these developments will be touched on briefly; a more detailed discussion of these elements is provided in Chapter 5. We will also introduce a methodology for designing a management system for energy storage systems. This system is particularly suitable for the management of complex systems, featuring elements of uncertainty regarding the production of variable renewable energy, consumption (which is also variable) and interactions with the wider grid. Our methodology, based on fuzzy logic, is designed to respond to a number of requirements involving real-time treatment.

Chapters 2–4 concern cases involving a single entity: a commercial building, a tertiary building connected to a powerful network and an energy producer in a zone which is not connected to a larger network. These entities may be consumers, producers and storers of electrical energy.

Chapter 2 concerns the development of an energy management system for a commercial building such as a supermarket, integrating photovoltaic solar energy production and energy storage. Fuzzy logic is used to design an energy management strategy for the storage system. The storage system regulates the power drawn from the electrical network during peak and off-peak periods in such a way as to reduce electric bills and CO_2 emissions, while promoting self-consumption through the use of solar panels. Energy may be stored in a dedicated system, as well as through the use of adjustable loads. We present the results of simulations and compare various topologies (with or without photovoltaic generation and a storage system) on the basis of economic and ecological indicators.

In Chapter 3, we discuss the combination of three different technologies – variable intensity LED lighting responding to external luminosity, photovoltaic energy production and batteries – operating in a tertiary sector building over a dedicated DC (direct current) network. This configuration creates an intermittent production/intermittent charge/storage system which is designed to reduce certain electronic conversion stages. By maintaining a connection to the AC (alternating current) distribution grid, the system must guarantee energy supply for lighting purposes and eventually may supply power to the AC grid. To ensure that the system is able to respond to user needs in terms of lighting and to support the operation of the AC grid, while favoring consumption of solar electricity produced on-site, a real-time energy management system is developed using a methodology based on fuzzy logic, applied to the case of a DC network architecture.

In Chapter 4, we present a photovoltaic system with hybrid storage combining two different technologies: electro-chemical batteries and super-capacitors. This hybrid approach aims to combine the advantages of each technology in order to increase the life expectancy of the storage system and to maximize overall yield. The system in question is designed to supply electricity to island or isolated habitats. A supervisory algorithm based on fuzzy logic is also presented. The main objective in this case is to monitor a projected production program while respecting the constraints operating on the electric network management system (power smoothing, frequency control, etc.). A comparative study of different storage configurations, particularly with regard to the life expectancy of storage elements and average energy cost, is also presented.

The full innovative potential of smart grids can only be released by promoting interaction between the different actors involved in the electric system (producers, consumers, storage and network operators), increasing their "electrical intelligence". These actors may have very different consumption and production profiles, with very varied economic and social objectives and/or constraints. New types of actor may emerge alongside new economic models, all of which may contribute to solving energy and climate issues, promoting the development of renewable energy sources. It is important that all actors should benefit in these cases, including those in situations of "energy poverty". These questions and issues will be discussed in Chapters 5–8, which present several case studies involving very different actors.

In Chapter 5, we highlight the diversity of actors involved in a smart grid, defining the rationale of individuals, which may vary and may impact a whole group of actors. We also address the issue of economic and sociological changes brought about by the use of smart grids, including changes to the value chain, contractual models, socio-economic profiles of consumers and governance. The social acceptability of mass participation in energy management is also discussed, particularly with regard to load management in multi-actor commercial buildings (e.g. shopping malls) and in a domestic context (households in residential buildings).

Chapter 6 concerns possible exchanges of electrical energy flow and services between a commercial building, such as a supermarket, and other actors such as renewable energy producers, network operators, third-party consumers (e.g. residential buildings), an electrochemical battery storage system and a diesel generator, all grouped together in a network for the purposes of self-consumption. We need to define ways of managing these exchanges, financial sustainability and acceptability for all of the actors involved, from energy professionals to consumers. Our study concerns a collective self-consumption system established between actors in a given geographical zone. First, we will present a case study concerning energy mutualization between commercial, tertiary and residential buildings, introducing the notion of an energy service aggregator. We will then present a method of energy management based on fuzzy logic, as applied to our case study. A specification will be established for each actor, drawing on expertise provided by a sociologist in order to assess the conditions of acceptability and the implication of each actor in the energy mutualization process. We then propose the introduction of a load management acceptability coefficient, to be integrated into the supervision strategy. Several different scenarios, with and without energy management, will be compared on the basis of economic, environmental, self-production and self-consumption indicators.

Chapter 7 concerns the management of a local energy community such as an eco-neighborhood. The objectives considered include increasing energy efficiency in the neighborhood and reducing CO_2 emissions by increasing the production and use of photovoltaic energy in the local energy network, which also includes energy storage systems and gas turbines, which must be used in an optimal manner, while guaranteeing the operation and stability of the neighborhood network. This can only be attained by achieving a balance between supply and demand. The aim here is to identify the best way of exploiting production capacity in response to an increase in new uses of electricity (such as electric vehicles), and also to develop evolutive energy management systems into which new production mechanisms can be integrated with ease. Our method aims to predetermine the production profile of generators so as to ensure global optimization of an objective function for the urban electrical network, then to adjust operating points over the course of a day to account for any differences identified through a communication network. There may be several possible solutions, so our two-level optimization approach is designed to identify the best option for any system in order to:

– maximize production from renewable sources by taking account of availability, which depends on weather conditions, and of their usage within the electric system in order to promote self-consumption;

– minimize the cost of energy production within the micro-grid;

– minimize the equivalent CO_2 quantities emitted by conventional generators.

The batteries of electric and rechargeable hybrid vehicles will, in future, represent a significant amount of storage capacity, and this may be exploited by the electric grid when vehicles are plugged in. It may also be used, more specifically, by buildings. A variety of different technologies will be presented in Chapter 8:

– Vehicle to Grid (V2G): the vehicle (via its battery) feeds into the electric grid through a charging point on a public thoroughfare or carpark (station, shopping mall, etc.);

– Vehicle to Home (V2H): the vehicle powers a home, generally one cut off from the electric network;

– Vehicle to Home and Grid (V2HG): the vehicle powers a home which is also connected to an external electric network, meaning that the vehicle can feed into or draw power from this network;

– Vehicle to Building (V2B): the vehicle powers an apartment or service building. Evidently, in this case, there would be several vehicles in a carpark;

– Vehicle to Station (V2S): the vehicle powers a railway station building or railway equipment, and vice versa. Again, this situation would involve multiple vehicles.

Finally, Chapter 8 introduces different configurations which may be used to exploit the reversibility of the charge in electric vehicles such as those described above. The potential services and energy management questions associated with a fleet of electric vehicles interacting with an electrical distribution network will be discussed in greater detail. We will also describe an energy supervision system based on fuzzy logic, and look more closely at the uses of reversible charge in the context of train stations.

1

Storing Electrical Energy in Habitat: Toward "Smart Buildings" and "Smart Cities"

1.1. Toward smarter electrical grids

1.1.1. *The move to decentralize electrical grids*

The traditional organization of an electrical grid is based on centralized management, at the level of the transport grid to which conventional nuclear, thermal or hydraulic production systems are connected. Originally, the distribution grid only supplied consumers, and only carried power flows from high voltage points, through connections to the transport grid, toward lower voltage points. The possibilities for adjustment at the distribution level are limited, and ancillary services (voltage and frequency control) are provided by production units connected to the transport grid [ROB 12c, ROB 15].

The development of decentralized production, generally low power, unplanned and not monitored by a central entity, has brought about significant changes. Producers are often connected to a distribution grid and dispersed across a territory, contrasting with the classic model of high-power production on a few, clearly defined sites. The effects of integrating this production, which generally comes from wind and solar sources, are becoming increasingly noticeable and bring in new constraints. The variable nature of wind and photovoltaic sources, which is difficult to predict, adds a further level of complexity to grid management issues.

The liberalization of the electricity market within the European Union, beginning in the early 21st Century, has resulted in a clear separation between the management of energy production, which is subject to competition, and the management of transport and distribution grids: evidently, the infrastructure involved cannot be duplicated. In France, the CRE (*commission de régulation de l'électricité*, Electricity Regulation Commission) [CRE] is charged with ensuring that the new competition mechanisms are respected, that competition does not have a negative effect for consumers and that there is no danger to an infrastructure crucial to both the economy and security of the country. Liberalization has led to a need for new approaches to managing the electricity system, alongside new market mechanisms integrating the characteristics of new decentralized sources. Given that the electricity grids themselves cannot be rebuilt, development is needed at three specific levels:

– at the source level, using the possibilities offered by power electronics to develop new control and supervision strategies and provide ancillary services, notably through the implementation of energy storage; to develop multi-source systems (integrating intermittent renewable production, classic, predictable sources and storage) featuring integrated and optimized energy management [ROB 15];

– at the grid level, rolling out smart grids and developing new grid architectures, such as micro-grids, in order to increase the efficiency, security and availability of electricity grids, and increasing energy storage capacity, either at a central point or dispersed across these grids [ROB 15];

– at the consumer level, in industrial processes, tertiary buildings and homes, through electric and rechargeable hybrid vehicles, and in guided transport systems (trains, subway systems and trams) [ROB 16], with the aim of modulating energy demands to correspond to consumption, renewable production availability and the constraints inherent in electric grids.

Interactions between these different aspects need to be coordinated to some extent, and this raises questions regarding the optimal and most acceptable level of decentralization; a system for communication between components is also required. These issues are not purely technological in nature, including economic and sociological aspects, and requiring new developments on the judicial stage.

1.1.2. *Smart grids*

It is thus essential that we install and use new communication technologies as part of advanced management mechanisms. The level of intelligence in a grid depends on two factors. The first corresponds to the installation of a telecommunications network, mechanisms and equipment for remote control and automated network management within transport and distribution grids. The second involves advanced management of production (centralized and/or decentralized) and of loads, notably via the development of new products and services by producers and distributors, including network managers which increase the level of freedom available in piloting a grid. Final consumers may also benefit from special services and pricing offers, allowing the adoption of ambitious approaches to mastering instant demands for electricity and the integration of renewable energy sources (Figure 1.1).

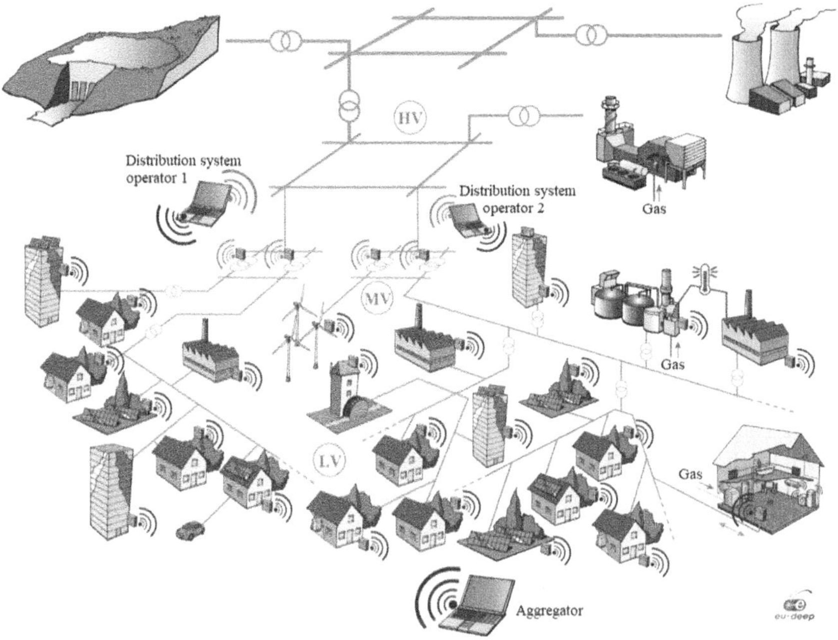

Figure 1.1. *Example of a future smart grid, including the distribution of regulation capacities across multiple sites via the Internet. HV = high voltage grid, MV = medium voltage grid, LV = low voltage grid (EU-Deep project). For a color version of the figures in this chapter see www.iste.co.uk/robyns/buildings.zip*

There are important issues to consider in relation to the infrastructure and reliability of communication grids, the "top layer" of infrastructure management software, the normalization of communication processes, and the security and confidentiality of data. Rapid and efficient management of extremely large quantities of data is essential for an electric system of this type to function effectively. For example, grid topology may need to be altered in response to an accident, or customer erasure may be decided upon in accordance with their contract conditions, in response to an unexpected change in local consumption. The rollout of large numbers of captors and measurement instruments (such as the Linky connected meters in France) means that the volume of information produced and used to manage the electric system is constantly increasing. A modular, evolutive and extendable grid architecture is therefore necessary.

In France, in 2016, buildings absorbed 45% of total final energy consumption (across all energy types). It is thus crucial to increase their energy efficiency and to develop their capacity to produce and store energy, integrating mechanisms for efficient energy management in connection with the existing distribution grid.

1.2. Storage requirements in buildings

A priori, buildings which are directly powered by the grid have no need to store electrical energy, with the exception of certain critical buildings which have their own backup supply for safety and security reasons, maintaining services such as lighting in public buildings, ensuring equipment continues to function in hospitals or guaranteeing that certain business systems continue to operate to avoid economic losses (e.g. data servers or sites devoted to specific sensitive industries). Renewable energy, which may be produced locally using solar panels, for example, creates different requirements. The inherent variability of production, uncertainty in predictions and the priority given to local consumption, reducing transportation losses, create a greater need for local electrical storage. Unlike onboard systems, storage solutions in buildings are not subject to weight constraints as they are not carried by the system; however, volume remains a significant consideration. Another point to consider is that the grid used should correspond to the application, for example in terms of DC voltage and current.

The increase in the use of electrical energy in buildings is due to the flexibility it offers, as well as to the fact that pollution may be avoided at the point of use. If electricity is produced by burning fossil fuels at a power station, for example, the pollution emitted – including greenhouse gases – will not be released in the building itself, but elsewhere, i.e. at the power station. In order to reduce emissions, electricity needs to be produced using non-polluting renewable resources; furthermore, construction and de-construction phases of the production unit need to minimize consumption of energy from non-renewable sources, and, more generally, polluting emissions need to be minimized across the whole lifecycle.

Storage systems, which will become increasingly present in buildings in coming years, respond to the needs of these applications, compensating for the variation in renewable energy production while potentially providing services for other actors in the electrical system. Although prices are likely to decrease somewhat, storage systems remain expensive; the provision of services to other actors is a means of financial valorization, as long as the implications of aging are also taken into account. In this regard, work is currently ongoing to identify ways in which electric vehicles may provide services to the electrical distribution grid, or, more locally, to the buildings to which they are connected for charging purposes; this is known as Vehicle to Grid (V2G) [SAR 13, SAR 16a] or Vehicle to Home (V2H) [VEN 16, VEN 17]. The same considerations apply to storage systems included in commercial, tertiary (office) buildings and residential neighborhoods.

1.3. Difficulties in storing electrical energy

The main drawback of electric power is the high cost of storage. While electrostatic energy and magnetic energy can be stored (in capacitors and superconducting magnetic energy storage, SMES, respectively), these solutions only provide a very partial response with regard to the timescale under consideration. To obtain high-capacity storage at an acceptable price, electrical energy must be transformed into another form of energy. Electrochemical storage, using lead batteries, has long been used for onboard applications and emergency power supplies. Storage in the form of kinetic energy, using flywheels, has been used over the last few decades in fixed applications such as emergency power supplies and for certain onboard applications, for example in satellites.

Electrochemical batteries store electrical energy supplied in continuous form. Flywheel storage involves electrical machines which must be able to operate at variable speeds, i.e. variable frequency. Since the grid provides electricity in the form of alternating voltage and current at a fixed frequency, these storage techniques were little used until the development of power electronics in the 1970s. We now possess the capacity to transform the form and characteristics of currents and voltages as required.

Ragone plots, showing power density and energy density, are often used to compare different technologies and highlight the specific energy/power balance of each [ROB 15]. Figure 1.2 shows a simplified example, comparing several electrochemical technologies and supercondensers [MUL 13].

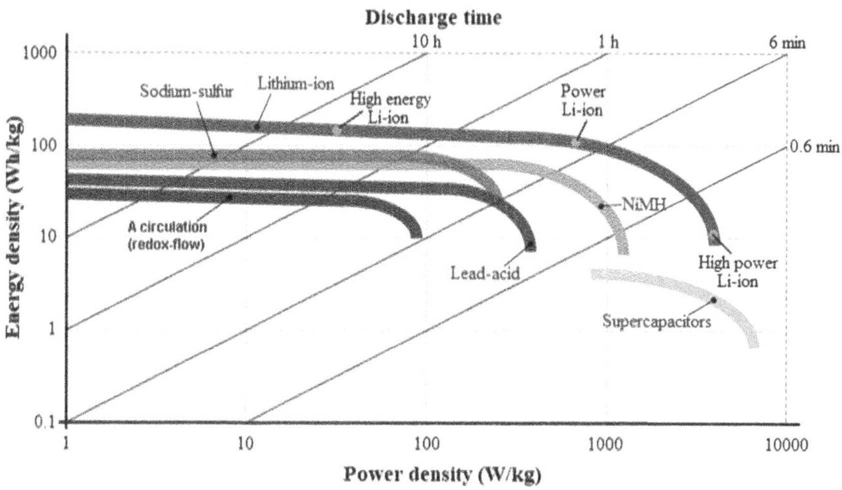

Figure 1.2. *Example of a Ragone diagram showing several electrochemical technologies and supercapacitors [MUL 13]*

In addition to power and energy, we need to take account of factors including life expectancy, response time and yield.

Life expectancy still represents a major technological constraint with regard to battery usage. It depends on the temperature of the battery, which should be neither too high nor too low, the frequency of charge–discharge cycles and the depth of discharge (DoD). The manufacturers of electrochemical solutions give figures of between 1000 and 10,000 charge–discharge cycles, with a maximum DoD and for a specified range of operating temperatures. Based on daily charge–discharge cycles, life expectancy thus varies from 3 to around 20 years. Life expectancy may be increased by reducing the operating temperature (e.g. via air conditioning) and DoD.

The energy capacity of supercapacitors is considerably lower than that of batteries, but they provide far higher power dynamics and a longer life expectancy in terms of charge–discharge cycles, in excess of 10,000. The combination of supercapacitors with Li-ion batteries is a useful solution for dynamic global storage systems, providing high storage capacities with high life expectancy. The supercapacitors handle rapid electrical variations, while the electrochemical batteries respond to regular energy needs. Note that flywheel storage systems can also provide high dynamic levels with a far higher number of charge–discharge cycles than that which is possible with electrochemical batteries [ROB 15].

Hydrogen is another possibility, enabling electricity to be produced via a fuel cell, and it can be produced from electricity (from renewable sources, for example) using an electrolyzer. However, the yield of the charge–discharge cycle is relatively low at approximately 25–30%; this means that the cost over a lifecycle remains excessive for the moment.

In the context of buildings, direct energy consumption presents the advantage of a better overall energy yield. The energy conversion required for storage results in losses, which differ widely depending on the storage technology in question. Over a full cycle, these losses may vary from less than 10% to 50%, or even more in the case of hydrogen. Nevertheless, this notion of yield needs to be relativized if stored energy comes from a source where energy shedding is used in the case of overproduction, for example in wind or solar power. We need to look at the overall balance in order to identify the best strategy (shedding or storage with a certain rate of loss) in response to economic or even environmental criteria.

Finally, note that electrical energy may be stored as a different form of energy before being used. This is the case for hot water tanks in domestic networks, where energy is finally used in thermal form, and for hydrogen produced through electrolysis, which may then be used for combustion. Certain loads include storage capacities which may be exploited to give flexibility by modulating their power supply from the grid. This is the case for cold storage in supermarket freezers, for example, and for storage in the batteries of electric vehicles.

1.4. Electricity supply in buildings

1.4.1. *Building supply and consumption*

The electrical energy consumed by buildings may be produced locally or supplied by the distribution grid. Buildings are generally powered by the grid, unless they are isolated (e.g. a mountain chalet) or have their own power supply. This situation is increasingly common with the development of renewable wind and photovoltaic solar supplies.

Figures 1.3 and 1.4 show typical profiles for domestic and commercial consumers. They show the way in which consumption varies depending on the time of day, the season and the load type.

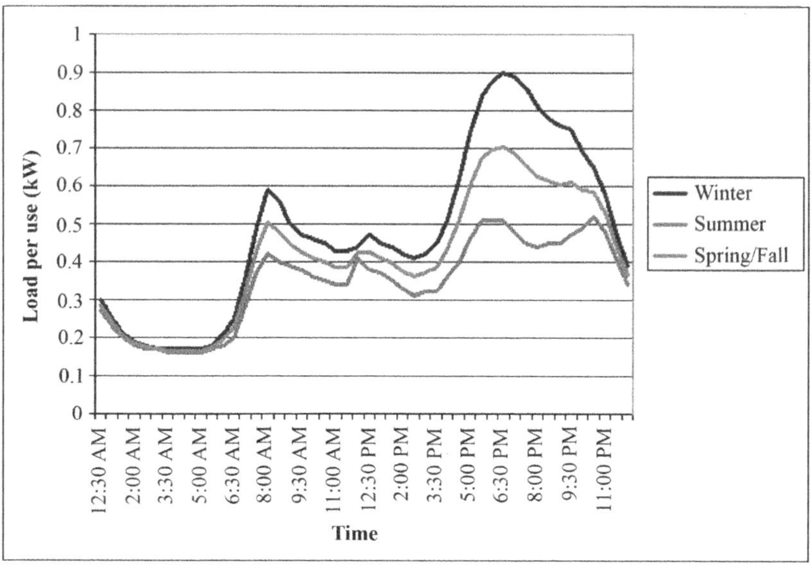

Figure 1.3. *Typical profiles for domestic consumers excluding electric heating (RTE)*

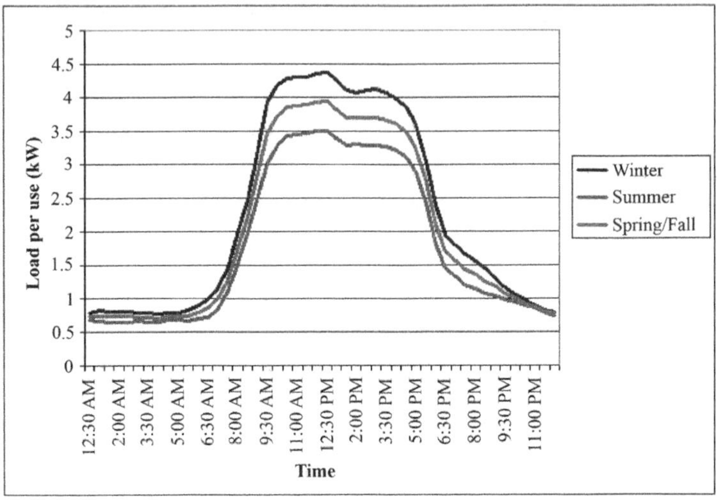

Figure 1.4. *Typical profiles for tertiary and small-scale manufacturing consumers (RTE)*

Figure 1.5 shows the power consumption profile for a large supermarket over the course of a week. The subscribed power in the example shown is 1200 kW. This is a non-optimized value, ensuring that this limit is never exceeded; going over this threshold would mean paying expensive penalties to the network operator. The addition of local storage and production would enable the optimization of subscribed power, reducing the cost of energy drawn from the grid.

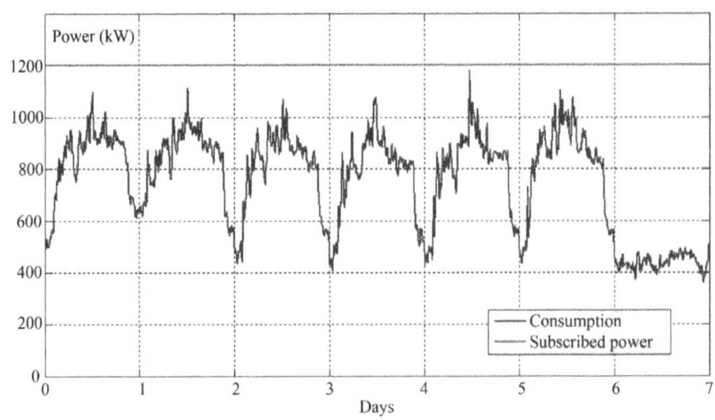

Figure 1.5. *Power consumption profile for a large supermarket over the course of a week*

In a supermarket, positive and negative cold storage may account for between one-third and half of electrical consumption. Negative cold storage can be adjusted within the limits imposed for food conservation, and this may be exploited as part of an energy management approach. Figure 1.6 shows a possible evolution of freezer temperature:

– phase 1, descending temperature, corresponds to positive consumption, and is equivalent to a cold storage system absorbing electrical energy;

– phase 2, increasing temperature, corresponds to zero consumption, but cold is being lost.

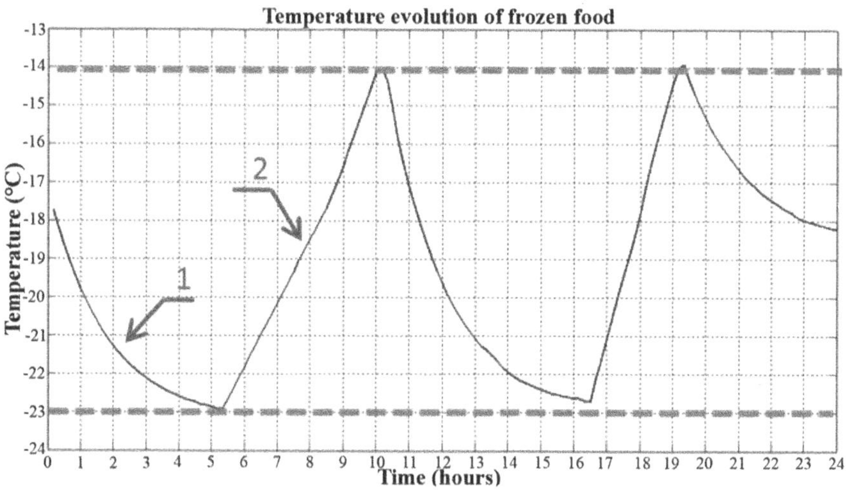

Figure 1.6. *Cold storage*

Load consumption management is nothing new. As early as the 1970s, a system was implemented to manage hot water sanitation tanks in order to avoid electrical peaks every morning and evening. Two approaches were used to achieve this: first, a pricing incentive with different tariffs for peak and off-peak production, and second, the option to control hot water tanks remotely at times of low consumption. Figure 1.7 shows the positive impact of these measures on the power profile for hot water tanks over several years. Note that these solutions were only possible because the tanks possess the capacity to retain heat over a 24-hour period.

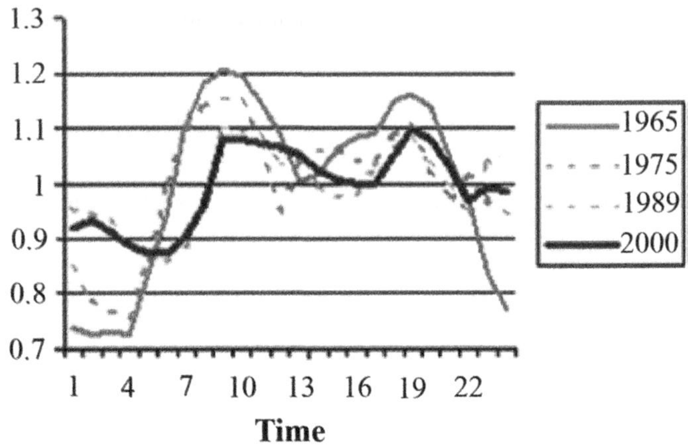

Figure 1.7. *Influence of power consumption of electric water heaters on power profiles (normalized) for mainland France [DOB 13]*

These examples highlight the variability of power demands made on the network by different types of loads and show the value of smoothing these variations, using an optimized pricing strategy and exploiting the existing storage capacity of charges (particularly water heaters, in this case), or, now, using additional storage systems for electrical energy. The integration of fluctuating renewable energy, which is difficult to predict at a local level, also justifies the use of storage systems. Storage capacity may be further exploited to provide additional services to electricity distribution or transmission grids, increasing their potential to generate profit [ROB 15]. Charging electrical vehicles using a building's electrical network offers another means of load modulation, storage and service provision for the electrical grid [BOU 15, BOU 16, ROB 16].

1.4.2. *Self-production and self-consumption*

The development of renewable energy sources has encouraged local production of electricity in proximity to the power-consuming load, i.e. local consumption. This is particularly true for energy obtained from a primary source which varies independently of demand and does not have a natural form of storage (hydraulics or combustible, for instance), and for which production is difficult to predict – particularly solar and wind power. In this context, self-production and self-consumption of renewable energy are defined as follows, with reference to the illustration in Figure 1.8:

– self-production is the portion of total consumption which is supplied by local renewable energy sources, for example over the course of the day, i.e. surface C/surface B (generally calculated over the course of a year);

– self-consumption is the portion of renewable production which is consumed in real–time, for example over the course of a day, i.e. surface C/surface A (generally calculated over the course of a year).

Using Figure 1.8, we can also define the conditions for zero-energy or positive-energy buildings, extending our calculations over a full year:

– a building can be considered zero-energy if surface B is equivalent to surface A, calculated over the whole year;

– a building can be considered positive-energy if surface B is smaller than surface A, calculated over the whole year.

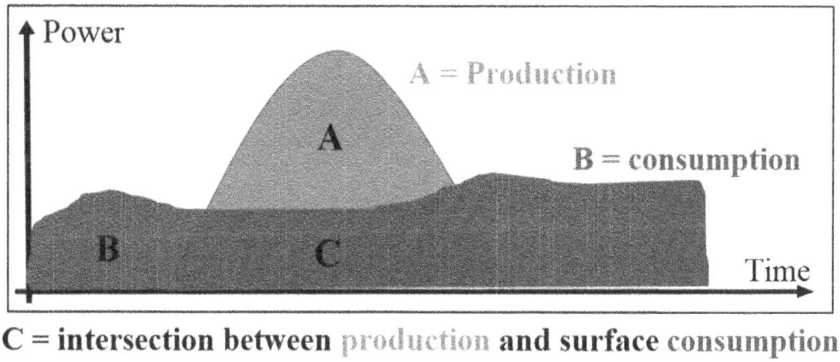

Figure 1.8. *Self-production and self-consumption over a 24-hour day*

In the absence of an energy storage system, a local network can function autonomously, or in isolation from the main grid, only if production A is greater than or equal to consumption C at all times.

1.4.3. *Micro-grids*

Micro smart grids are small-scale electrical networks designed to provide a reliable and high-quality power supply to a small number of consumers. An example of a micro-grid structure is presented in Figure 1.9. Micro-grids operate by aggregating multiple local production systems of varying types (gas micro-turbines, fuel cells, diesel generators, solar generators, wind turbines, small hydraulic systems, etc.), consumer installations, storage systems and supervisory and demand management tools. They may be directly connected to the distribution grid or operate as an island. This concept is now being extended to heating and gas networks. The micro-grid concept can also be applied to multi-fluid systems and on differing scales (building, neighborhood, industrial or manufacturing group, village, etc.) [CRE].

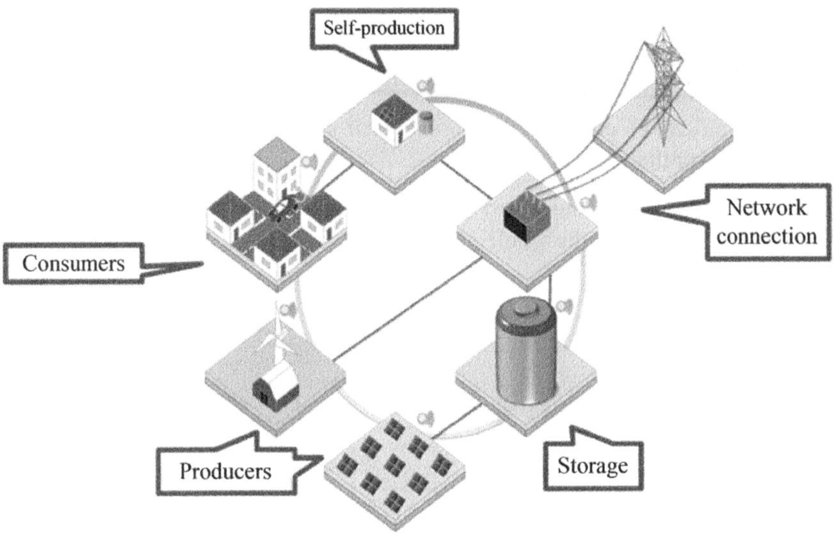

Figure 1.9. *Example of a micro-grid structure [CRE]*

Electric micro-grids may be categorized by size, as well as by utility (reliability, resilience and effectiveness of the network, difficulty of access to energy, operation in poor weather conditions, emergence of eco-neighborhoods, multi-energy possibilities, energy saving, etc.), falling into five main categories [CRE]:

– micro-grids for commercial, small manufacturing or industrial zones: these zones consume large amounts of electricity and include companies and industries carrying out a variety of different activities, all with different energy requirements. The aim is to optimize energy management to maximize control of the zone profile in relation to the distribution grid;

– university campus micro-grids: the aim in this case is to improve energy management on campus in order to reduce energy consumption;

– micro-grids for isolated zones with little or no connection to the grid, or which are temporarily cut off due to poor weather conditions: micro-grids are a way of exploiting local renewable energy and avoiding dependence on polluting, costly diesel generators. Micro-grids can also prevent a total loss of electrical power in towns affected by severe weather events;

– eco-neighborhoods: operating along very similar lines to micro-grids for commercial or industrial zones;

– "life base" micro-grids (military bases or hospitals), with their own means of production and storage and their own distribution infrastructure: the micro-grid model guarantees energy autonomy, enabling continued operation even during power outages in the main grid. This is essential for installations such as military bases or hospital, which must be able to fulfill their functions even though the main supply is down.

Micro-grids present several advantages [CRE]:

– in technical terms, micro-grids allow optimal management of renewably produced electricity at the local level. They may act as an ancillary service alongside the public distribution grid, assisting in maintaining a stable voltage and "lightening" the load in cases where they are cut off from the distribution grid;

– in economic terms, depending on size, a micro-grid may be used as an aggregator, adjusting to markets (spot market, adjustment market and capacity market). Micro-grids also make it possible to delay network investment, as the proximity between production and consumption sites enables the optimization of energy transportation. They also reduce the volume of technical losses;

– in social terms, micro-grids provide a response to evolutions in the basic energy needs of a territory. They are notably more secure and more reliable in case of incident. As local projects, they also promote initiatives and the development of new partnerships between local actors;

– in environmental terms, they permit better integration of renewable energy sources into networks and avoid the need to build thermal power stations in zones which are only weakly interconnected [ROB 12c].

Moreover, the infrastructure needed for a smart electrical grid is complex and can take several years to install; micro-grids are simpler in terms of implementation, and can act as a catalyst to the creation of smart grids. Responding to many of the issues associated with smart grids and the integration of renewable energy sources on a smaller scale, they provide an illustration of the ways in which larger grids may operate.

An example of a micro-grid can be found on the island of Kythnos, in the Cyclades archipelago in the Aegean Sea (Figure 1.10). Installed as part of the European “More Micro-Grids” program, it provides power to 12 houses. It is made up of photovoltaic solar panels with a power of 10 kWc, a 5 kW diesel generator, battery banks with a capacity of 53 kWh, and monitoring and communication systems, powered by a 2 kWc photovoltaic system. This micro-grid has been used to test both centralized and decentralized control strategies in island mode and to test communication protocols.

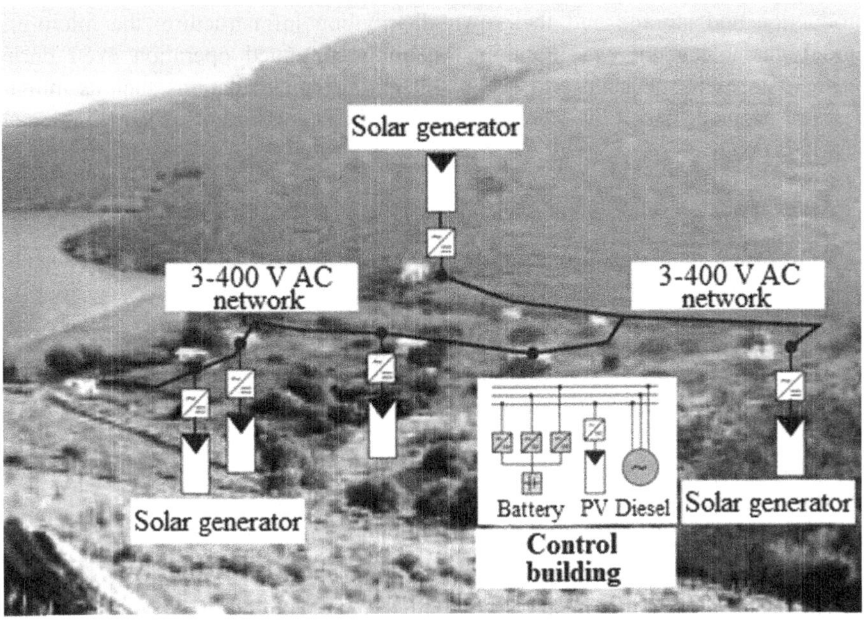

Figure 1.10. *Micro-grid on the island of Kythnos (source: ABB)*

1.5. Smart buildings

Buildings have a key role to play in the development of smart energy grids, micro-grids, eco-neighborhoods and smart cities. Figure 1.11 shows different characteristics of a smart building, including local energy production and storage and controllable loads with modulable consumption (lighting, heating, electric vehicles, etc.), which may be connected to the electricity distribution grid and to external sources or in islanded mode, i.e. cut off from the main grid.

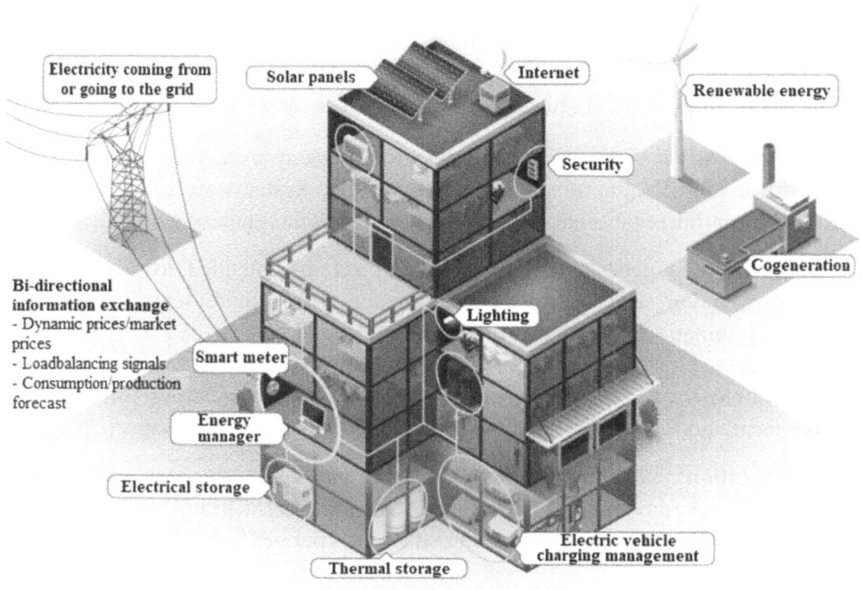

Figure 1.11. *Smart building [http://www.objetconnecte.com/batiments-intelligents-marche-iot/]*

Smart buildings include a variety of technical equipment, providing an interface between:

– some or all weather conditions (heat, luminosity, air quality, wind);

– some or all of the energy-consuming equipment needed for the building to function (heating, air conditioning, ventilation, lighting, electric vehicles, etc.), for production and for storage of local energy;

– some or all areas of the building and some or all periods of use, whether the building is residential or used for other commercial purposes.

The "smartness" of a building can vary according to the range of services provided by technical equipment. This "smartness" involves a number of different aspects, and will ideally aim to maximize energy efficiency and user comfort. These may be grouped into four areas:

– **Automation**: typically, following a measurement or time setting, a technical tool (such as a captor) triggers an adjustment process (e.g. variations in temperature, lighting or ventilation). For example, luminosity sensors may trigger the activation of lighting or of blinds, depending on the strength of the sun and the time of day. In a domestic setting, automation of this type is generally known as domotics. Actions may be triggered by dialog between devices, with or without human intervention.

– **Centralization**: some or all of the devices present are controlled by a single system, for example a thermostat for home heating. Systems are often in place to override the centralized system, for example thermostatic regulation on radiators.

– **Regulation**: a generic term covering all actions intended to adjust energy consumption and/or user comfort in accordance with requirements, technical or otherwise, automatic or manual. For example, a window that can be opened is a regulation device in its own right. Similarly, rules, consumption information or raising awareness of green behaviors are all means of regulating (or, to use a different term, modifying) the sociotechnical conditions of energy usage.

– **Information**: consumption must be closely monitored, over the course of the year, month, week, day, hour, by device, etc. The aim here is to identify connections between technical devices, consuming devices, periods of use and effective consumption. This information is not sufficient to develop a full picture of energy usage practices, but provides information regarding timing and, in some cases, the intensity of use (e.g. oven temperature settings).

Connected electric meters, such as the Linky meter in France, play an essential role in smart buildings, acting as a communication interface between the electricity grid and the consumer's equipment. A concentrator installed in a distribution station collects information from connected meters using power-line communication (PLC), acquiring data from the electrical devices in the local environment (transformers, switches, etc.) and sending it to the grid operator. The grid operator's computer system can then be accessed by energy suppliers, who use regular metering data for their clients to calculate bills. Figure 1.12 shows the relationships between consumers, suppliers and grid operators which will develop over the coming years. Smart meters provide consumers with precise and frequent updates regarding their electricity consumption via a dedicated site. They improve the quality of service offered by grid operators by generalizing remote meter reading, enabling bills to be systematically established on the basis of real consumption, and encouraging the development of varied offers tailored to specific requirements in order to improve the response to demand in periods of high consumption [CRE]. Connected meters are also being developed for use in gas networks (e.g. Gazpar in France).

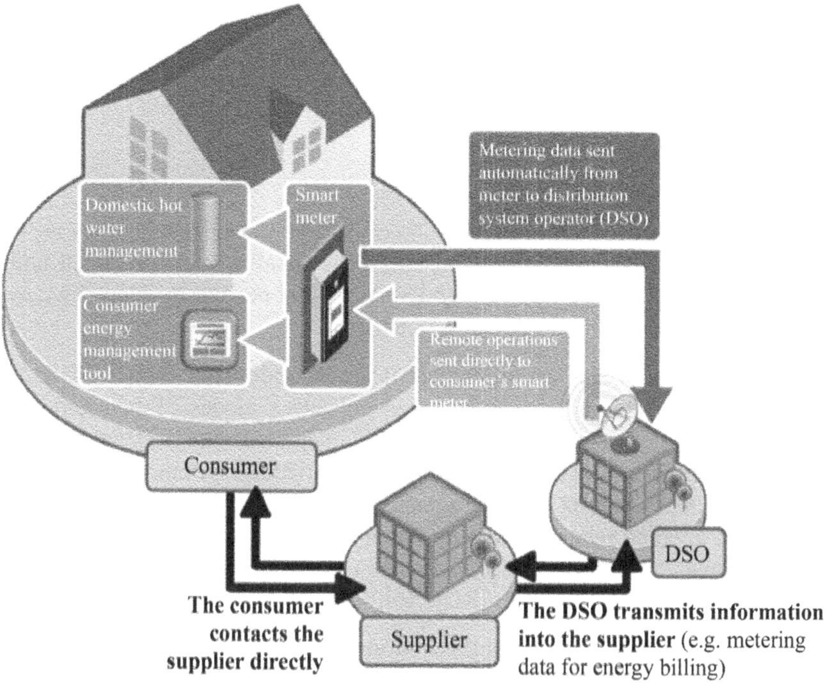

Figure 1.12. *Connected meter principle [CRE]*

The final consumer is thus at the heart of these future electric grids. Their behavior contributes to maintaining the balance between supply and demand at all times and to limiting peaks in consumption. To this end, consumers may adopt two different behavior types; these individuals are known as flexi-consumers [CRE]:

– downward modulation (balancing): in response to a specific request, the consumer temporarily reduces their electricity consumption below the usual level. This may notably include temporarily cutting off power to some of the most powerful electrical equipment (heating, washing machine, an electric vehicle being charged), by delaying consumption (the washing machine or vehicle charging may start a few minutes or a few hours later than usual, activated manually or automatically) or by stopping certain uses entirely (e.g. acting on a domestic water heater or on production machines in an industrial setting);

– upward modulation (consumption shift): in response to a request from actors in the electric system, a consumer temporarily delays consumption, either manually or automatically. This may be carried out in anticipation of the use of certain equipment. The modulation does not increase electricity consumption, but encourages the user to use power at more helpful times (e.g. when renewable power sources are producing large amounts of electricity or to exploit dedicated or pre-existing storage capacity which may be present in certain loads).

1.6. Smart cities

Towns and cities account for barely 2% of the Earth's surface, but they are home to 50% of the global population, consume 75% of all energy produced and are responsible for 80% of CO_2 emissions. Urban centers need to develop new, energy efficient services in a range of domains [CRE]:

– Smart transportation and mobility: different individual modes of transport (cars, motorcycles, bicycles, walking) and collective modes of transport (buses, metro systems, trams, taxis, etc.) must be integrated into a single, efficient, accessible, affordable, safe and environmentally friendly system. This will reduce the environmental footprint of the city, optimize the use of urban space and provide citizens with a varied range of mobility options to respond to all of their needs. The cities and towns of tomorrow will also make use of the latest public transportation and electric mobility technologies [ROB 16].

– Development of a sustainable environment: towns and cities will become increasingly involved in waste and energy management, for example via the construction of eco-neighborhoods. With regard to waste, towns and cities will aim to reduce or even eliminate waste production, and will implement efficient systems for retrieving and exploiting garbage. In terms of energy, towns and cities will make greater efforts to increase energy efficiency (developing low-energy public lighting, higher-performance transportation solution, etc.) and will implement local energy production systems (solar panels on the roofs of public buildings, producing energy through waste processing, etc.).

– Development of responsible urbanization and smart habitat: the high value of real estate in city centers combined with limited availability of land means that urbanization is a complex issue. The urban spread model which has dominated up to the present day is costly in terms of space, public amenities and energy, and is no longer tenable. Towns and cities will move toward solutions which respect the essential need for privacy, ensure sufficient sunlight, are able to evolve easily and facilitate community life. Buildings will need to become "smarter" to facilitate and improve energy management, and, potentially, to reduce consumption.

Citizens will have a central role to play in the towns and cities of tomorrow. No longer simply consumers of services, they will be seen as partners and stakeholders in urban development. They will be able to take on this role due to a democratization of the means of information, enabling enhanced participation.

The implementation of smart cities will be a progressive process. Demonstrators will be used to test different functions made possible by smart grids and smart buildings, to identify the new economic models essential to the viability of these functions, and to examine questions of acceptability and the implication of different actors (producers, consumers, energy grid operators, aggregators, etc.). In Lille, France, for example, demonstrator programs have already been established, including Live Tree (Lille Vauban en Transition Energétique, Ecologique et Economique: Energy, Environmental and Economic Transition in the Vauban neighborhood of Lille) [LIV] and "So MEL, So Connected" (supported by the ADEME, Agence de l'Environnement et de Maîtrise de l'Energie en France: the French Environment and Energy Management Agency) [SOM].

1.7. Socio-economic questions

1.7.1. *Toward new economic models*

The notion of an economic model includes multiple dimensions: a macroscopic (top-down) dimension, whereby public authorities may establish frameworks to encourage the development of smart electricity grids, and a microscopic (bottom-up) dimension, which relates to companies' ability to establish business models specific to their activities. Company business models describe the distribution of activities and, notably, the source of revenues, highlighting profitability. To develop a business model, actors need access to a number of basic elements [CRE]:

– identification of the actors in question and an understanding of their interest/risk in relation to smart electricity grids;

– cost–benefit analysis within a coherent framework. For example, the conclusions drawn from cost–benefit analysis of connected meters are different depending on the perspective taken, which may be limited to the distributor or take account of all costs and benefits from producers to final consumers;

– once a calculation framework has been defined, expenditure items and sources of profit should be defined along the new value chain;

– an estimation of the financial values associated with these profit and expenditure items;

– identification of possible funding sources.

In order to develop these business models, we need to identify the way in which smart electric grids modify the value chain. In the new value chain, final users will take on a particularly important role, as the balance of the electrical system will be managed primarily by directing consumption; in the existing system, balance is based on controlling production. The integration of renewable and intermittent energy sources makes electricity production harder to control and to manage. Final users will thus play an active role in the energy system, and will become the central focus for other actors in the value chain.

For final users to take an active part in the definition of future economic models for smart grids, a certain number of crucial steps need to be taken [CRE]:

– provide better information regarding the economic and environmental benefits to consumers and communities obtained through the use of new ICT technologies in managing the electrical system (improved public electricity system, increased quality and frequency of consumption information to enable greater control, etc.);

– provide easy access to information via a variety of simple interfaces (smartphones, computers, etc.), enabling reactivity and normalizing the use of services linked to smart networks;

– guarantee the protection of personal data and manage the way in which individuals' information is shared.

The increasing interdependence between different actors (producers, consumers, storers, network operators, aggregators, etc.) inherent in the development of smart grids makes the development of new economic models increasingly complex, while offering new economic possibilities.

1.7.2. *Social acceptability*

The introduction of technical developments which are seen as overwhelmingly positive in terms of problem-solving or progress may be met with resistance from the intended users. Innovations are slow to take root and can face opposition for reasons other than price. These forms of resistance are always baffling as they contradict the presupposition that progress (improvements, solutions, additional services) will always be welcome in a rationale based on utility and profit maximization (cost–benefit approach). Sociology can cast some light on the issues of resistance and on the conditions of overcoming this resistance to reach the level of support needed for a change which is *a priori* beneficial to be adopted.

Social acceptability refers to the set of individual judgments relating to the acceptance (or non-acceptance) of a practice or condition, whereby individuals compare this solution with possible alternatives in order to determine desirability. It is expressed by groups, active in the political life of a society, who share a judgment with regard to the practice in question [YEL 13].

This constitutes a collective process insofar as acceptability becomes "social" once a minimum level of support is attained. This minimum reflects a threshold which varies from subject to subject, and which results in the diffusion of acceptance throughout the target social group via a number of different mechanisms (conformity, normality, innovation, etc.).

Acceptability is the result of a reasoning or rationality which structures the mode of comparison between different possible scenarios. Rationality may be modeled as the interaction between four notions:

– an objective, i.e. the aims of the actor (money, pleasure, recognition, distinction, information, conservation, comfort, etc.);

– resources, i.e. which is required for the objective to be attained (skills, information, allies, money, network, etc.);

– constraints, i.e. which acts against the achievement of the objective (rivals, competition, lack of skills, isolation, weaknesses, etc.);

– strategy, i.e. the use of resources with regard to the objective and the constraints to overcome. Rationality is thus an interplay between cognitive and reflexive elements, resulting in a way of thinking and acting in relation to a short-, medium- or long-term goal.

Acceptability becomes a social and influential concern when an opinion is shared by a sufficient number of individuals (threshold effect) and/or by a group with sufficient weight within the social milieu. Not all social groups have the same influence within their social environment. Groups with the highest capital (economic, cultural or relational) have more influence in prescriptive terms and their opinions may have a knock-on effect.

The reasoning involved in acceptability develops within a multi-layered context. This explains the systemic nature of the social acceptability process: as the issue is not limited to personal preference, it is useful to understand social psychology.

In terms of context, the parameters to consider include:

– individual reasoning, following the structure described above (objective, resources, constraints, strategy);

– social context, which is made up of three aspects:

- macro: regulations, market, price, incentives, social culture, national politics, etc.,

- meso: geography, territory, operators, technical constraints in the immediate area, etc.,

- micro: organization, operation, requirements, resources, values and image of the actor, etc.

In this context, values, image and attitudes carry a heavy weight. Beliefs, worldviews and preconceived ideas all contribute to the formulation of a positive or negative attitude toward a proposition which may or may not be accepted. The meso and macro levels are equally important; however, their influence is subtler in that it is less immediately obvious and more debatable.

1.8. Storage management

Different time frames need to be taken into account when establishing an energy system management strategy, with or without one or more energy storage solutions (Figure 1.13):

– long-term supervision, over the course of a day or more;

– medium-term supervision, corresponding to durations from approximately 30 min to 1 hour;

– real-time supervision, which corresponds to the shortest timescale used in ensuring the system functions correctly, ensuring stability, goal attainment, hazard response, etc. This timescale may be anywhere from a few hundredths of a second to several minutes.

Longer-term planning (for storage over several days, weeks, months or years) may also be needed to ensure efficient storage management and to guarantee economic viability.

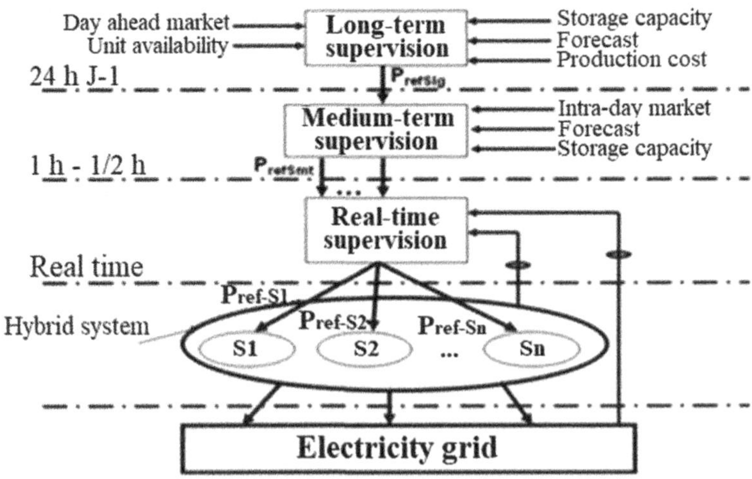

Figure 1.13. *Different timescales to consider when managing an energy system, with or without energy storage systems*

Managing energy storage represents a significant challenge due to the complexity of the problems involved, the economic and ecological objectives, and the fact that there is no single solution for achieving these goals [NEH 11, ROB 12b, ROB 13a, ROB 13b, ROB 15, ROB 16]. Three broad groups of tools for supervising hybrid systems with storage have been proposed in the literature:

– **causal formalization tools** [ALL 10, FAK 11, ZHO 11, DEL 12]. This approach consists of identifying power flows which may be inverted in order to identify reference power values. It requires the use of a detailed mathematical model of sources and storage systems, alongside excellent real-time data concerning different flows and the associated losses;

– **explicit optimization tools** with objective functions [ROB 12a, SAR 13, SAR 16, ROB 16]. This approach is necessary to guarantee that optimal choices have been made to ensure maximization, for example of energy produced from renewable sources. However, minimization of a well-formulated cost function is no easy matter, particularly in real-time;

– **implicit optimization tools**, for example using fuzzy logic [CHE 00, LEC 03, LAG 09, COU 10, ZHA 12, MAR 12, KAD 13, ROB 13a, ROB 13b, ROB 15, BOU 15, BOU 16, ROB 16]. This type of tool is eminently suitable for the management of complex systems, involving quantities or states which are difficult to predict and not particularly well known in real-time (wind, sunshine, network frequency and state, variation in consumption, etc.).

A variety of approaches may be considered and combined for storage management purposes, including filters, correctors and artificial intelligence tools.

A methodology for the design of supervisors for managing hybrid energy production systems including storage is developed in this book [ROB 13a, ROB 13b, ROB 15, ROB 16]. Our method is an extension of those commonly used in designing control systems for industrial processes: Petri networks [ZUR 94, LU 10] and GRAFCET [GUI 99]. These approaches allow a graphical and "step by step" approach to the construction of system control, making analysis and implantation considerably easier. They are particularly well suited to sequential logic systems. However, in the case of hybrid production units including random variables and continuous states, tools of this type encounter certain limits. The method we propose is an extension of this graphical approach to include fuzzy and poorly known quantities.

This methodology does not require the use of mathematical models, as it is based on system expertise, represented by fuzzy rules. Input may be random, and supervision may aim to achieve several objectives simultaneously. Transitions between operating modes are progressive as they are determined by fuzzy variables.

Finally, our methodology enables storage management by convergence toward a load level, along with control of complexity for the purposes of real-time processing.

It is made up of eight stages, each assisting in supervisor design (discussed at length in Chapter 1 of [ROB 16] and implemented in several chapters of the present book):

– determination of a system specification, with objectives, constraints and resources clearly laid out;

– definition of a supervisor structure, including determination of the required input and output;

– determination of operating modes using functional graphs, with a graphical representation of modes of operation developed based on knowledge of the system. These modes are expressed verbally, which makes it easier to integrate socio-economic considerations;

– definition of the membership functions of fuzzy variables;

– determination of fuzzy modes using operational graphs;

– extraction of fuzzy rules, characteristics of the fuzzy supervisor, from operational graphs;

– definition of indicators to use in assessing whether or not objectives have been met. For example, these may include an indication of power, energy, voltage quality, yield, or concern economic or environmental aspects;

– optimization of supervisor parameters using design of experiments (DoE) and/or genetic algorithms, for example.

This methodology will be developed progressively over the course of this book, considering energy storage based on one or more technologies in the context of different buildings (supermarkets, tertiary sector buildings, micro-grids, residential neighborhoods, and with regard to the integration of reversible-charge electric vehicles). Causal and explicit optimization methods will also be used.

1.9. Methodologies used in developing energy management for storage systems

In this book, we will progressively implement several different methods which may be used to develop a management approach for storage systems, based on one or more storage technologies and in different building-related contexts. Table 1.1 shows a summary of the different contexts which will be considered, along with our objectives, the tools and disciplines used, and the methods to design an energy management system which will be discussed in later chapters.

Chapter/theme	Objectives	Methods, tools and disciplines used in designing an energy management system
2. Commercial buildings consuming and producing electricity	Reduce the electricity bill Reduce CO_2 emissions	Fuzzy logic
3. Tertiary buildings consuming and producing electricity. Implementation of a dedicated DC electrical architecture	Promote self-consumption of local production Reduce CO_2 emissions	Fuzzy logic
4. Photovoltaic solar producer with hybrid storage in the context of a non-interconnected grid	Production injected into an electricity grid according to a planned profile. Participate in grid frequency control Increase life expectancy of storage systems	Fuzzy logic Electrochemical battery degradation model Economic analysis
5. System including multiple tertiary or residential actors within a smart grid	Economic and sociological conditions to promote energy load balancing	Economic openness to contract theory Openness to the notion of social acceptability
6. Energy pooling between buildings of varying natures, consumers and producers	Promote self-consumption of local production and load balancing with respect to the public distribution grid. Reduce electricity bill and CO_2 emissions	Fuzzy logic Integration of an acceptability coefficient
7. Residential community, e.g. eco-neighborhood	Promote self-consumption of local production of renewable energy	Hierarchical and causal control
8. Reversible charge electric vehicles	Minimize cost of transporting electrical energy through the distribution grid. Maximize local consumption of renewable energy	Fuzzy logic with optimization via genetic algorithm

Table 1.1. *Different aims and construction methods for energy management systems described later in this book*

2

Energy Storage in a Commercial Building

2.1. Introduction

Over the last few years, European energy authorities have been seeking to reduce CO_2 emissions and fossil fuel consumption through the development of renewable energy [EUR 01, EUR 02, EUR 03, ROB 12c]. The increasing demand for buildings to meet high energy performance standards and the development of ultra-low energy buildings (passive homes) are key priorities, as the greatest potential for energy savings is found in the residential sector (i.e. homes) and in commercial buildings (tertiary sector), estimated at approximately 27% and 30% of energy consumption, respectively [EUR 01].

In this chapter, we will develop an energy management system for a commercial building, such as a supermarket, including photovoltaic solar production and energy storage.

2.2. Managing energy storage in a supermarket

2.2.1. *Introduction*

In this section, we will describe a large supermarket connected to the electric grid, with a solar power installation on the roof and a storage system. A general methodology for designing an energy supervision approach for the system was developed in [COU 10, ROB 15]. The aim of this study is to devise an electrical energy management system for a supermarket-type building by applying this methodology.

An energy management strategy for the storage system will be developed using fuzzy logic [BOR 98, BOU 98, BUH 94]. The storage system enables the power provided by the network to be regulated across peak and off-peak periods in order to reduce electricity costs and CO_2 emissions. Storage functions may be carried out by a dedicated system, as well as, in part, through the use of loads whose power can be modulated.

We will begin by defining the connection configuration of the supermarket. The objectives, constraints and means of action in this supervision will also be presented, along with the principles of electric billing and of the energy management strategy. Next, we will develop a supervisor based on fuzzy logic. We will present the results of simulations, and different topologies (with or without photovoltaic power (PV)) will be compared using economic and ecological indicators.

The study presented in this chapter takes no account of investment and maintenance costs, nor of aging in dedicated storage systems. This latter consideration will be discussed in Chapter 4.

2.2.2. *System characteristics*

Several topologies for connecting a commercial building to the electric grid are presented in [RIC 11, SEC 10, LOC 09]. Figure 2.1 shows the electrical configuration of the supermarket-type commercial building considered in this study. It includes a photovoltaic system [ROB 12c] and a storage system, installed parallel to the supermarket load [ZHA 12]. All of this equipment is installed downstream of the electric meter. Photovoltaic production should be consumed, as a priority, by the load (self-consumption).

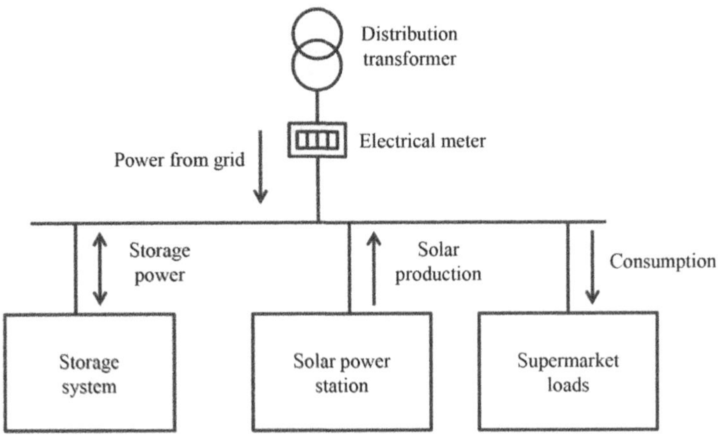

Figure 2.1. *Electrical configuration of a supermarket-type building including photovoltaic production and storage*

The PV system and the supermarket load are modeled by the production profile and load profile, respectively, shown in Figure 2.2.

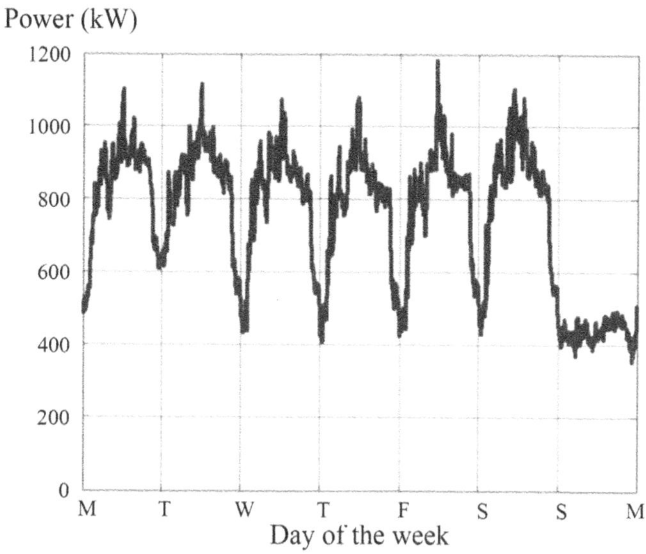

a) Consumption profile over a week (1: Monday; 7: Sunday)[1]

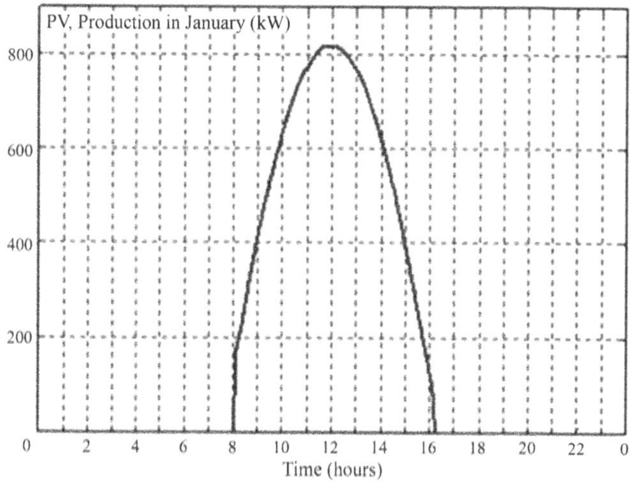

b) PV production profile over a 24-hour day

Figure 2.2. *Consumption and production profiles for a fictional supermarket. For a color version of the figures in this chapter see www.iste.co.uk/robyns/buildings.zip*

1 Translator's note: as a general rule, French supermarkets do not open on Sundays – clearly, this fictional supermarket is no exception.

There are several distinct types of electrical load in a supermarket: refrigeration and freezing apparatus, lighting systems, air handling systems, computers, IT servers, cash management systems, etc. This profile is shown in Figure 2.2a.

The PV is modeled using a production profile based on irradiance data provided by the Photovoltaic Geographical Information System (PVGIS) [PVG 17]. The values used are those for the month of January in Lille, northern France. No account is taken of weather conditions or the specific hours of sunlight for each day of the week. An example of an idealized profile is shown in Figure 2.2b. An example of a real profile can be found in Chapter 2 of [ROB 12c].

The generic model used for the storage system [COU 07] is shown in Figure 2.3. P_{max} is the output power limit, τ_{stor} is the time constant for the storage system and W_{stor} is the stored energy in the system.

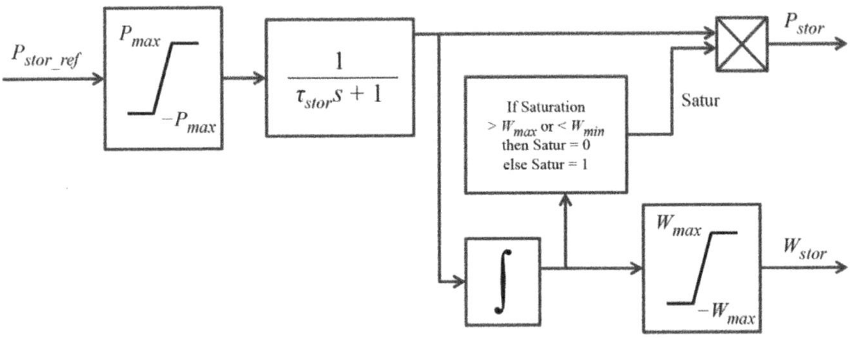

Figure 2.3. *Simplified model of the storage system*

In practice, the storage system may consist of a specific system (e.g. batteries), but this may be supplemented by exploiting the storage capacity of certain loads, such as the negative cold storage capacity, given that refrigeration (positive cold) and freezing (negative cold) represent between one-third and one half of the energy consumption of a supermarket.

The first step in the supervisor design methodology, as described in [COU 10, ROB 15] and in Chapter 1, consists of determining specifications and identifying the characteristics and objectives of the system. Table 2.1 shows a summary of the objectives, constraints and means of action involved in supervision. Electricity prices vary between pricing periods. These price variations are used to establish energy management strategies, thus reducing the electricity bill and CO_2 emissions by balancing consumption across peak and off-peak times.

Subscription power is a constraint which limits power consumption. The load must not exceed this limit on pain of financial penalties. The means of action is the storage reference power.

Objectives	Constraints	Means of action
– Reduce the electricity bill. – Balance consumption between peak and off-peak periods in order to reduce CO_2 emissions – Ensure energy is available	– Price of electricity during different periods – Limitation of subscription power – Limitation of storage capacity	– Reference power of storage

Table 2.1. *Objectives, constraints and means of action involved in the energy management strategy*

2.2.3. *Electricity billing*

An electricity bill is made up of three components: an annual premium, based on the subscription power and selected options, active energy consumption costs (per kWh) and reactive energy consumption above an authorized limitation during the winter season, established by a regulatory tangent (0.4, i.e. cosφ = 0.93). There may be an additional penalty if the supplied power is greater than the subscribed power. Figure 2.4 shows the calculation process.

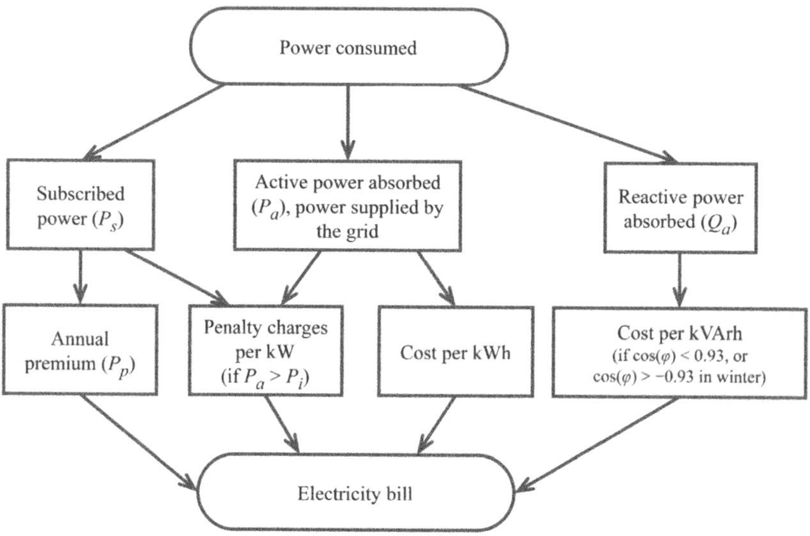

Figure 2.4. *Composition of an electricity bill*

The annual premium is calculated from the subscribed power P_s. This power is calculated as a function of the subscribed power defined over each pricing period P_i. Equation [2.1] shows how the annual premium is calculated, where P_p is the total annual premium (€), P_{kW} is the price per kW consumed (€/kW), k_i is a period coefficient defined by the electricity supplier, P_i is the subscribed power (kW) which varies as a function of the pricing period and n is the number of the pricing period:

$$\boldsymbol{P_s = \sum_{i=1}^{n} k_i (P_i - P_{i-1}) \text{ where } P_0 = 0} \quad [2.1]$$

$$\boldsymbol{P_p = P_{kW} P_s}$$

Table 2.2 shows an example of the pricing table for the three winter pricing periods.

	Peak period in winter (PPW)	Shoulder period in winter (SPW)	Off-peak period in winter (OPW)
k_i	k_1=1	k_2=0.77	k_3=0.38
P_{kWh}(€/kWh)	0.1151	0.07662	0.04641
P_{kW} (€/kW)	66.12		

Table 2.2. *Example of a price table for different pricing periods in winter [EDF 11]*

The different pricing periods are presented in section 2.2.5.

2.2.4. *Objectives of the energy management strategy*

Based on the objectives and constraints set out in Table 2.1 and the prices shown in Table 2.2, the objectives of the energy management strategy are as follows:

– do not exceed subscribed power;

– reduce power drawn from the grid during peak period in order to reduce the electricity bill;

– ensure the availability of the storage system for the next pricing period. Charge the storage system during off-peak or peak periods only if absolutely necessary;

– PV production should be consumed or stored in the storage system as far as possible. If there is still excess power, this should be sent to the grid if the purchase agreement permits negative supply (i.e. supply from the installation to the grid).

The architecture of the supervision system is shown in Figure 2.5. The inputs are the subscribed power P_s, the absorbed power P_a (power supplied by the grid), the SOC (state of charge of the storage system) and the time t, in accordance with the objectives listed above.

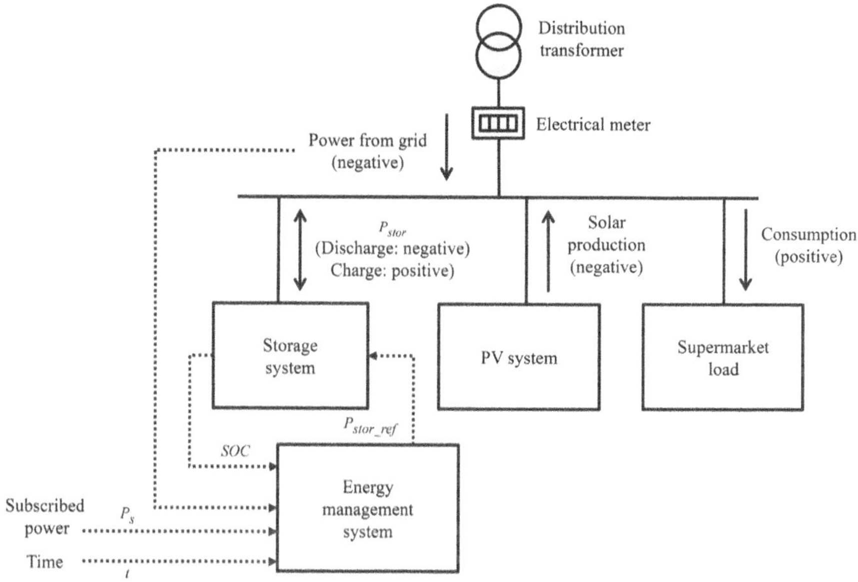

Figure 2.5. *Architecture of the supervision system*

2.2.5. *Fuzzy logic supervisor*

As we have seen, the supervision system must respond to several objectives. Fuzzy logic is a helpful tool for solving problems of this type, as it has the capacity to take account of multiple objectives and constraints [ROB 15], and identify compromises linked to real situations [COU 10, SPR 09, SIE 07, ZHA 10].

The objectives for energy management described in the previous sections can be expressed as a series of rules relating to different pricing periods.

As shown in Figure 2.5, the architecture of the supervision system features four inputs. There is only one output, corresponding to the means of action mentioned in Table 2.1.

The objectives of the supervision strategy differ according to the pricing period. These pricing periods are broken down into six different operating modes. Table 2.3 shows the different pricing periods for the winter season (the most important in terms of cost) and the corresponding operating modes.

Pricing periods	Description	Operating mode	Description
PPW	0900–1100 and 1800–2000 hours, Mon–Fri	PPW1	0900–1100 when PV production is high
		PPW2	1800–2000 when PV production ceases
SPW	0700–0900, 1100–1800 and 2000–0100 hours, Mon–Fri	SPW1	0700–0900 and 1100–1800 which are followed by a PPW period
		SPW2	2000–0100 which is followed by an off-peak period
OPW	0100–0700 hours, Mon–Fri, weekends and public holidays	OPW1	0100–0700, Mon–Fri and all day Saturday, when the supermarket is open
		OPW2	Sunday or public holidays when the supermarket is closed

Table 2.3. *Pricing periods for the winter season and the corresponding operating modes. PPW: peak period in winter, SPW: shoulder period in winter, OPW: off-peak period in winter*

The supervision strategy will thus be made up of several operating modes. Figure 2.6 shows a graphical representation of supervision using fuzzy logic. This graphics tool, which is presented in Chapter 5 of [ROB 15], assists in the design of different operating modes. The modes mentioned above are represented by rectangles with rounded corners and system states are represented by transitions. The three pricing periods are represented by blocks N1.1, N1.2 and N1.3. There are different operating modes for each pricing period, and the transition conditions are specified in Table 2.3. Figures 2.7, 2.8 and 2.9 respectively show the details of operating modes for the peak period, the shoulder period and the off-peak period. Different objectives for each operating mode are shown in blocks N1.1.1 to N1.3.2. The order of the elements shows the priority given to objectives.

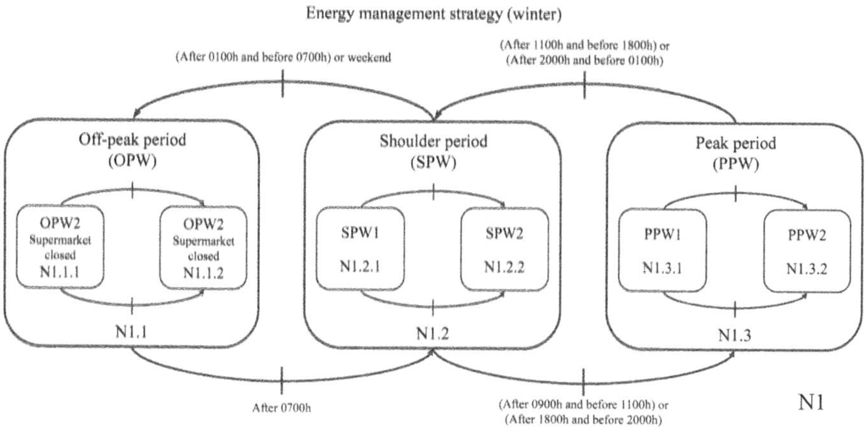

Figure 2.6. *Graphical representation of the supervision strategy*

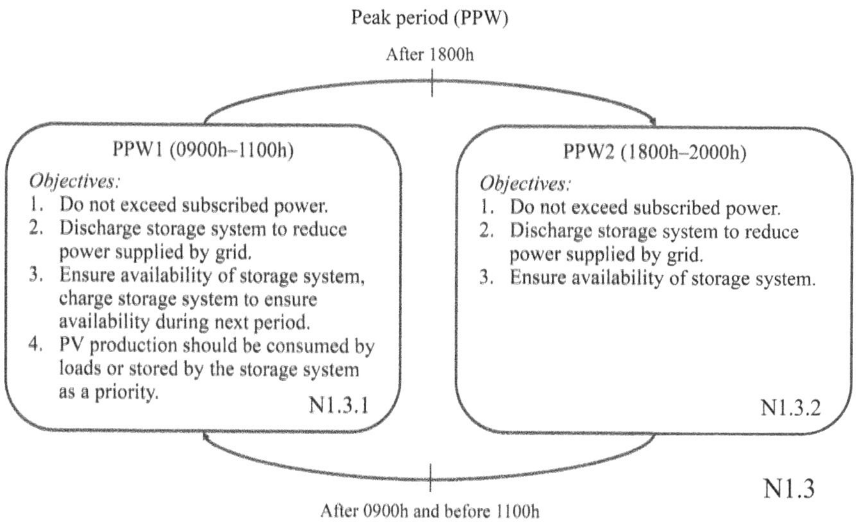

Figure 2.7. *Details of the peak period in winter*

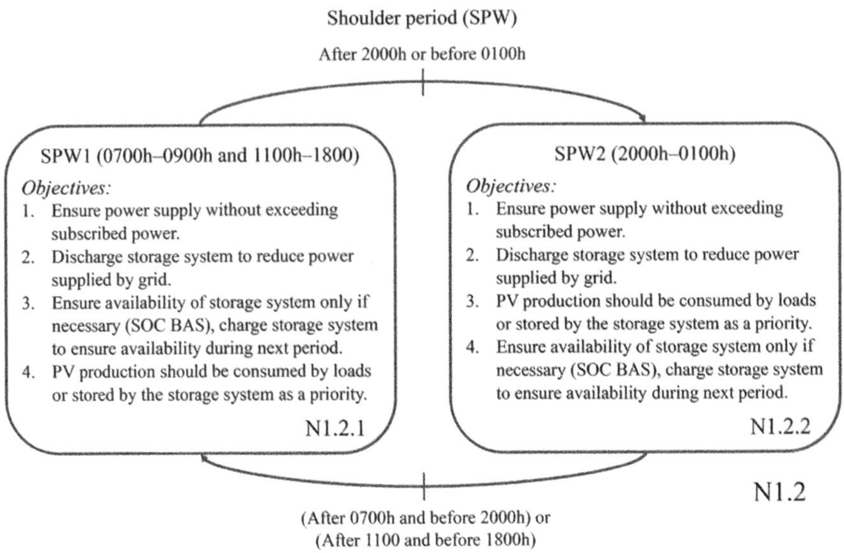

Figure 2.8. *Details of the shoulder period in winter*

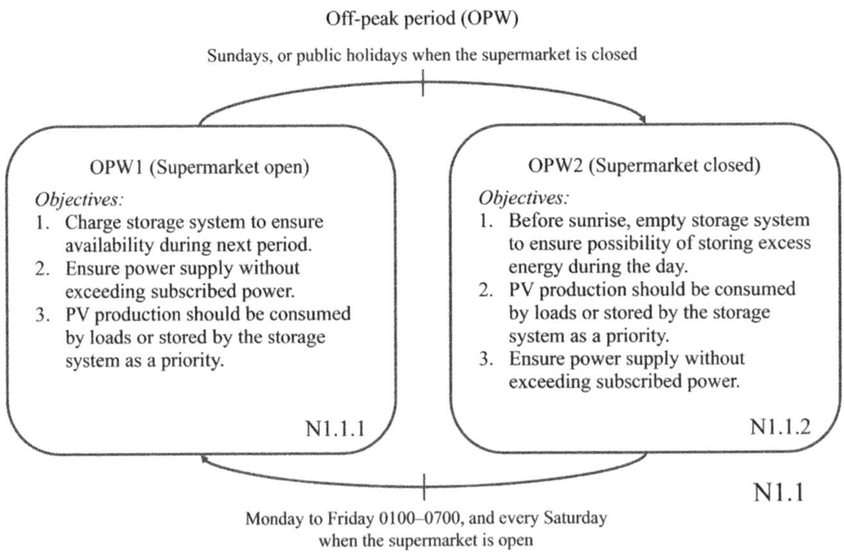

Figure 2.9. *Details of the off-peak period in winter*

These six operating modes will be described in greater detail in the following sections.

2.2.5.1. *PPW1 operating mode*

During this period, there is significant PV production and the price of electricity is high. The objectives of supervision strategy N1.3.1 are shown in Figure 2.7.

The power difference ΔP is defined by equation [2.2], where P_i is the subscribed power for each pricing period, as shown in [EDF 11], and P_a the active absorbed power, as shown in Figure 2.4:

$$\Delta P=(P_i-P_a)/P_i \qquad [2.2]$$

Figure 2.10 shows the convention selected for ΔP and the different fuzzy levels to be used in the fuzzy rules.

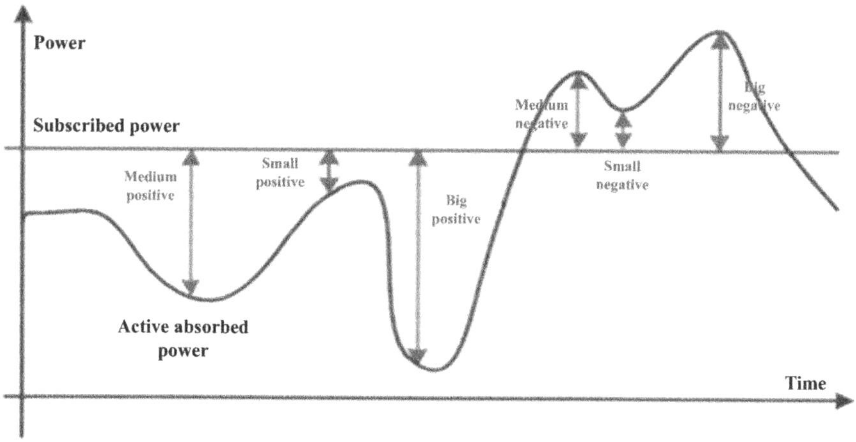

Figure 2.10. *Convention on power and different fuzzy levels*

The fuzzy algorithm is made up of three parts: fuzzification, inference and defuzzification [VAS 99].

Fuzzification

The two membership functions for the input variables are shown in Figure 2.11. As we have seen, one of the objectives of energy management is to ensure that the power drawn from the grid does not exceed the subscribed power. For this reason, the membership function of ΔP is not symmetrical about 0. In Figure 2.11, the names of the fuzzy subsets are as follows: Z = Zero, M = Medium, B = Big,

BN = Big Negative, MN = Medium Negative, MP = Medium Positive, BP = Big Positive. The Small Positive and Small Negative states of the ΔP input variable shown in Figure 2.10 are concatenated into a single Zero subset.

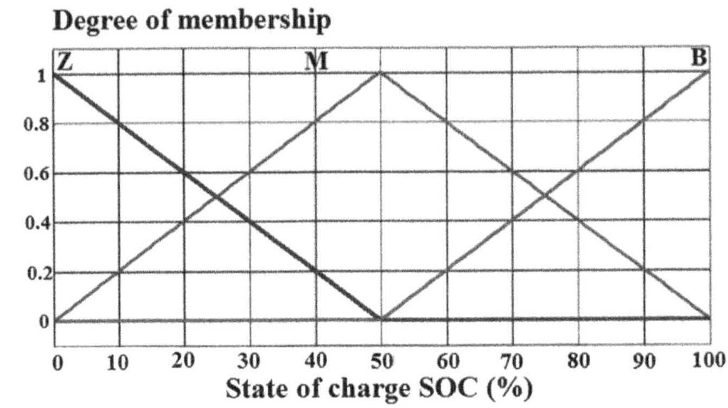

a) State of charge of the storage system

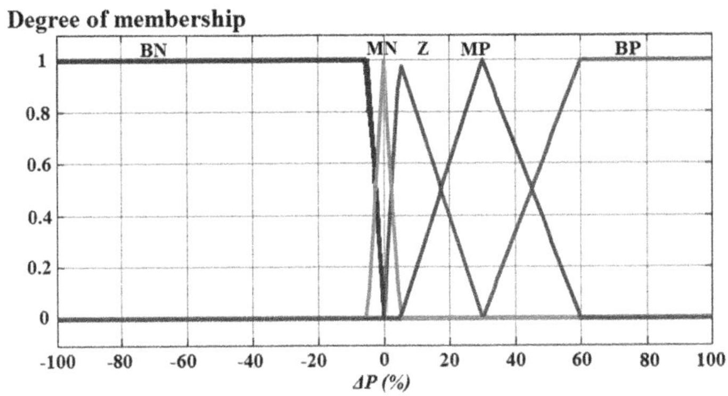

b) Power difference between subscribed power and supplied power

Figure 2.11. *Membership functions of input variables*

Inference

The fuzzy rules are expressed as follows:

IF ΔP is Medium Positive **AND** SOC is Medium, **THEN** P_{ref} of the storage system is Medium Negative.

IF ΔP is Big Negative **AND** SOC is Medium, **THEN** P_{ref} of the storage system is Big Negative.

These rules are defined as a function of the objectives for N1.3.1 indicated in Figure 2.7. As a function of the real situation (price of electricity, consumption, subscribed power, state of charge of the storage system, etc.), these rules enable the supervision system to behave in the most appropriate manner to attain the desired objectives. Table 2.4 shows the fuzzy rules corresponding to strategy PPW1 (SN = Small Negative, MN = Medium Negative, BN = Big Negative, Z = Zero, SP = Small Positive, MP = Medium Positive, BP = Big Positive).

		ΔP				
	P_{ref}	BP	MP	Z	MN	BN
SOC	B	SP	BN	BN	BN	BN
	M	MP	MN	MN	BN	BN
	Z	BP	SN	MN	BN	BN

Table 2.4. *Fuzzy rules: power reference for the storage system in strategy PPW1*

Defuzzification

The membership functions of the output quantity are shown in Figure 2.12.

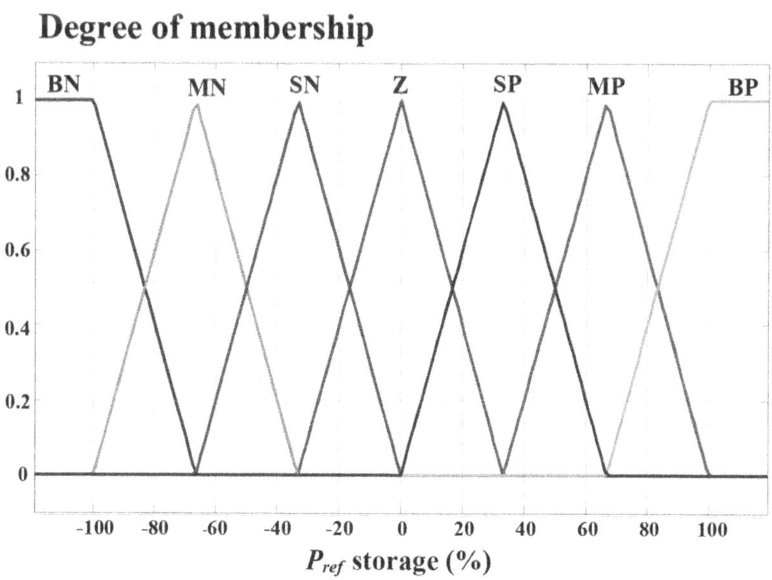

Figure 2.12. *Membership functions of the output variable*

Figure 2.13 represents the variation of P_{ref} as a function of the SOC and ΔP. It shows the nonlinear relationship between input and output variables obtained using fuzzy logic. Note that, in most cases, the storage system is discharged. When ΔP is close to 100% (power consumption close to 0), the storage system will be responsible for supplying the power consumed by the supermarket.

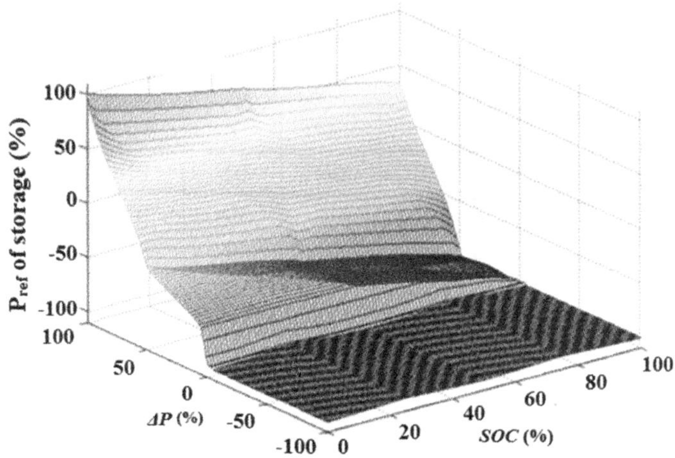

Figure 2.13. *Surface generated by strategy PPW1*

2.2.5.2. *PPW2 operating mode*

This mode covers the period between 1800 and 2000 hours in the winter. During this time, there is no PV production. The storage system must be discharged in order to reduce electricity costs during this peak period.

The objectives for N1.3.2 are shown in Figure 2.7. The membership functions of the inputs and output are the same as those shown in Figures 2.11 and 2.12, but the fuzzy rules are different as the objectives are different. The fuzzy rules for strategy PPW2 are shown in Table 2.5.

		ΔP				
SOC	P_{ref}	BP	MP	Z	MN	BN
	B	BN	BN	BN	BN	BN
	M	MN	MN	MN	BN	BN
	Z	SN	SN	SN	BN	BN

Table 2.5. *Fuzzy rules: power reference of the storage system using strategy PPW2*

The relationship between input and output is different to that obtained using strategy PPW1, as shown in Figure 2.14. Note that the difference between PPW2 and PPW1 is most evident when ΔP is close to 100%.

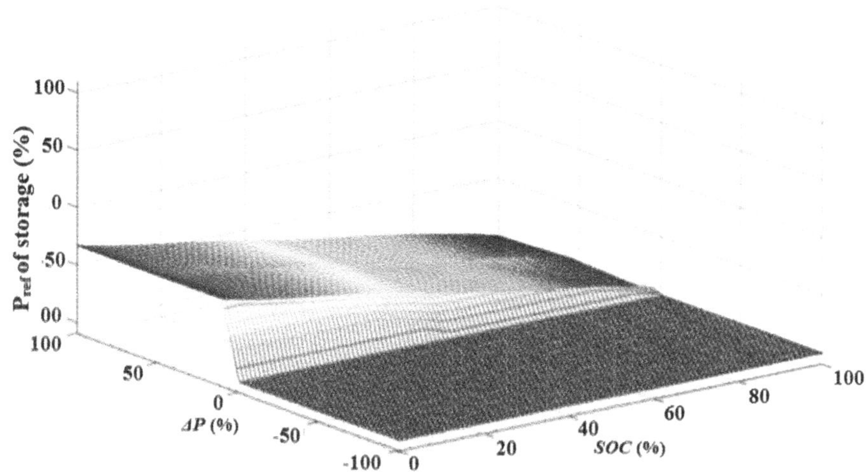

Figure 2.14. *Surface generated by strategy PPW2*

2.2.5.3. *SPW1 operating mode*

This period runs from 0700–0900 and 1100–1800 hours, Monday to Friday. PV production is high during this period, but it fluctuates a lot and is difficult to predict (due to passing clouds, cloud cover, etc.). For this reason, the management strategy needs to be flexible. The fuzzy rules for this period, following the objectives for N1.2.1 shown in Figure 2.8, are presented in Table 2.6.

		ΔP				
	P_{ref}	BP	MP	Z	MN	BN
SOC	B	SP	Z	MN	BN	BN
	M	MP	SP	MN	BN	BN
	Z	BP	SP	Z	MN	BN

Table 2.6. *Fuzzy rules: power reference of the storage system using strategy SPW1*

The surface generated by this strategy is shown in Figure 2.15. Comparing Figures 2.13 and 2.15, we see that the two strategies produce very similar results: when ΔP is close to 1, the storage system is charged. This similarity is also present in the fuzzy rules for strategies PPW1 and SPW1. They differ in the volume contained within the surface and in the values attained using SPW1, which are higher than the values for PPW1. This indicates that during period SPW1, the storage system will be charged more when the level of consumption is low (ΔP close to 1), and will be discharged less when the level of consumption is high (ΔP close to 0). This is because electricity is cheaper during SPW periods than in PPW periods, and to the fact that one of the objectives of SPW1 is to ensure the availability of energy from the storage system.

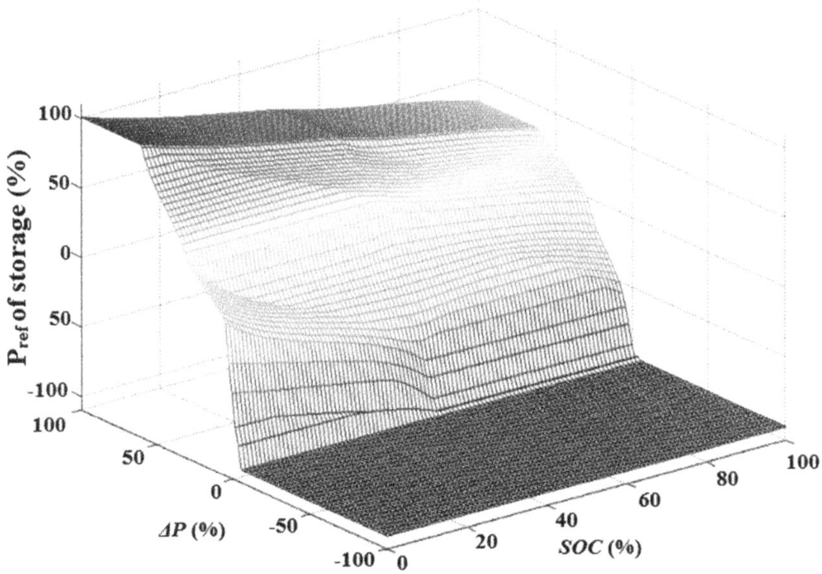

Figure 2.15. *Surface generated by strategy SPW1*

2.2.5.4. *SPW2 operating mode*

This period runs from 2000 to 0100 hours, when there is no PV production. It is followed by an off-peak period. During the off-peak period, the storage system will be charged and consumption will always be well below the subscribed power. The availability of the storage system is therefore less important during SPW2. The fuzzy rules for this period, following the objectives for N1.2.2 shown in Figure 2.8, are presented in Table 2.7.

		ΔP				
	P_{ref}	BP	MP	Z	MN	BN
SOC	B	Z	Z	Z	MN	BN
	M	Z	Z	Z	MN	BN
	Z	SP	SP	Z	MN	BN

Table 2.7. *Fuzzy rules: power reference of the storage system using strategy SPW2*

Figure 2.16 shows the surface generated by strategy SPW2. Note that in comparison with strategy SPW1 (Figure 2.15), the storage system is less charged in SPW2. It is only charged when the SOC is low (e.g. below 50%), because the availability of stored power is not a main requirement for this period, given that it is followed by an off-peak period.

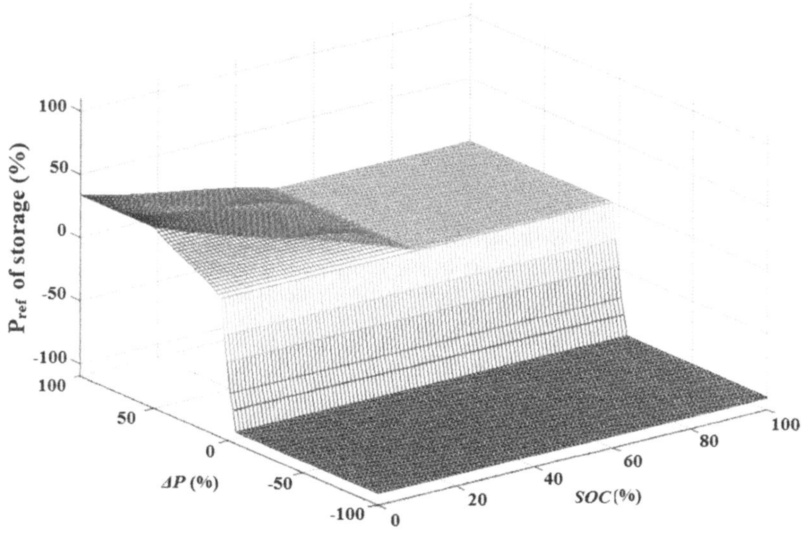

Figure 2.16. *Surface generated by strategy SPW2*

2.2.5.5. *OPW1 operating mode*

OPW1 is one of the off-peak periods, running from 0100 to 0700 hours, Monday to Friday. During this period, the annual premium and the price of electricity are cheaper than at other times. The storage system may be charged during this period due to the low cost of electricity. The fuzzy rules for this period, following the objectives for N1.1.1 shown in Figure 2.9, are presented in Table 2.8.

		ΔP				
	P_{ref}	BP	MP	Z	MN	BN
SOC	B	BP	BP	Z	BN	BN
	M	BP	BP	MP	MN	BN
	Z	BP	BP	BP	MN	BN

Table 2.8. *Fuzzy rules: power reference of the storage system using strategy OPW1*

Figure 2.17 shows the surface generated by strategy OPW1. Note that in most cases, when ΔP is positive, the storage system is charged to maximum, except when ΔP is close to 0 (the power supplied by the grid is close to the subscribed power P_s), as the main objective for this period is to charge the storage system to ensure that energy will be available during the following period.

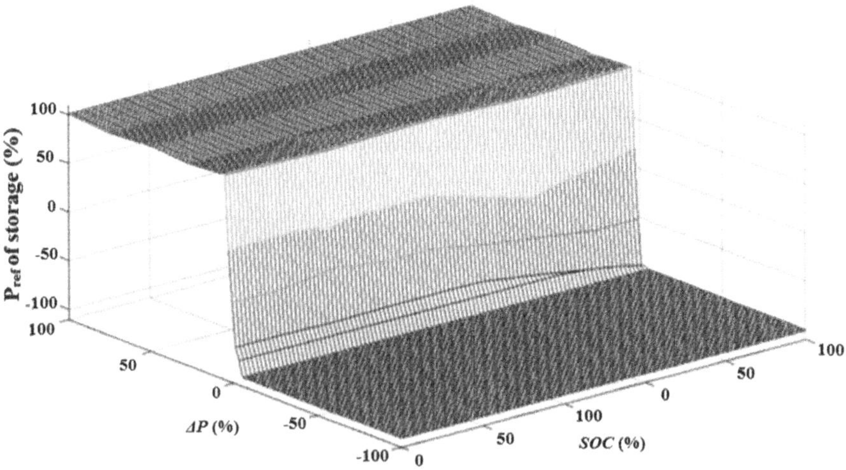

Figure 2.17. *Surface generated by strategy OPW1*

2.2.5.6. *OPW2 operating mode*

This period is rather different, covering Sundays or public holidays when the supermarket is closed. During this time, the level of consumption by the supermarket is low. PV production during the day is likely to exceed consumption. In this case, if the storage system is full, excess power will be sent to the grid. As we stated in section 2.2.4, photovoltaic production should be consumed by loads or stored as a priority. Thus, if the storage system is full, it must be partially discharged before sunrise. The storage system will then have sufficient room to store energy during the day. At the end of the day, if the storage system is not fully charged, it may be charged from the grid during the night in preparation for the following working day.

During this period, SOC is no longer the variable for consideration. According to the objectives of N.1.1.2 (Figure 2.9), this strategy can be expressed as a simple input–output relationship. This relationship is shown in Figure 2.18.

Using this strategy, the storage system is discharged most of the time. It will only be charged when ΔP is close to 100% (PV production is close to consumption). In this case, it is better to charge the storage system to prevent PV production from being sent to the network: consumption or storage of this energy takes priority.

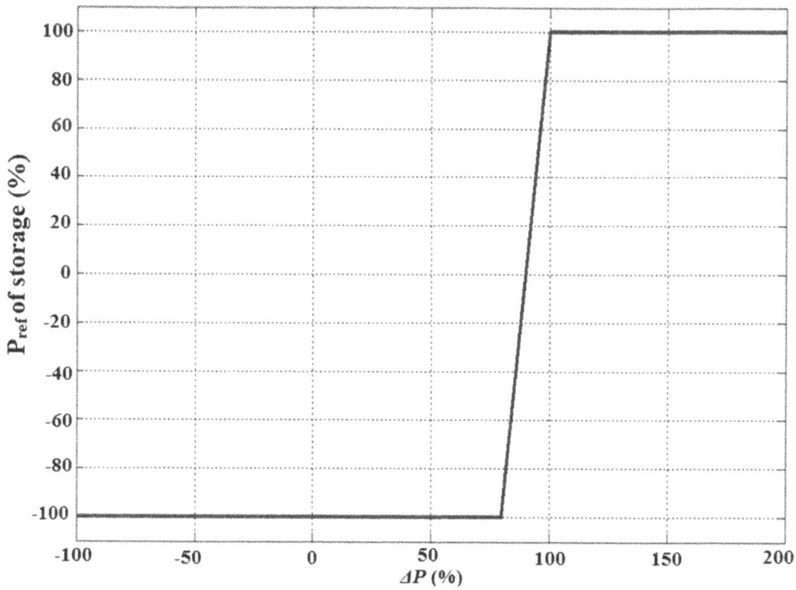

Figure 2.18. *Graph of strategy OPW2*

2.2.6. *Simulation*

For our digital simulation, we considered a supermarket with a surface of 13058 m^2 and a PV system with a peak power rating of 1.16 MW. The storage system has a capacity of 1700 kWh and a maximum power of 400 kW.

Figure 2.19 shows simulation results for this supermarket over the course of a week. These results were obtained using data for the month of January when the pricing period system is at its most complex. In the first figure, the continuous line shows subscribed power and the dashed line shows power supplied by the grid. In the second figure, the continuous line shows load power and the dashed line represents PV production. In the third figure, the continuous line shows the state of charge of the storage system.

Note that, for this week in January, PV production was lower than consumption from Monday to Saturday, which *a priori* indicates 100% self-consumption. On Sundays, however, PV production was greater than consumption. Simulations for other seasons were carried out using different PV and consumption figures. During the summer, PV production is often greater than consumption.

From Figure 2.19, we note that the supplied power does not exceed the subscribed power, even when the load power exceeded this subscribed power. This is made possible by the use of the storage system.

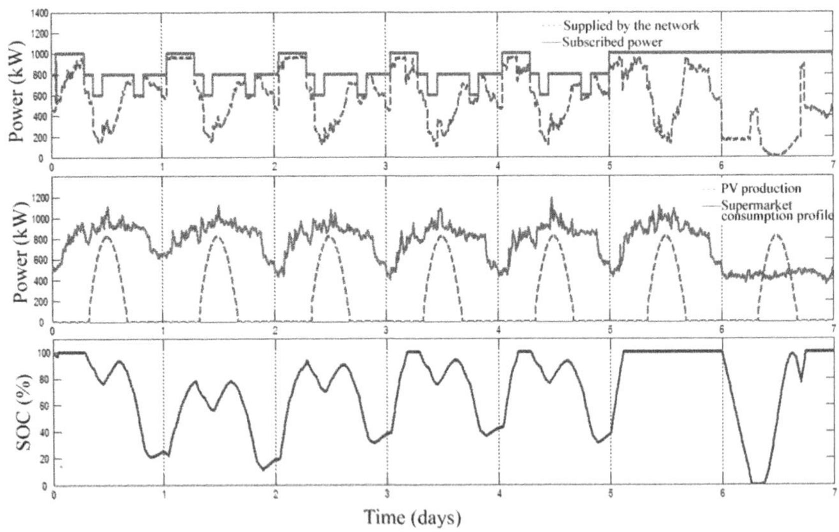

Figure 2.19. *Simulation results for a week in January*

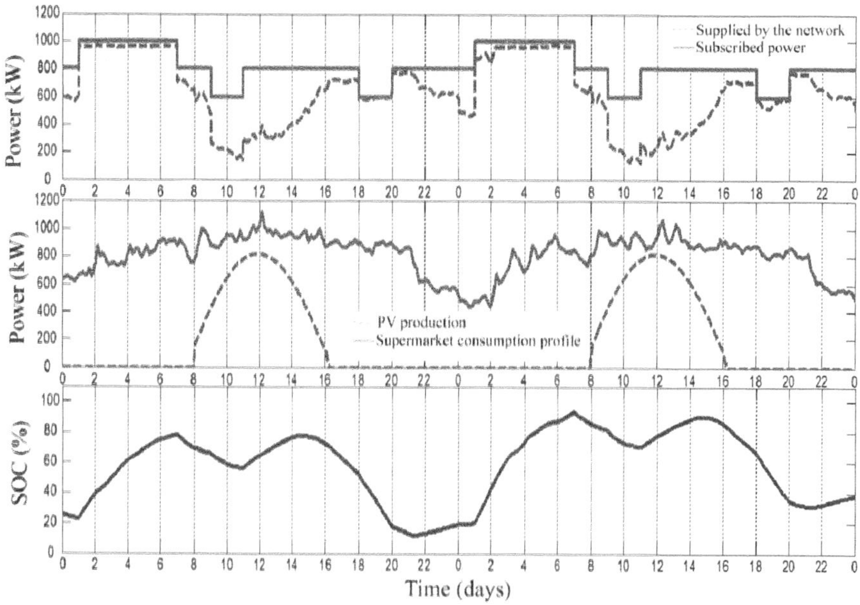

Figure 2.20. *Zoom on simulation results for two working days (days 2 and 3)*

Figure 2.20 shows a closer view of the simulation results for two working days. We see that:

– during the periods from 0700 to 0900 hours and 1600 to 1800 hours, the storage system slowly discharges to ensure that consumed power does not exceed subscribed power;

– during the peak period from 0900 to 1100 hours, PV production is used to power the load, and the storage system is slightly discharged;

– during the shoulder period from 1100 to 1300 hours, PV production is consumed by the load or stored as a priority;

– during the peak period from 1800 to 2000 hours, there is no further PV production, and the storage system is discharged at maximum power (steeper slope in SOC) as the price of electricity is highest during this period;

– during the period from 2000 to 2200 hours, the storage system is slightly discharged to ensure that consumed power does not exceed subscribed power;

– during the period from 2200 to 0100 hours, the storage system is slightly recharged as ΔP is high and the SOC is low;

– during the period from 0100 to 0700 hours, the storage system is charged at maximum power as the price of electricity is lowest during this period.

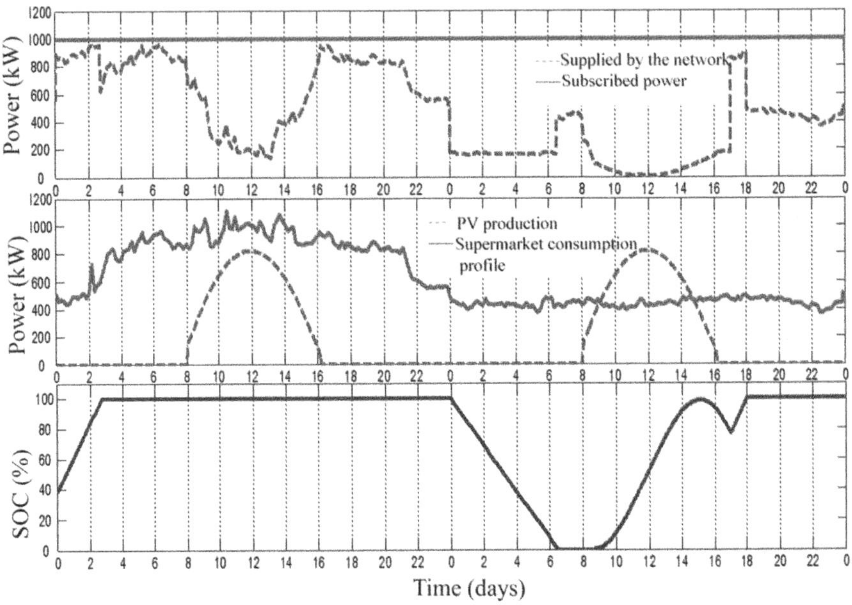

Figure 2.21. *Zoom on simulation results for the two weekend days (days 6 and 7)*

Figure 2.21 shows a closer view of the simulation results for two weekend days, including one day when the supermarket is closed. We see that:

– during the period from midnight to 0300 hours on Saturday, the storage system is charged, but then is not used for the rest of the day;

– during the period from midnight to 0700 hours on Sunday, the storage system is completely discharged in order to ensure that storage is available to absorb PV production during the day;

– during the period from 0900 to 1500 hours, when PV production is greater than consumption (the dashed line is higher than the continuous line), the storage system begins to store excess energy production until PV production falls below consumption (the dashed line falls below the continuous line);

– from 1700 hours onward, after the sun has set, strategy OPW1 is applied. The storage system is charged until full.

2.2.7. *Performance analysis using indicators*

In this section, we will analyze our simulation results using economic and ecological criteria to evaluate conformity with the objectives defined in Table 2.1. The economic criteria used are the annual premium (Pp in Figure 2.4.) and the cost of electricity consumption for a week (cost in kWh in Figure 2.4). The ecological criterion used is CO_2 emissions, estimated using production data supplied by the operator of the French transportation network (RTE France).

Table 2.9 shows the values for these indicators obtained from our simulation results for different electrical configurations.

Case 1 represents a configuration with no storage system or PV system. The subscribed power is 1200 kW for all pricing periods.

Case 2 represents a configuration with storage, PV system and the associated supervision system. The subscribed power is as follows: OPW 1000 kW, SPW 800 kW and PPW 600 kW. The addition of storage and a PV system allows a reduction in subscribed power, with a significant reduction in the subscription cost paid to the electricity supplier and smaller reduction in the cost of electrical energy consumption.

	Annual premium (€)	**Consumption for a week (€)**	**CO_2 emissions for a week (T)**
Case 1	79344	8782	11.604
Case 2	54747	6066	8.541
Difference	–24597 (–31.00%)	–2716 (–30.93%)	–3.063 (–26.40%)

Table 2.9. *Comparison of results with or without storage and PV systems*

As shown in Table 2.9, the use of storage and PV systems results in a significant reduction in electricity bills and CO_2 emissions between case 1 and case 2. The annual premium falls by 21594 € (31%) over the course of a year, and the electricity bill falls by 2716 € (30.93%) per week. The CO_2 emissions from energy production fall by 3.063 metric tons (26.04%) per week.

This estimation of CO_2 emissions produced by the French transportation network operator RTE France is based on overall production [RTE 17a] and an estimation of CO_2 emissions for each type of production [RTE 17b]. Figure 2.22 shows a comparison of CO_2 emissions over a period of one week. The dashed line shows CO_2 emissions for case 1 and the continuous line represents CO_2 emissions for case 2.

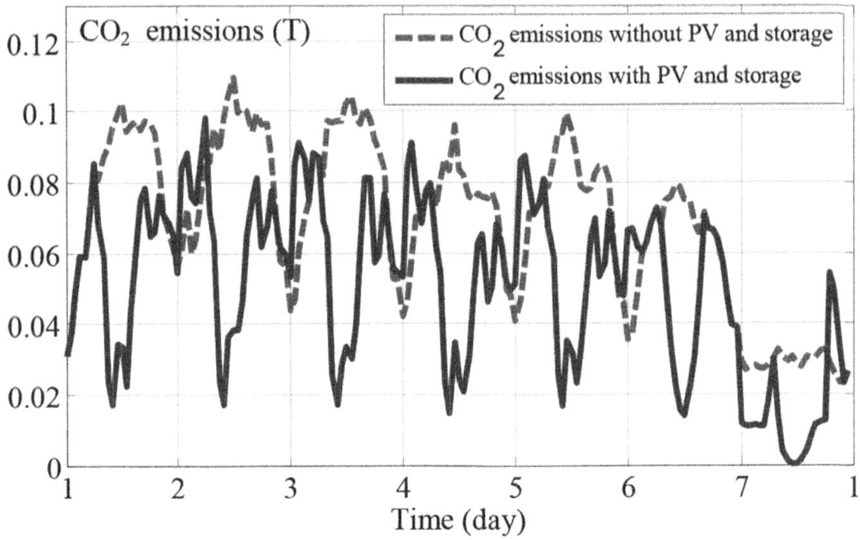

Figure 2.22. *Comparison of CO_2 emissions resulting from energy production*

We see that during the daytime, the continuous line is lower than the dashed line. During this period, PV production permits a reduction in the power supplied by the network and in CO_2 emissions generated by thermal electricity production. During the night, the continuous line is higher than the dashed line. During this period, the storage system is charged using power from the grid, meaning that CO_2 emissions are higher due to increased use of conventional energy sources. However, this increase is more than compensated by the reduction in daytime CO_2 emissions. Overall, CO_2 emissions are lower in case 2 than in case 1. This highlights the compromise between economic and environmental objectives attained through the use of a supervision strategy.

As CO_2 emissions fluctuate considerably according to the season, consumption and production type, we used CO_2 emissions data for a whole year in order to evaluate the impact of the energy management strategy. Table 2.10 shows the differing CO_2 emissions for our two cases. Note that CO_2 emissions are significantly lower in case 2, with a reduction of 136.2 T (52.38%) in CO_2 emissions.

We should note that the calculation of CO_2 emissions is based on French data. Much of the electricity produced in France comes from nuclear and hydraulic sources, which emit very little CO_2. The unit of CO_2 emissions per MWh is thus relatively low compared to other European countries. The reduction in CO_2 emissions would be higher if we considered data for the whole of Europe or other countries.

	CO_2 emissions per year (T)
Case 1	260
Case 2	123.8
Difference	–136.2 (–52.38%)

Table 2.10. *Comparison of CO_2 emissions with and without PV and storage systems over a year*

2.3. Conclusion

Energy management in a commercial building including PV production and a storage system is a complex problem, due to variations in the price of electricity, different types of consumption, types of purchase contract, etc. Energy management must take account of these constraints and permit the simultaneous attainment of multiple objectives. A storage system may be constructed using a variety of equipment, such as electrochemical batteries, negative cold chambers (refrigeration), etc. The power indications given by the supervision system not only control the storage system, but also provide a reference for scheduling the use of certain loads.

Our supervision strategy based on fuzzy logic is a valuable tool for solving problems of this type. This strategy enables systems to be controlled in a way that responds to both economic and ecological aims.

A graphical method was used in designing the supervision strategy. This supervision strategy was then evaluated using economic and ecological indicators. Energy bills and CO_2 emissions were reduced by the application of the proposed solution.

This study does not take account of investment and maintenance costs, nor of the aging of new apparatus (storage and PV). In order for this type of solution to be profitable, the economic gains must at least cover costs over the lifetime of the equipment.

2.4. Acknowledgments

The authors wish to thank the ADEME (*Agence française de l'Environnement et de la Maîtrise de l'Energie* – French Environment and Energy Management Agency) and the Nord – Pas de Calais region for their financial backing for the GISEP project (*Gestion Intelligente des Sources d'Energie électrique intégrant du Photovoltaïque des charges de bâtiments commerciaux et des procédés de stockage d'énergie* – Smart Management for Electrical Energy Including Photovoltaic Sources in Commercial Buildings and Energy Storage Procedures), in which context the case study presented in this chapter was developed.

3

Energy Storage in a Tertiary Building, Combining Photovoltaic Panels and LED Lighting

3.1. Introduction

According to figures published by the *Agence De l'Environnement et de la Maîtrise de l'Energie* (ADEME, Environment and Energy Mastery Agency) in 2017 [AFE 18], worldwide energy consumption for lighting was 2700 TWh, representing approximately 19% of total worldwide electricity consumption in 2009. Consumption for lighting in France was 56 TWh, approximately 12% of total electricity consumption. Of this, 37 TWh was consumed by non-residential buildings (commercial buildings, offices, etc.). Lighting for buildings in this category thus accounts for approximately 66% of electrical consumption for lighting in France.

Lighting technologies have progressed considerably in recent years, aiming to reduce consumption while maintaining lighting levels. LEDs (Light Emitting Diodes) offer the most promising solutions in terms of energy efficiency. These products are now readily available and increasingly popular, due to the quality of lighting which they offer (with adjustable warmth and color), ease of integration in buildings (notably in confined spaces, where their low level of thermal dissipation is particularly valuable) and their price, which is now competitive. The long-life expectancy of these elements, counted in tens of thousands of hours, also contributes to offering increased value for money. Public policies encouraging the use of energy-efficient devices in order to reduce overall electricity consumption have contributed to the rapid implementation of LED-based lighting solutions.

In parallel, energy transition policies have led to an increase in local electricity production, near to the site of use, for example with the installation of photovoltaic systems in buildings. This raises issues with regard to the intermittent nature of this production; electricity production needs to equal consumption at all times in order to ensure the correct operation of the grid to which these buildings are connected, and to guarantee user satisfaction. Electrical storage systems may be used to compensate for the high variability and low predictability of photovoltaic production, attenuating these constraints. Electrochemical batteries are currently the preferred solution.

The three technologies mentioned above – LED lighting, photovoltaic production and batteries – all function, *a priori*, using direct current (DC). By grouping these devices into a dedicated DC electrical network (at least initially), it becomes possible to associate intermittent production, intermittent loads and storage, while minimizing the number of electronic conversion processes. By maintaining a connection with the AC distribution grid, the system offers the possibility of supplying electrical power to the AC grid itself, in addition to guaranteeing electric power for lighting systems (Figure 3.1).

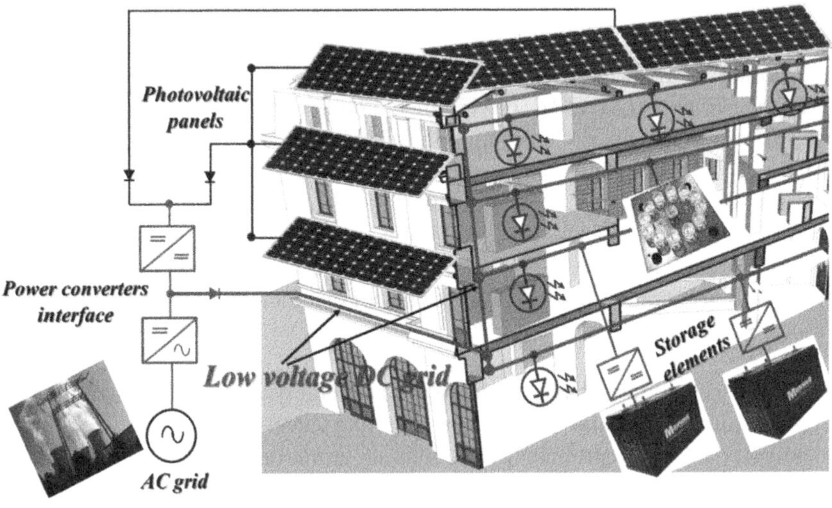

Figure 3.1. *Association of three technologies in a building: PV, batteries and LEDs. For a color version of the figures in this chapter see www.iste.co.uk/robyns/buildings.zip*

Evidently, a real-time energy management system is needed in order for this assembly to satisfy lighting requirements and respond to the needs of the electric grid.

In this chapter, we will describe an energy management strategy based on fuzzy logic [ROB 13a], applied to a DC network architecture described below [KAD 13].

3.2. DC network architecture

Figure 3.2 shows the proposed architecture connecting various power devices. A DC bus, with a minimum variable voltage of 200V, is used to connect photovoltaic panels (PV), LED lighting systems running on 48VDC and a battery storage system running on 48VDC, while maintaining a connection to the AC electrical network.

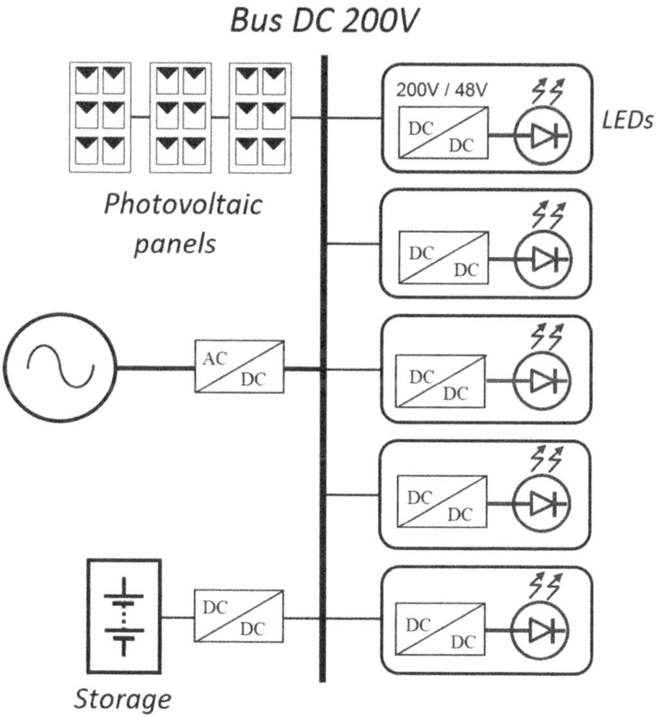

Figure 3.2. *Connection architecture*

The connection between the PV system and the DC bus is direct, with no DC/DC converter; in this, the architecture differs from the solutions which are usually envisaged. The aim here is to reduce conversion steps. The DC/DC converter usually used allows MPPT (Maximum Power Point Tracking) control, which is a means of extracting maximum power from panels as a function of solar radiation. In the absence of this device, the same function may be carried out by the AC/DC converter which connects the DC bus to the AC electrical grid; the voltage of the DC bus is set at a value which ensures maximum power to be extracted from the PV panels as a whole. The choice of voltage for the DC bus may have implications, including either the addition of a 50Hz transformer or of an AC/DC converter architecture with integrated transformer (this solution is interesting in terms of protection equipment [HAS 14]). LEDs powered using 48VDC must be connected to the DC bus via a unidirectional power DC/DC interface. Finally, the storage system, made up of electrochemical batteries with a nominal voltage of 48VDC, is also connected via a DC/DC converter, with, in this case, bidirectional power.

3.3. Energy management

Real-time energy management of the system is based on the first six steps of the methodology set out in Chapter 1 (section 1.8), using fuzzy logic.

These steps are:

1) determination of a system specification, with objectives, constraints and resources clearly laid out;

2) determination of a supervisor structure;

3) determination of a functional graph, presenting and organizing the different operating modes for the system;

4) determination of membership functions for input and output variables, enabling the translation of system variables into fuzzy variables;

5) determination of the operating graph, used to obtain the fuzzy algorithm introduced into the computer processor;

6) creation of fuzzy rules based on these operating graphs.

3.3.1. *Specification*

3.3.1.1. *Objectives*

The objectives of the system are:

– to ensure maximum power extraction from the PV panels (MPPT function);

– to maximize local use of energy from the PV generator, prioritizing use over injection into the electric grid, promoting self-consumption of the electricity produced by the building: this should reduce the consumption of power from the grid and consequently the bill for electricity supplied by the distributor. It should also reduce the equivalent CO_2 per kWh balance drawn from the network;

– to ensure the availability of energy in the storage system, ensuring the quality of the energy supply, and reduce the use of grid power as much as possible during peak consumption periods.

3.3.1.2. *Constraints*

The principal constraints are:

– maximum charge and discharge power of the batteries;

– the state of charge of the batteries, which has both minimum and maximum limitations, determined by the dimensions of the storage system;

– the power generated by the PV system and that consumed by the LED lighting system, which are highly variable and difficult to forecast, increasing the complexity of energy management for the system.

3.3.1.3. *Means of action*

The means of action are the reference power exchanged with the grid ($P_{\_GRID\_ref}$) and the reference power of the storage system ($P_{\_STO\_ref}$), as shown in Figure 3.3.

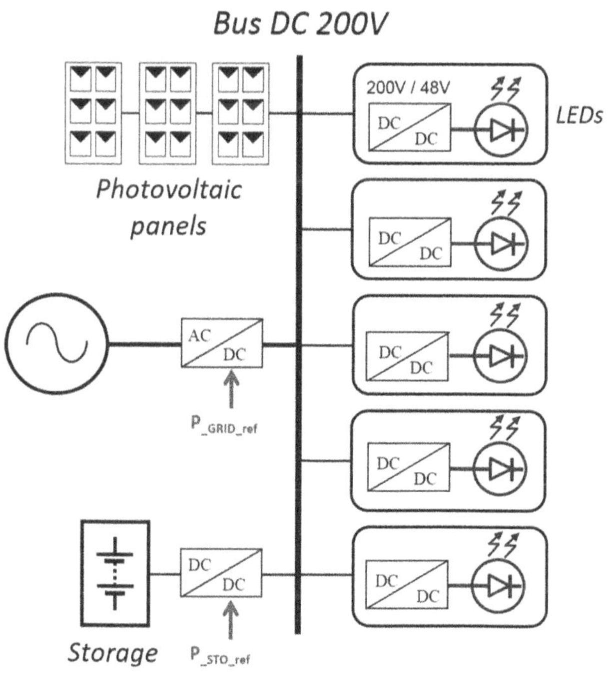

Figure 3.3. *Connection architecture and means of action ($P_{\_GRID\_ref}$ and $P_{\_STO\_ref}$)*

3.3.2. *System inputs/outputs*

In this step, certain system inputs are identified in relation to the objectives mentioned above. First, in order to fulfill the MPPT function, the error (ΔV) between the measured voltage of the DC bus V_{bus} and the reference voltage V_{MPP} supplied by the MPPT algorithm as a function of environmental conditions must be entered into the system to allow maximization of the point of operation of PV panels, close to the MPP (Maximum Power Point).

The second input is the SOC (State Of Charge) of the storage device, which is required to ensure the availability of energy in the storage system. This input is directly obtained from the battery management system or via a model based on battery power integration and its capacity.

Finally, in cases where the SOC is low, the supervisor will order electricity to be drawn from the grid. Two other input values are therefore needed: the equivalent CO_2 per kWh for electricity from the grid (*CO2*) and the price of electricity (*Ep*). *CO2* variable data for periods of a few minutes may be retrieved from specialist websites (such as *ECO$_2$mix*, for France, which uses 15 min time steps [RTE 18]). The evolution of electricity prices over time, *Ep*, depends on the pricing contract. In the long term, it seems reasonable to suggest that consumers may be able to access this data at regular time intervals in order to take account of the frequent price variations which occur over the course of a day. These values are used to define grid access conditions (GAC), economic indicators (electricity prices) and environmental indicators (equivalent CO_2 per kWh drawn from the network). GAC values are normalized to give a figure between 0 and 1 (0 for the least favorable grid access conditions, 1 for the most favorable conditions).

An operating diagram for the supervisor is shown in Figure 3.4. The part of the system constructed using fuzzy logic should offer the best compromise regarding the reference power of the storage system, $P_{\_STO\_ref}$, the desired objectives and the existing constraints.

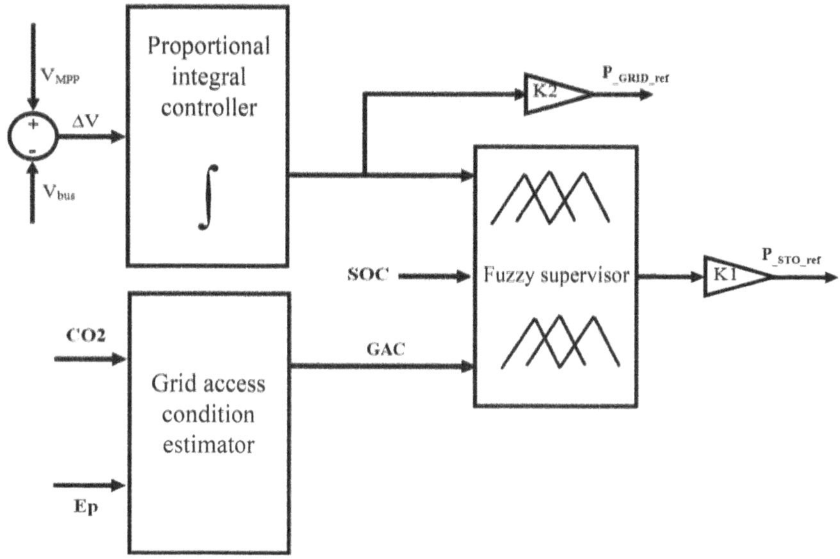

Figure 3.4. *Supervisor structure*

3.3.3. *Functional graph*

A functional graph can be defined based on knowledge of the system (Figure 3.5). Operating modes are represented by rectangles with rounded corners, and system states are represented by transitions. As the figure shows, the supervision system is broken down into three main modes of operation, depending on the state of charge of the storage system: Small (N3), Medium (N1) and Big (N2). For each operating mode, an initial test is then carried out to determine the sign of the power exchanged with the grid. Power at the point of connection to the grid is positive when it is drawn from the grid (levels N1.2, N2.2 and N3.2) and negative when electricity is being injected into the grid (N1.3, N2.3 and N3.3). Levels N1.1, N2.1 and N3.1 correspond to situations in which very low amounts of power are being exchanged with the grid in either direction. A final test is carried out to determine grid access conditions, GAC: Small denotes poor access conditions, and Big denotes the contrary. A low GAC value corresponds, for example, to periods where the price per kWh of electricity drawn from the network is high, and the equivalent CO_2 for this quantity of energy is also high. A high GAC value indicates favorable access conditions.

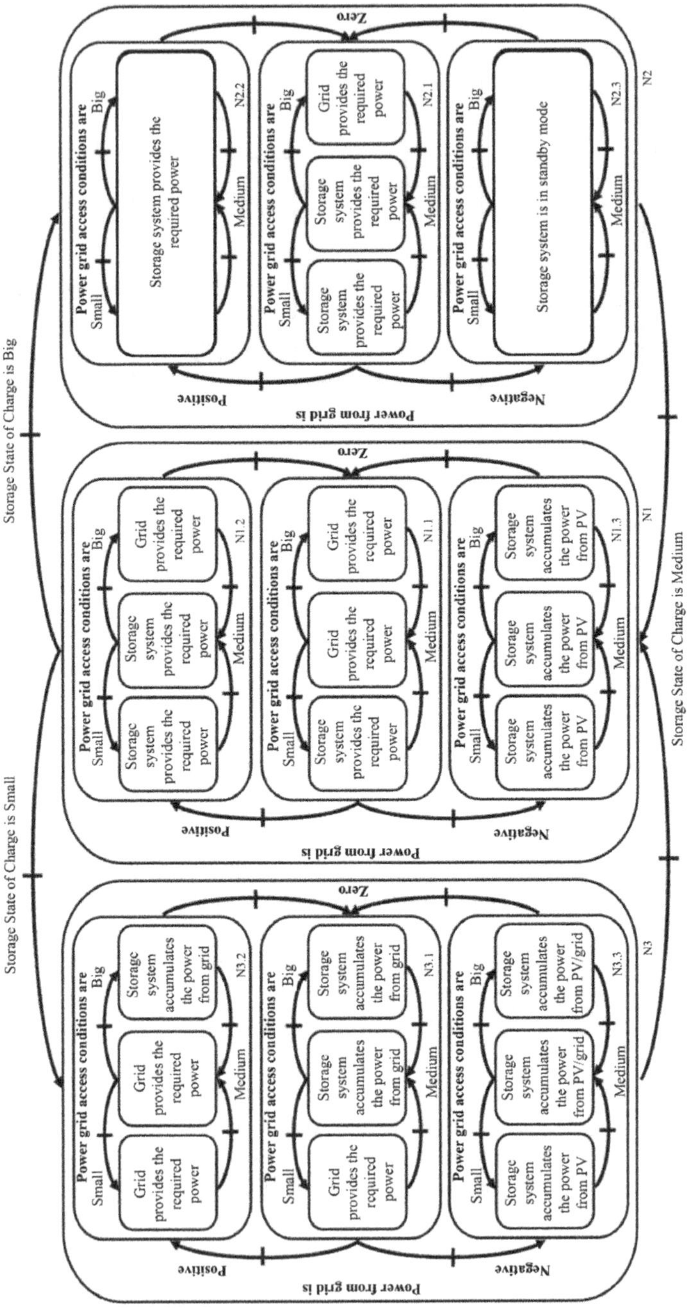

Figure 3.5. *Functional graph*

To illustrate this notion, consider level N3.2. This corresponds to a situation where the storage state of charge is Small, and where electrical power is being drawn from the network (power from grid is Positive). The behavior of the storage system will thus depend on the GAC: if these are good (Big), then additional power will be drawn from the grid in order to progressively recharge the storage system. If conditions are poor (Small), however, the power drawn from the grid will remain unchanged.

3.3.4. *Determination of membership functions*

The purpose of this step is to define membership functions, denoted by μ, for the input and output variables of the fuzzy supervisor. In this study, these functions will be defined empirically. Following the functional graph, the membership functions for each input and output value, in normalized form, are shown in Figures 3.6–3.9:

– the state of charge (*SOC*) is *Small (S)*, *Medium (M)* and *Big (B)* – Figure 3.6;

– the grid access conditions (*GAC*) are *Small (S)*, *Big (B)* or *Medium (M)* – Figure 3.7;

– the grid power (P_{GRID}) is *Positive (P)* when power is drawn from the grid and *Negative (N)* when power is being injected back into the network. When this value is low, whatever the sign, the label assigned is *Zero (Z)* – Figure 3.8;

– the storage reference power ($P_{\_STO\_ref}$) is *Positive (P)* when the storage is being used to supply energy, *Negative (N)* when the storage system is accumulating energy and *Zero (Z)* when the value is very low, whatever the sign – Figure 3.9.

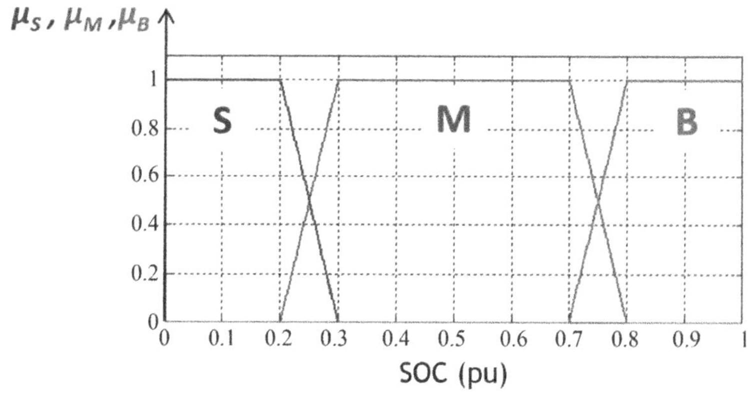

Figure 3.6. *Membership function of the SOC variable (state of charge of storage)*

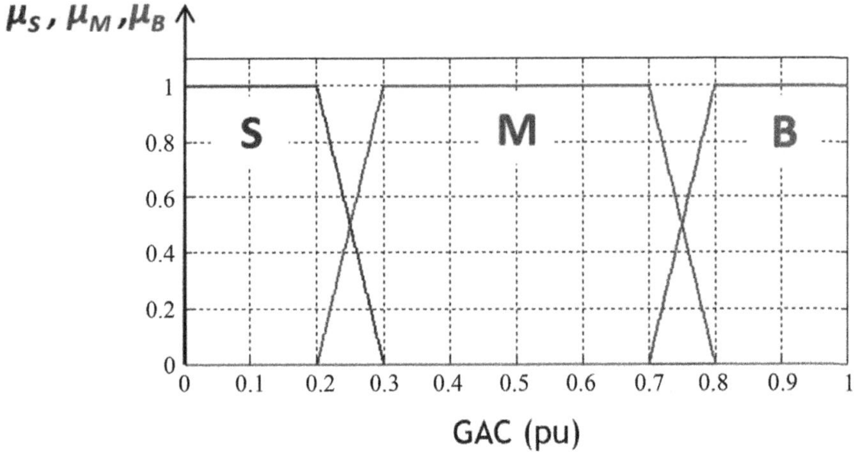

Figure 3.7. *Membership function of the GAC variable (grid access conditions)*

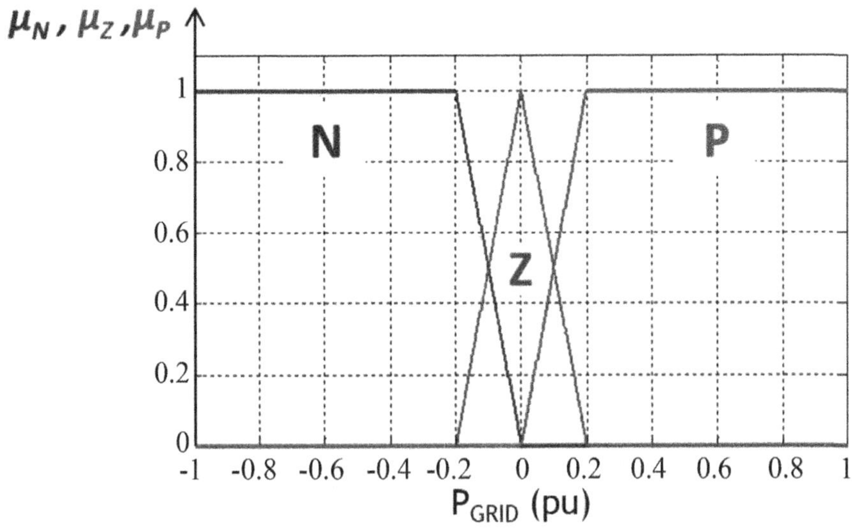

Figure 3.8. *Membership function of the P_{GRID} variable (electrical power exchanged with the electric grid)*

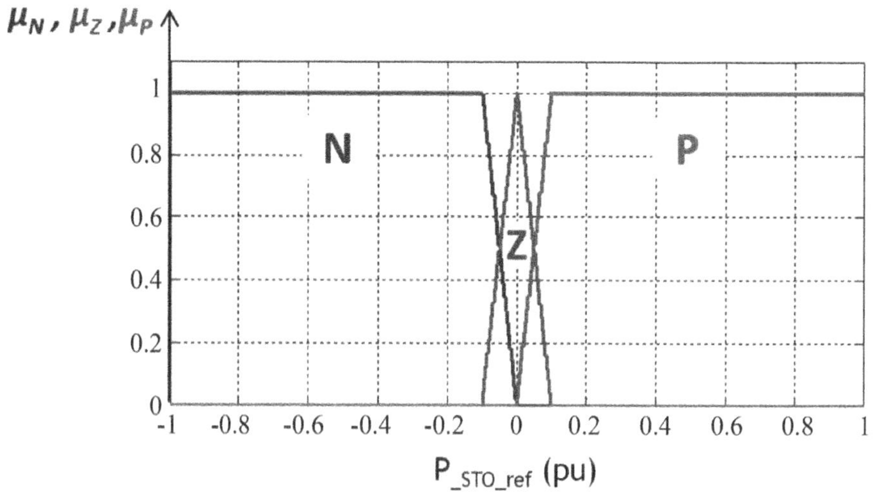

Figure 3.9. *Membership function of the $P_{\_STO\_ref}$ variable (power reference applied to the storage system)*

3.3.5. *Operational graph*

This graph results from the application of the membership functions from Figures 3.6 to 3.9 to the functional graph in Figure 3.5. It features the three test conditions (SOC, the power exchanged with the grid P_{GRID} and GAC) giving access to different possible system states. Each block in this graph provides information on the behavior to be adopted by the storage system.

As an illustration, let us return to level N3.2. This level is attained when *SOC is S*, then P_{GRID} *is P*. According to the grid access conditions (*S, M* and *B*), the storage reference will be *Z, Z* and *N*, respectively.

3.3.6. *Fuzzy rules*

The operational graph shown in Figure 3.10 results in 23 rules (out of a maximum of 27, deduced from the 3×3×3 fuzzy subsets in Figures 3.6–3.8), which are shown in Table 3.1.

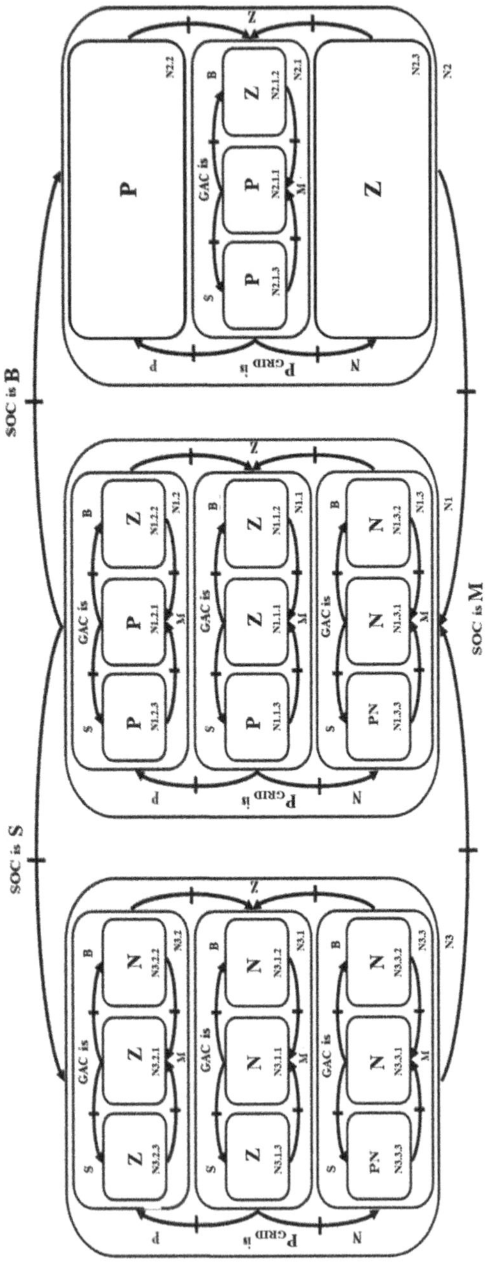

Figure 3.10. *Operational graph*

Level	Rules							
N1.1.1	IF	SOC **M**	IF	P_{GRID} **Z**	AND IF	GAC **M**	THEN	$P_{\_STO\_ref}$ **Z**
N1.1.2	IF	SOC **M**	IF	P_{GRID} **Z**	AND IF	GAC **B**	THEN	$P_{\_STO\_ref}$ **Z**
N1.1.3	IF	SOC **M**	IF	P_{GRID} **Z**	AND IF	GAC **S**	THEN	$P_{\_STO\_ref}$ **P**
N1.2.1	IF	SOC **M**	IF	P_{GRID} **P**	AND IF	GAC **M**	THEN	$P_{\_STO\_ref}$ **P**
N1.2.2	IF	SOC **M**	IF	P_{GRID} **P**	AND IF	GAC **B**	THEN	$P_{\_STO\_ref}$ **Z**
N1.2.3	IF	SOC **M**	IF	P_{GRID} **P**	AND IF	GAC **S**	THEN	$P_{\_STO\_ref}$ **P**
N1.3.1	IF	SOC **M**	IF	P_{GRID} **N**	AND IF	GAC **M**	THEN	$P_{\_STO\_ref}$ **N**
N1.3.2	IF	SOC **M**	IF	P_{GRID} **N**	AND IF	GAC **B**	THEN	$P_{\_STO\_ref}$ **N**
N1.3.3	IF	SOC **M**	IF	P_{GRID} **N**	AND IF	GAC **S**	THEN	$P_{\_STO\_ref}$ **PN**
N2.1.1	IF	SOC **B**	IF	P_{GRID} **Z**	AND IF	GAC **M**	THEN	$P_{\_STO\_ref}$ **P**
N2.1.2	IF	SOC **B**	IF	P_{GRID} **Z**	AND IF	GAC **B**	THEN	$P_{\_STO\_ref}$ **Z**
N2.1.3	IF	SOC **B**	IF	P_{GRID} **Z**	AND IF	GAC **S**	THEN	$P_{\_STO\_ref}$ **P**
N2.2	IF	SOC **B**	IF	P_{GRID} **P**	THEN			$P_{\_STO\_ref}$ **P**
N2.3	IF	SOC **B**	IF	P_{GRID} **N**	THEN			$P_{\_STO\_ref}$ **Z**
N3.1.1	IF	SOC **S**	IF	P_{GRID} **Z**	AND IF	GAC **M**	THEN	$P_{\_STO\_ref}$ **N**
N3.1.2	IF	SOC **S**	IF	P_{GRID} **Z**	AND IF	GAC **B**	THEN	$P_{\_STO\_ref}$ **N**
N3.1.3	IF	SOC **S**	IF	P_{GRID} **Z**	AND IF	GAC **S**	THEN	$P_{\_STO\_ref}$ **Z**
N3.2.1	IF	SOC **S**	IF	P_{GRID} **P**	AND IF	GAC **M**	THEN	$P_{\_STO\_ref}$ **Z**
N3.2.2	IF	SOC **S**	IF	P_{GRID} **P**	AND IF	GAC **B**	THEN	$P_{\_STO\_ref}$ **N**
N3.2.3	IF	SOC **S**	IF	P_{GRID} **P**	AND IF	GAC **S**	THEN	$P_{\_STO\_ref}$ **Z**
N3.3.1	IF	SOC **S**	IF	P_{GRID} **N**	AND IF	GAC **M**	THEN	$P_{\_STO\_ref}$ **N**
N3.3.2	IF	SOC **S**	IF	P_{GRID} **N**	AND IF	GAC **B**	THEN	$P_{\_STO\_ref}$ **N**
N3.3.3	IF	SOC **S**	IF	P_{GRID} **N**	AND IF	GAC **S**	THEN	$P_{\_STO\_ref}$ **PN**

Table 3.1. *Fuzzy rules for the supervisor*

3.4. Simulation results

The system presented in Figure 3.2 was modeled in order to test the proposed fuzzy supervisor by means of simulations.

The model of the photovoltaic generator was constructed using characteristics for polycrystalline PV modules. The LED lighting power was modeled using a simplified trapezoid load profile. A duration of 2 hours was selected for the increasing and decreasing slopes (from 0600 to 0800 hours and from 1800 to 2000 hours respectively) between 0W and 3000W. Between 0800 and 1800 hours, the power consumption for lighting is presumed constant and equal to 3000W. The power converters are presumed to be perfect, with 100% efficiency.

The capacity of the storage system is 120 Ah and the maximum power of the system is 5000 W.

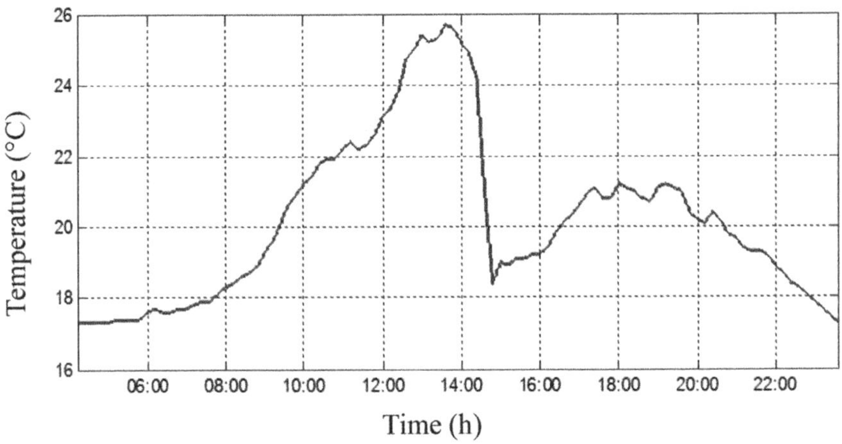

Figure 3.11. *Graph of ambient temperature*

As shown in Figure 3.4, the power reference for the AC/DC converter connecting the whole system to the electrical grid (Figure 3.3) is determined using a proportional-integral controller and is intended to regulate the voltage of the main DC bus to ensure maximum overall power extraction from the photovoltaic panels. The evolution of this voltage is shown in Figure 3.12.

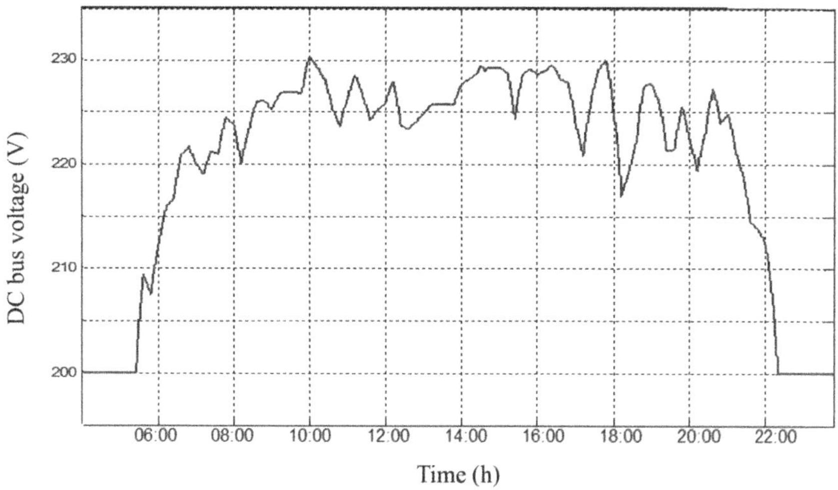

Figure 3.12. *DC bus voltage*

The behavior of the fuzzy supervisor will now be studied for three different levels of grid access conditions: purely favorable (low economic cost and equivalent CO_2 per kWh), purely unfavorable (high economic cost and equivalent CO_2 per kWh) and variable (intermediate economic cost and equivalent CO_2 per kWh).

In each case, the solar radiation and temperature conditions are shown in Figures 3.10 and 3.11. The voltage of the DC bus is shown in Figure 3.12.

The consumption curve for LED lighting has been simplified in order to allow us to analyze the performance of the supervisor. The maximum power consumed is 3kW. Daily energy consumption is 36kWh.

The initial state of charge of the storage system is 50%. We will observe its evolution over the course of 24 hours, along with different powers:

– P_{GRID}: power exchanged with the distribution grid, following the conventions below:

- positive: power drawn from the grid,

- negative: power supplied to the grid;

– P_{LED}: power consumed by LED lighting;

– P_{PV}: power supplied by photovoltaic panels;

– P_{STO}: power exchanged with the storage system, which may be:

- positive: power supplied by the storage system,

- negative: power absorbed by the storage system.

3.4.1. *Case 1: favorable grid access conditions (GAC)*

In this first test case, the GAC (cost per kWh, equivalent CO_2) are presumed to be as favorable as possible. This gives us the following values, taken from Figure 3.7: $\mu_S = 0$, $\mu_M = 0$, $\mu_B = 1$.

Figure 3.13 shows the different powers, while Figure 3.14 shows the evolution of the state of charge (SOC) of the storage system.

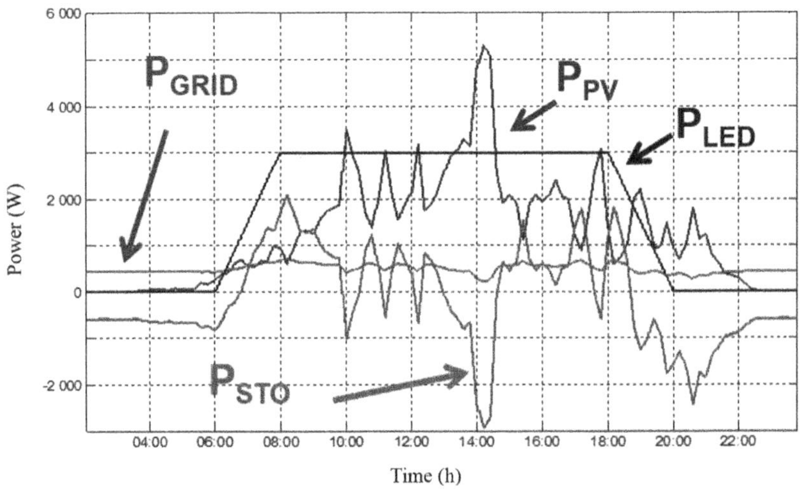

Figure 3.13. *Power profiles with favorable GAC*

First, note the complementarity of the three powers P_{PV}, P_{STO} and P_{GRID} in satisfying the electric load for lighting P_{LED}.

We see that the grid is used in a regular and moderate manner (averaging between 500W and 600W); the additional power required for the LED lighting system is obtained from PV production and storage. Figure 3.14 indicates that storage is refilled during peaks in PV production or at the end of the day, and in given lighting conditions will complete the cycle at a higher SOC than that observed at the beginning of the day. The grid is thus used to power lighting and to rebuild storage.

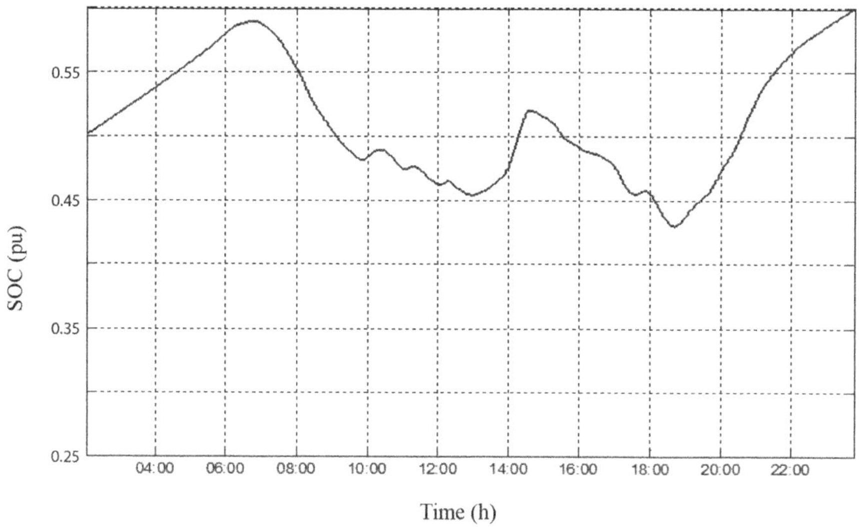

Figure 3.14. *SOC with favorable GAC*

3.4.2. *Case 2: unfavorable GACs*

For this second test case, the network access conditions (cost per kWh, equivalent CO_2) are presumed to be as unfavorable as possible. This is represented by values $\mu_S = 1$, $\mu_M = 0$ and $\mu_B = 0$ from Figure 3.7.

Figure 3.15 shows the different powers, while Figure 3.16 shows the evolution of the state of charge (SOC) of the storage system.

When the lighting load begins to consume power, this is initially supplied by the storage system, the photovoltaic generator progressively begins to contribute. This results in discharge of the storage system. When the SOC reaches approximately 30% (point (1) – Figures 3.15 and 3.16), the distribution grid comes into play to supplement PV production and power from storage. The storage system recharges slightly during peaks in solar radiation (points (2) and (3) – Figure 3.16), then during the period when consumption decreases but PV production is still ongoing. As electricity from the grid is used as little as possible due to poor GAC, the storage system struggles to rebuild its energy reserve at the end of the cycle. This situation would evidently not be tenable in the long term, given similar operating conditions and the same system dimensions.

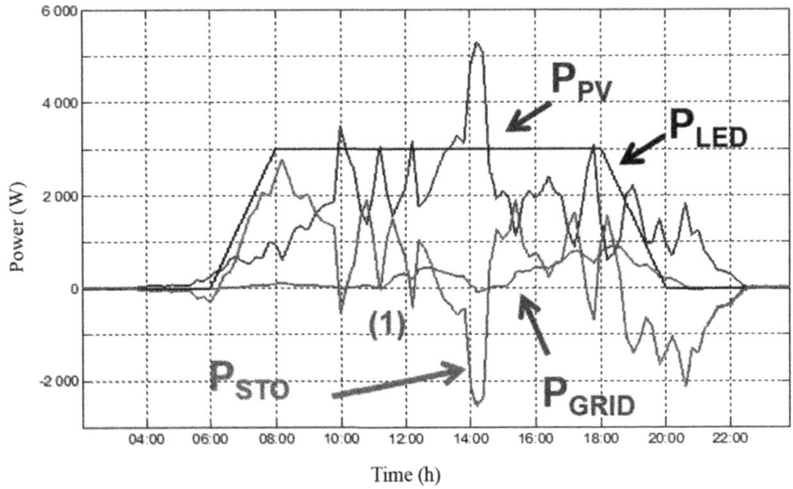

Figure 3.15. *Power profiles for unfavorable GAC*

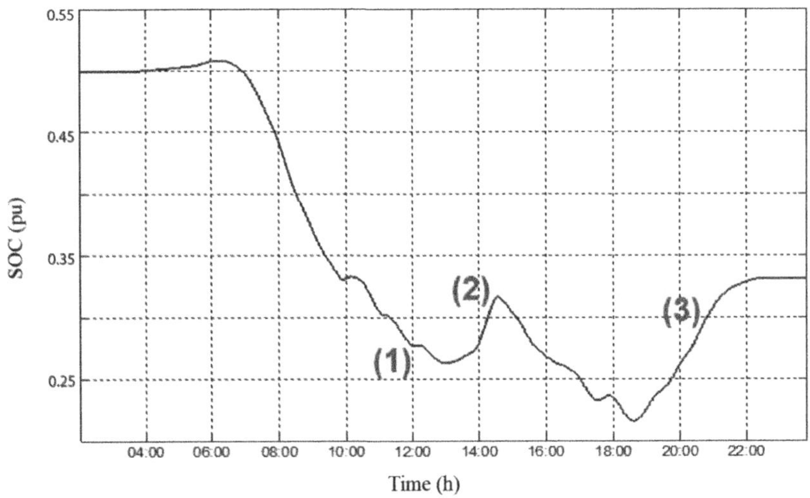

Figure 3.16. *SOC for unfavorable GAC*

3.4.3. *Case 3: variable GAC*

As shown in Figure 3.4, GAC (grid access conditions) are established based on the equivalent CO_2 for each kWh of electricity absorbed from the distribution grid (denoted by CO_2) and on the price of electricity (denoted by *Ep*). These two values are shown in Figures 3.17 and 3.18 respectively. The quantity of CO_2, in g/kWh (Figure 3.17), evolves in 15 min steps. This data may be obtained from dedicated website. Pricing data is dependent on the chosen subscription; in this case, the contract includes 8 hours of off-peak and 16 hours of more expensive peak rates. The evolution of the GAC variable is shown in Figure 3.19.

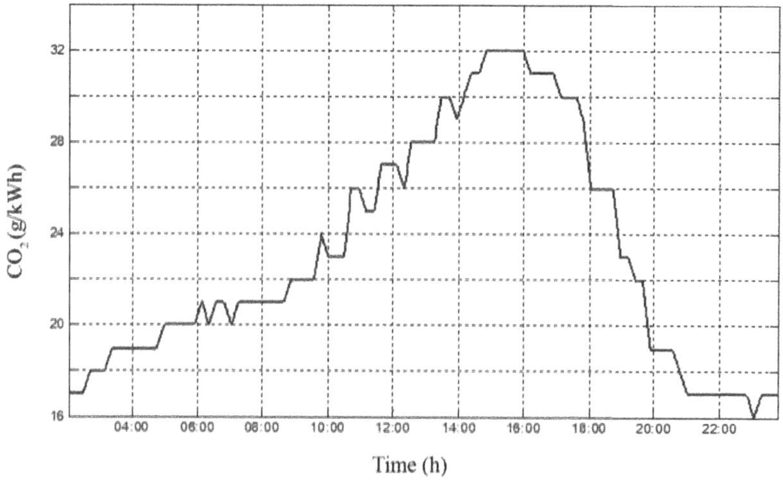

Figure 3.17. *Evolution of the equivalent C0₂ per kWh drawn from the grid*

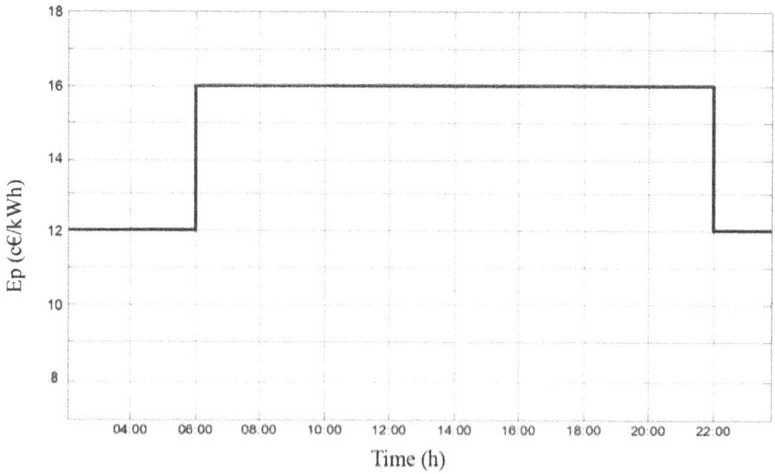

Figure 3.18. *Evolution of the price of electricity drawn from the network, Ep*

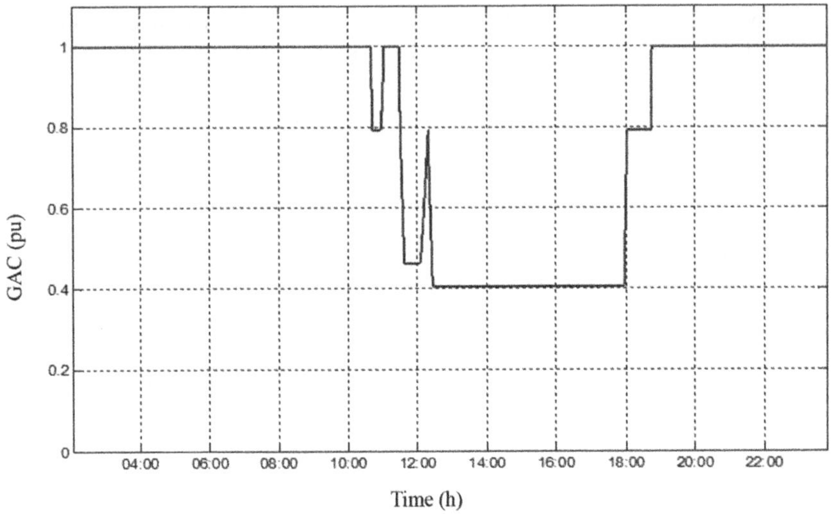

Figure 3.19. *Evolution of GAC*

Figure 3.20 shows the different power profiles, while Figure 3.21 shows the evolution of the state of charge of the storage system (*SOC*).

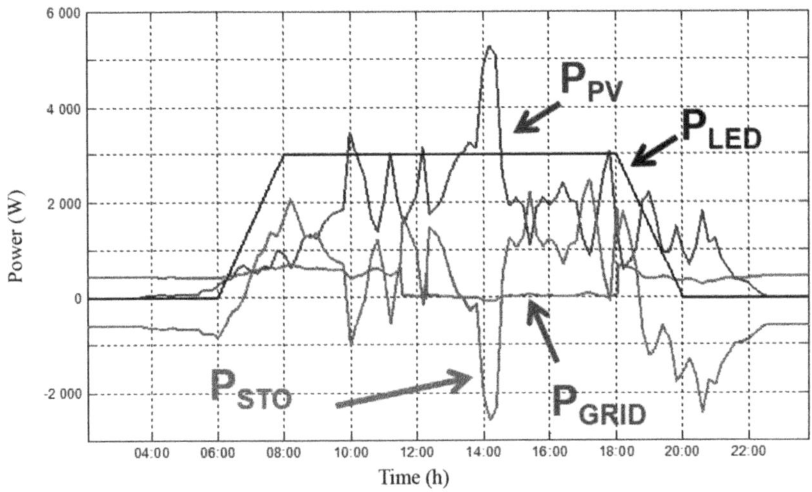

Figure 3.20. *Power profiles for variable GAC*

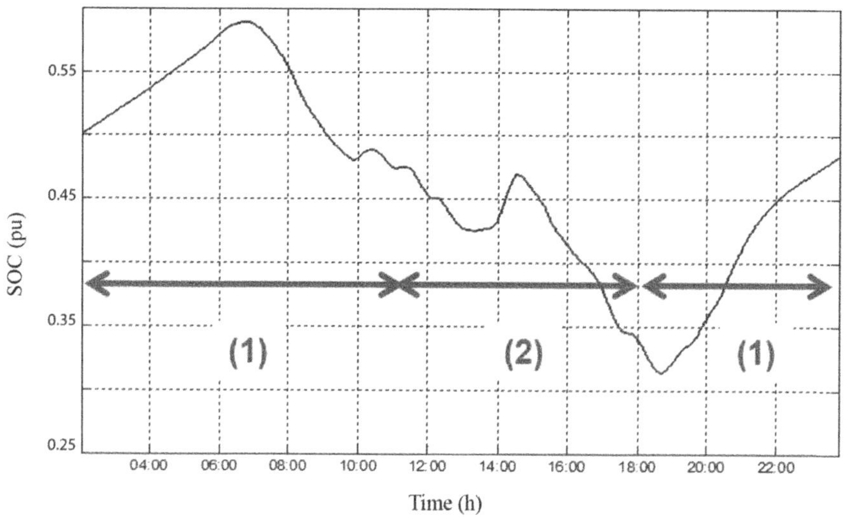

Figure 3.21. *SOC in the case of variable GAC*

The periods of the day labeled (1) correspond to phases where the GAC are favorable. In this case, energy from the grid is used to contribute to powering the load, on the one hand, and to recharging the storage system, on the other hand (particularly after 1800 hours). From around 1130 to 1800 hours, the GAC are unfavorable, and use of the grid to power the lighting system is strictly limited; the storage system is used to supplement photovoltaic production. In this case, the evolution of GAC toward the end of the day and overnight enables the storage system to be recharged.

3.4.4. *Comparison of results*

In our three test cases, we see that the LED lighting load is powered by a mixture of sources, varying according to grid access conditions (GAC). When the GAC are favorable, the electric grid will be used to contribute to powering the LED system, supplementing the power supplied from storage and by the PV panels. It will also contribute to recharging the storage system. When the GAC are strongly unfavorable, storage is prioritized as a source of power to supplement PV production. Grid power is only used when the state of charge of the storage system descends below a certain level, near to the lower limit. The case involving variable GAC is more realistic, and shows the interest of using a supervision system to regulate the use of electricity from the grid in accordance with GAC.

Note that the state of charge (SOC) of the storage system at the end of the cycle is not the same in each case, and tends to deviate from the initial SOC of 50%. In the case of favorable GAC, the final *SOC* stands at approximately 60%; in the opposite case, the final SOC is only 34%. In the first case, it would thus be possible to consider reducing the quantity of energy drawn from the grid. In the second case, more power would need to be drawn from the grid. These changes could be made by optimizing the membership functions of the different variables involved, which were defined empirically for the purposes of this study.

3.5. Conclusion

In this chapter, we have considered the development of a fuzzy supervisor, applied to a power system which combines photovoltaic production, LED-based lighting and a battery-based electrical storage system, interconnected via a DC bus and with a connection to the AC grid. Our aim was to prioritize local consumption of energy produced by the photovoltaic panels and to draw as little energy as possible from the distribution grid.

We began by proposing a DC architecture that reduces the number of electronic power converters.

We then developed our supervision system following the methodology set out in Chapter 1. This step consisted of defining the following elements for the supervisor:

– the specification (objectives, constraints and means of action);

– inputs and outputs;

– the functional graph;

– membership functions of variables;

– the operational graph;

– fuzzy rules.

To validate our supervisor, we simulated three different test cases, altering the grid access conditions (favorable, unfavorable and variable). The results demonstrated that the supervisor permits regulation of the electrical power drawn from the network, which is adjusted as a function of these access conditions.

An additional step to optimize membership functions would further refine the results obtained using our supervisor, notably by ensuring that the storage system is sufficiently recharged at the end of the cycle, allowing it to return to the initial SOC.

3.6. Acknowledgments

The authors wish to thank the ADEME (*Agence française de l'Environnement et de la Maîtrise de l'Energie* – French Environment and Energy Management Agency) and the Nord – Pas de Calais region for their financial backing for the OCESE project (*Optimisation du Couplage Énergie Solaire et réseau d'Éclairage LED au sein de bâtiments tertiaires* – Optimization of Coupling Between Solar Energy and LED Lighting in Tertiary Buildings), in which context the case study presented in this chapter was developed.

4

Hybrid Storage Associated with Photovoltaic Technology for Buildings in Non-interconnected Zones

4.1. Introduction

Renewable energy sources, such as photovoltaic solar power, offer a promising alternative to traditional solutions in combatting climate change, notably by reducing the greenhouse gas emissions generated by electricity and heat production. In addition to being clean and widely available, these solutions reduce the energy dependency of countries where they are adopted. They will play a major role in the transition from a centralized to a distributed production structure, closely linked to the notion of smart grids. Progressively, in relation to specific economic and geographical contexts, they will come to compete more and more with conventional methods. In many cases, PV production is the best available solution in micro-grid contexts in isolated or poorly connected areas, thanks to the ease with which small, modular generator units can be integrated.

Renewable energy sources can thus contribute to improving the security of supply while reducing CO_2 emissions. They are also valuable tools for the development of rural zones, leading to the creation of jobs and the re-evaluation of local resources. Furthermore, in regions with isolated communities, they significantly reduce the costs involved in producing electricity.

Nevertheless, these renewable sources present certain characteristics which pose major problems in terms of balance in the grid, and in the supply/demand balance as a whole. In the case of photovoltaic solar production (as in the case of wind power), generation is variable and relatively unpredictable, as it is dependent on weather conditions which are naturally stochastic.

One solution to this issue is the integration of storage systems, along with smart management of the whole system. This results in a complex energy system which must be designed and managed in an optimal and effective manner.

In this chapter, we will present a photovoltaic system with hybrid storage combining two different technologies: batteries and supercapacitors. This hybrid approach combines the advantages of each technology in order to increase the life expectancy of batteries and to improve yield from the whole system. The system discussed here is intended to supply electricity to residential habitat in insular or isolated environments. We will develop a smart supervision algorithm based on fuzzy logic, following the methodology laid out in Chapter 1 and in [ROB 15, ROB 16]. Our simulations will show that the desired objectives, in terms of conformity to a production program and respect of the different constraints affecting the electric grid technical management (power balancing, frequency regulation, etc.) are met.

We also present a comparative study of different storage configurations, paying particular attention to the life expectancy of storage elements and to the levelized cost of energy (LCOE).

This chapter is organized as follows. Section 4.2 provides a detailed overview of photovoltaic systems for use in buildings, with or without a connection to the grid. Section 4.3 describes the importance of storage in these systems. In section 4.4, we present a photovoltaic generator combined with a hybrid storage solution, and in section 4.5, we give a detailed description of the steps and indicators used in evaluating energy management. Finally, we analyze the results obtained for different storage configurations.

4.2. Photovoltaic systems in buildings and integration into the grid

4.2.1. *Context and economic issues*

In recent years, the world has been forced to face unprecedented energy and climate problems requiring a strong response. In confronting these issues, we will pass through a transitional stage in which the increased use of renewable energy, improvements in energy efficiency and the reduction in consumption behaviors will play a crucial role. The challenge in terms of renewable energy is significant: in 2015, only 16% of total electrical production in the European Union came from these sources, and the aim is to attain a figure of 27% by 2030. The majority of this energy currently comes from hydroelectric sources. This fact shows even more clearly how much work is still needed. A variety of climatic, energy-related and economic factors, such as global warming, the increasing scarcity of fossil fuels and the liberalization of the energy market, have increased awareness of renewable energy among users and politicians. Local authorities will have a key role to play in increasing the use of renewable energy in their own areas, through actions such as renovating public buildings to reduce consumption, prioritizing local sources, including sustainability considerations in their public actions, and/or supporting those involved in developing new approaches. In recent years, a number of climate and energy plans have been adopted to promote the development of decentralized energy production and a reduction in energy consumption. Currently, almost all electricity produced from renewable sources in France is sold to the grid. However, with the reduction of feed-in tariffs, this approach is set to become less profitable in the near future (notably due to the removal of medium/long-term purchase tariffs and to an increase in self-consumption, reducing connections to the public grid). In economic terms, renewable energy solutions are not always the most competitive option, but it seems reasonable to expect that they will become increasingly affordable over the coming years, particularly through attempts to reduce purchasing costs and via the increase in the price of fossil fuels. There is also a growing trend toward local consumption of energy production.

In this context, photovoltaic systems in buildings represent a valuable means of reducing the part played by traditional sources in overall electric production. Energy from renewable sources may be directly reinjected into the grid, using either a storage system (such as batteries) or a smart management system to balance energy production and local usage. There are several possible configurations for photovoltaic systems [DAN 14]:

– Autonomous photovoltaic systems are not connected to the grid, and are used in cases where grid access is not possible or prohibitively expensive.

– Grid-connected photovoltaic systems are the most widespread, and, in many countries, users can sell production back to the grid operator for an attractive price.

– Photovoltaic systems may also be integrated into new smart grid configurations known as micro-grids, in which decentralized production sources are combined with storage systems, other means of production where applicable, and consumers in a local area. At present, these micro-grids are only found in a few isolated areas; in the long term, however, it should become possible to connect them to the main grid so that they may participate in ancillary services. This type of micro-grid will be considered in this chapter, in the context of a photovoltaic generation system for a non-connected area, contributing to electricity provision for a number of households. There are also constraints imposed by the local grid operator, including a planned production program and frequency regulation.

4.2.2. *Examples of projects*

4.2.2.1. *Autonomous photovoltaic systems*

The role of autonomous systems is to provide power to one or more consumers in an area cut off from the electric grid. Figure 4.1 shows an autonomous PV system powering a mountain cabin. A storage system is used alongside PV generators to ensure constant power supply over several days, irrespective of fluctuations in production and decorrelation between production and demand. The storage system acts as a buffer, storing surplus production and providing electricity when demand outstrips production. Photovoltaic production can also be scaled back in case of excess (i.e. when the battery is fully charged and demand is insufficient).

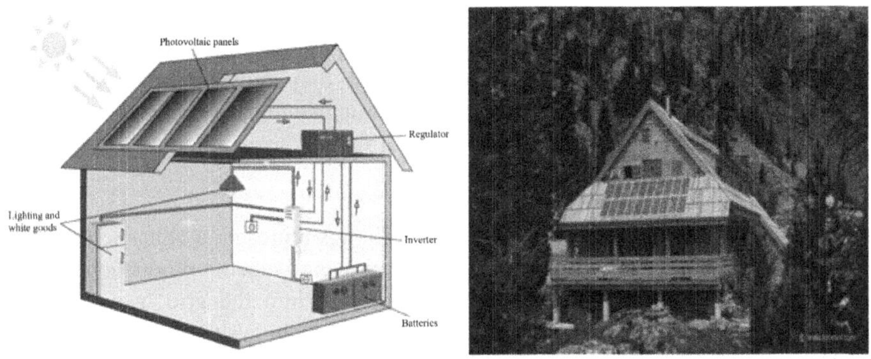

Figure 4.1. *Example of an autonomous photovoltaic system for a mountain cabin [EOS, PHO]. For a color version of the figures in this chapter see www.iste.co.uk/robyns/buildings.zip*

4.2.2.2. *Photovoltaic system connected to the grid*

Figure 4.2 shows a PV system connected to the electric grid, intended to contribute to the production of renewable energy within the grid. The energy produced by the panels is directly consumed by local loads in the habitat. Any surplus production which is not consumed instantly is injected into the network.

If grid access is cut off, the inverter ceases injecting production into the grid and production switches over to a parallel backup circuit, including a group of batteries. This disconnection protection system removes the risk of electrocution for maintenance workers in case of power cuts, and is managed by the inverter, which stops automatically if the grid is switched off.

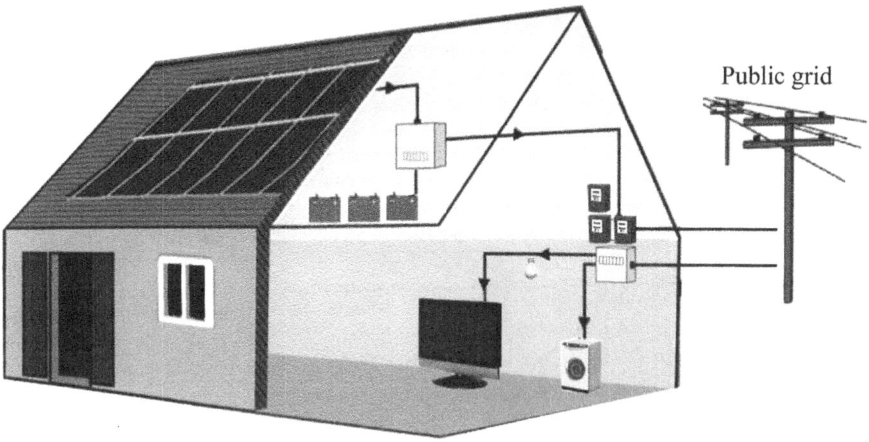

Figure 4.2. *Grid-connected photovoltaic system with self-consumption and isolation capacity [PIC 18]*

The photovoltaic panels are connected to a direct voltage bus through a converter, which converts power from DC to AC and ensures that the PV generator is always operating at optimal levels (maximum power). As the electrical characteristics of PV panels are weather-dependent, this form of management increases the profitability of the overall system.

In the context of the Live Tree program [LIV], the Université Catholique in Lille, France, installed an integrated photovoltaic generator in the RIZOMM building with a capacity of 90 kWc, 140 MWh/year. The generator is used alongside an Eaton type storage system (200 kWh/40 kW absorption, 80 kW discharge) and is connected to the grid. This is an example of self-consumption in an "island" format within the city of Lille. The island is particularly interesting in that it is a single electrical unit, i.e. the electricity supplied to each building comes from a single distribution point, following a star configuration. Electricity produced by the PV panels on the RIZOMM is self-consumed, with excess energy distributed to neighboring buildings connected to the same HTA-BT transformer station (Figure 4.3). The storage system contributes to maximizing self-consumption, among other things.

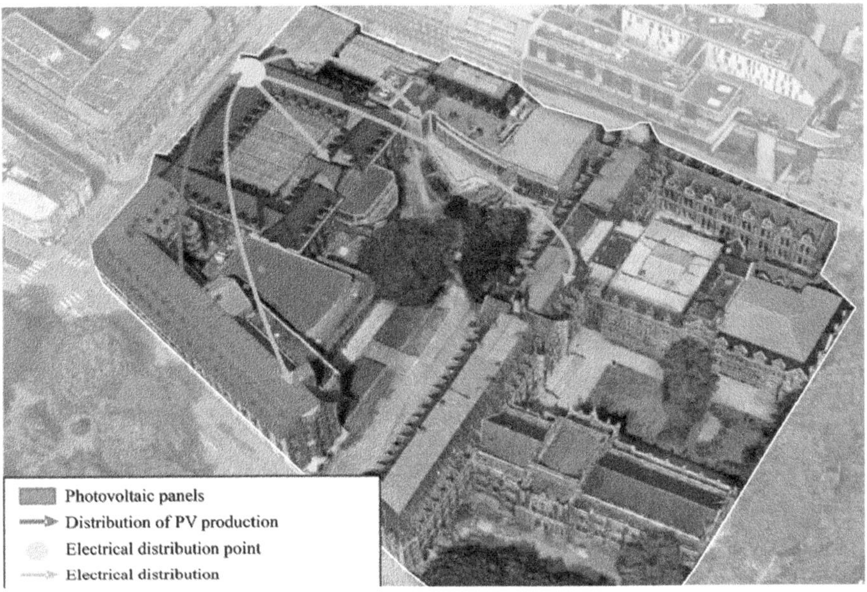

Figure 4.3. *Photovoltaic electrical generator for self-consumption in the RIZOMM building at the Université Catholique, Lille, delivering surplus power to neighbors*

4.2.2.3. *Photovoltaic systems as part of a mini-grid*

A mini-grid, such as the one shown in Figure 4.4 [THI 10], is made up of a photovoltaic generator, a storage system with either a set of lead batteries or a set of lithium batteries, a main source such as a generator, and converters to control the energy produced by these sources and supply consumers. This type of hybrid system is particularly suitable for isolated areas and rural supply.

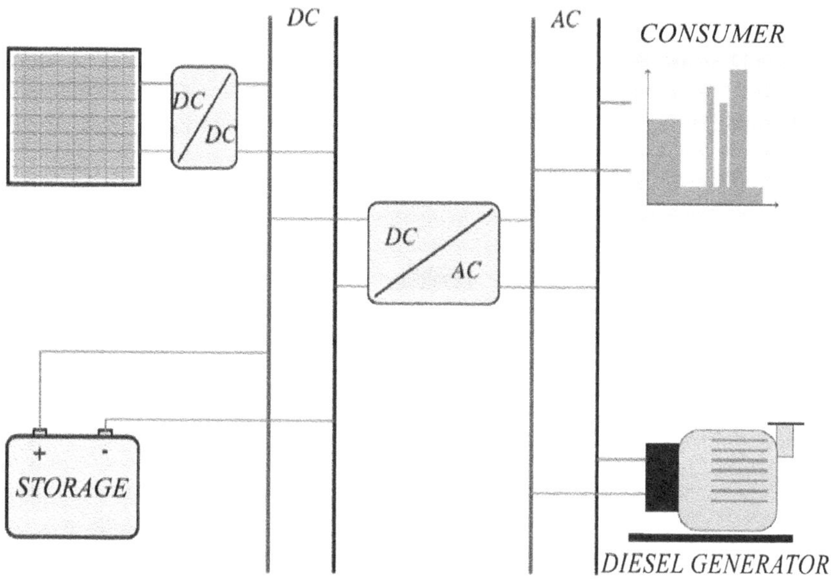

Figure 4.4. *Electrical architecture of an autonomous hybrid photovoltaic system [THI 10]*

The SOLEDO project (Solution globale pour un mini réseau avec Energie Décentralisée photovoltaïque en source principale et gestion Optimisée des flux de production et de consommation – Global Solution for a Micro-grid with Decentralized Photovoltaic Energy as a Principal Source and Optimized Management of Production and Consumption Flows), funded by the Agence Nationale de Recherche (ANR, National Research Agency) was carried out by the SATIE laboratory (at the ENS Cachan, Brittany site, CNRS).

The project consisted of developing a solution for a mini electrical distribution grid, including a photovoltaic system, with a power of between 10 and 100 kW. Production/consumption flows were optimized using a smart energy management and monitoring unit, with the option to adjust consumption by balancing, modulating the magnitude or offloading energy. This approach aims to reduce the dimensions of the photovoltaic area and thus to reduce the cost of the installation, minimizing the consumption of the generator and wear on the battery. Energy generation and management were monitored via a new communication and metering system using power-line communication (PLC) technology, eliminating the need for bulky solutions within a configuration in which equipment is not all housed in the same place. PLC technology enables the transportation of high-frequency signals by "piggybacking" the standard 50 Hz electric current delivered by conventional lines.

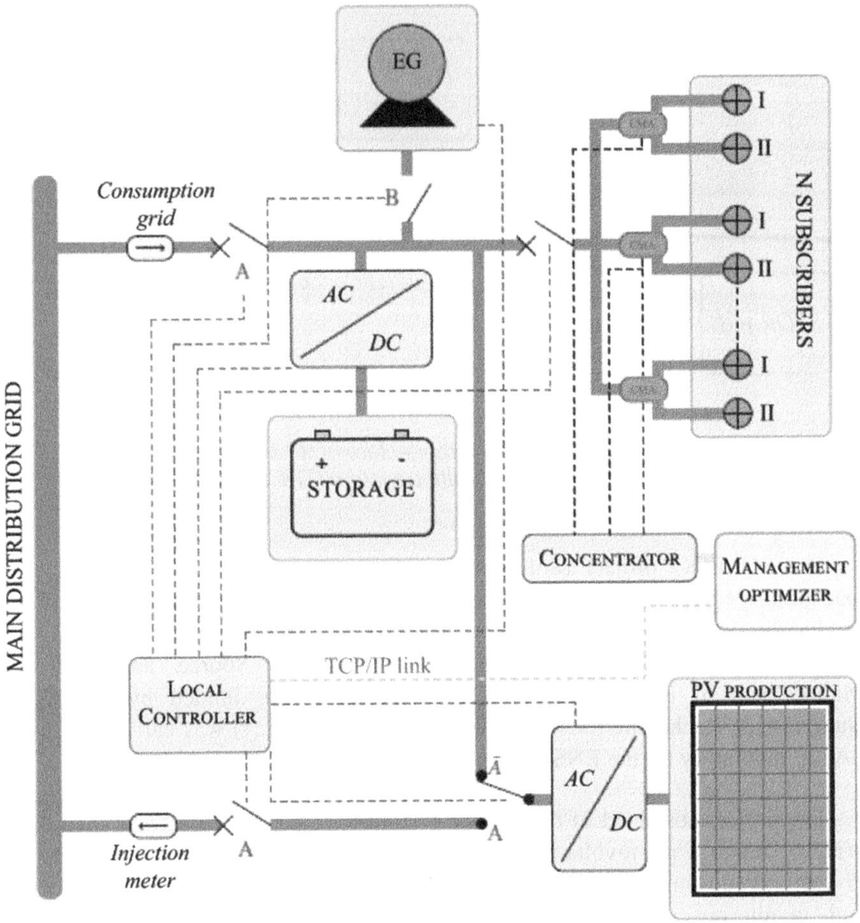

Figure 4.5. *Main diagram produced by the SOLEDO project [THI 10]*

In terms of general structure and management objectives, the study presented in this chapter follows the same logic as the SOLEDO project.

4.3. Importance of storage in photovoltaic systems

One of the main drawbacks of solar energy is the variability and difficulty of predicting production. To create a stable supply and enable permanent use of these systems, in which production and consumption are often decorrelated, some of the energy produced needs to be stored. Several different methods may be used for these purposes, using electrochemical technology (lead or lithium batteries), electricity (supercapacitors), in potential form (STEP pump systems, limited to large-scale storage), kinetic form (flywheel) or in the form of hydrogen (extremely expensive and not yet mature) [SIN 10, MUL 07, MUL 13, ROB 15].

4.3.1. *Photovoltaic systems for isolated sites*

Electrochemical storage is the most widely used option for photovoltaic systems [MUL 07, MUL 13]. In the case of isolated sites, storage systems must respect certain constraints, notably in terms of cost/performance, reliability, security, etc. [SYM 03]. Given these requirements, lead-acid batteries have generally been selected as they provide the best compromise in terms of investment cost, performance and maintenance [SIN 10, CON 16].

Lead-acid batteries are a tried-and-tested means of storing energy off-grid, making them the safest option. The solar charge controllers and inverters currently available on the market are designed to work with lead-acid batteries, and have been used for many years in real-world situations, so any teething problems have already been resolved. The initial investment is currently lower than that of lithium-ion batteries. With effective maintenance, lead-acid batteries operate at approximately 80–90% efficiency, but they do require regular efforts, including the addition of water to open batteries [CON 16].

However, in our opinion, a transition from the older lead-acid technology to higher-density lithium-ion batteries, offering greater resilience, longer life expectancy and a reduced cost over the whole of the lifecycle, is imminent [CON 16].

4.3.2. *Photovoltaic systems connected to the grid*

The electrical energy storage systems used in cases where a photovoltaic generator is connected to the main grid respond to the need to balance demand and production, and to meet the energy needs of houses, companies, municipalities, administrative entities and public services, compensating for the variability inherent in renewable energy supplies.

The objectives of the studied storage system are:

– to maintain energy supply during planned or unplanned breaks in supply from a generator, for example;

– to compensate for the power variabilities inherent in some renewable energy production;

– to balance supply and demand, provide ancillary services (primary and secondary reserves) and thus contribute to the stability of the electric grid;

– to enable renewable energy sources to play a greater part in electrical production, making it possible to attain ambitious sustainability goals.

These systems are particularly valuable in isolated zones which are not connected to the national grid, as in the French overseas territories (*Départements et Régions d'Outre-Mer*, DROM), which have significant renewable resources but are limited by the lack of a connection to the powerful transmission national grid, which is itself connected to the wider European network. This particularity means that all of the electricity consumed in these areas must be produced locally, and thermal production is often dominant. Because of this, the grids in question are more fragile and harder to balance and stabilize. Given the need to increase the part played by inherently variable renewable energy sources (above and beyond the oft-cited goal of 30%), it is crucial that we find a means of maintaining the stability of these networks; historically, management practices have relied on the flexibility offered by thermal production. The use of storage systems alongside variable renewable sources – in this case, photovoltaic solar power – is generally considered to offer a viable solution. The DROM are thus prime candidates for developing stationary electrical storage systems [LEN].

4.3.3. *Hybrid storage*

A number of studies [MAC 07, GLA 08, HAD 09] have discussed the implementation of various combinations of storage solutions. For isolated sites (autonomous habitat), the possibility of immediate access to high power enables changes in energy usage habits. Unlike conventional systems, which can only support the use of very low-consumption devices, these hybrid options have the capacity to supply high-power apparatus, for example cold storage systems which are particularly energy-hungry in the start-up phase.

Several companies offer storage solutions for use in grid-connected installations. These are able to provide support for the existing grid and are suitable for integration into smart grids. Products include:

– Li-ion battery containers (Saft, Mitsubishi Heavy Industries, Tesla);

– sodium-sulfur (NaS) batteries (NGK);

– hydrogen production and storage systems (CETH2 and McPhy);

– flywheel systems (Vycon, Beacon Power, Piller);

– superconductor systems (Maxwell).

However, there are very few solutions for electric grids including a combination of these different technologies; those which do exist are still in development. A number of examples are shown in Table 4.1:

Project/country	Storage technology	PV power installed
New Energy and Industrial Technology Development Organization (NEDO), Wakkanai City (Japan) [MOR 07, BLU 15]	– NaS (Sodium-Sulfur) batteries: 1.5 MW/11.8MWh – Ultracapacitors: 1.5 MW–25 kWh	5 MW
PNM New Mexico's Prosperity Energy Storage Project – New Mexico – USA, East Penn Manufacturing Co., Inc. [PNM 11]	– Batteries: Advanced Carbon VRLA (sealed lead batteries) – 1 MWh – Lead Acid ultra-batteries – 500 kW	500 kW

Table 4.1. *Examples of hybrid storage systems for electrical networks*

From Figure 4.6, we see that the best combination, in terms of satisfying both energy and power requirements, includes batteries and supercapacitors. The system must offer high levels of dynamism, high specific power and the capacity to support deep discharge in order to cope with peaks in power consumption. In addition to these characteristics, supercapacitors have the ability to withstand a high number of cycles compared to other storage technologies. By combining supercapacitors and batteries, it should be possible to optimize the size of storage systems and increase their life expectancy, while providing good dynamics and increased yield.

Finally, note that the management of hybrid storage systems is essential in order to attain desired objectives and to maximize the life expectancy of the most sensitive components (particularly batteries) [ROB 13a].

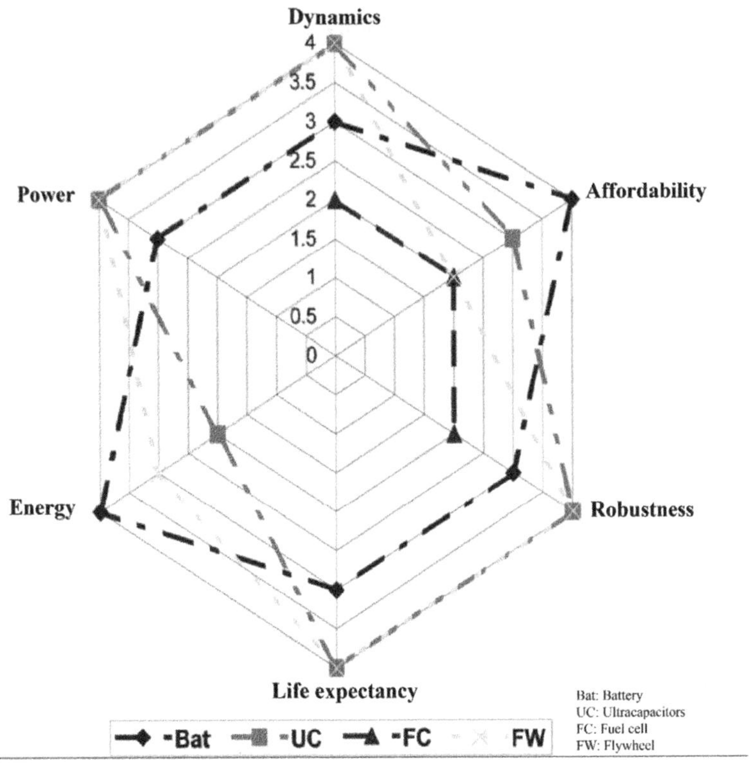

Figure 4.6. *Multi-criterion comparison of different storage technologies for the purposes of hybridization [SIN 10]*

4.3.4. *Electronic conversion structures for hybrid storage*

Photovoltaic cells produce direct current (constant sign) from solar radiation via a photoelectric effect. The majority of PV installations are used for low-voltage loads with a sine waveform. Electronic power converters are therefore needed to convert the waveforms of electrical quantities [ROB 12c].

PV systems may either be connected to an AC network or operate in isolated networks.

Autonomous systems are used to power loads directly. The solutions currently available on the market are battery-based uninterruptible power supplies (UPS), connected to a charge regulator and PV panels (Figure 4.7).

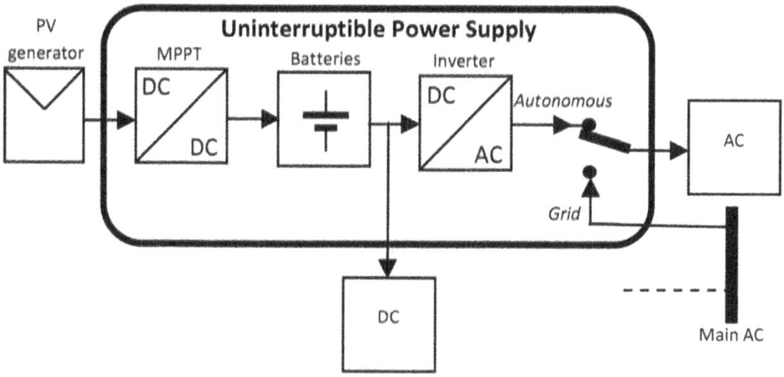

Figure 4.7. *Example of an autonomous photovoltaic system with storage*

Storage accounts for a large proportion of the cost of a system of this type, and the integration of PV production results in significant usage constraints due to its intermittent nature, with charge–discharge cycles taking place almost every day, often involving strong currents. An electronic power converter needs to be installed at the battery terminals in order to control charge and discharge power flows, ensuring the optimal use of the battery.

Currently, there are no storage units which combine high energy levels, the capacity to transport high powers and rapid dynamics within a small space ([ROB 15], Chapter 2). For this reason, a combination of technologies is needed to meet these requirements. The use of batteries for long-term energy storage alongside supercapacitors for rapid regulation of dynamic power is an economically viable option, but other solutions may also be considered, such as fuel cells, Redox batteries or flywheels.

Storage units may be used to store or generate electrical energy. They thus help to respond to the problems inherent in the use of intermittent renewable energy generators which result in fast power transients. A PV generator system including a combination of storage technologies may be considered as an active generator from the perspective of the grid: the availability of stored energy enables it to provide auxiliary services in the same way as a conventional generator. Furthermore, they may be assigned power references by the network operator, and supply this required power.

Various power electronic conversion structures have been designed to control energy exchanges between all of the elements involved (PV generator, first level of storage, second level of storage, grid, etc.) in an independent manner.

It is possible to create a hybrid AC generator with a shared bus (Figure 4.8(a)) by connecting all equipment to the AC electrical network. A communication network is needed in order to coordinate the operation of each element via a central controller. The advantage of this approach is that high-power energy sources and storage devices may be used, and may be located at a considerable distance from one another. However, this structure has two significant drawbacks:

– the voltage and frequency of the AC grid must be kept within a very small window in order to maintain operational stability in the generator;

– an inverter must be installed at the interface between each element and the grid, increasing the overall cost of the system.

In hybrid generators with a shared internal DC bus, only one inverter is needed, at the single point of connection to the AC grid, and all system elements are connected to a DC bus via power electronic converters (Figure 4.8(b)). This configuration is useful for low-power applications, and is less complex as an additional communication network is not required [LU 08].

Most of the PV systems on the market come with integrated power electronics. In this case, a mixed structure for the hybrid generator should also be considered. This non-invasive approach includes a complete PV system alongside storage elements connected to the system via their own inverters (Figure 4.8(c)).

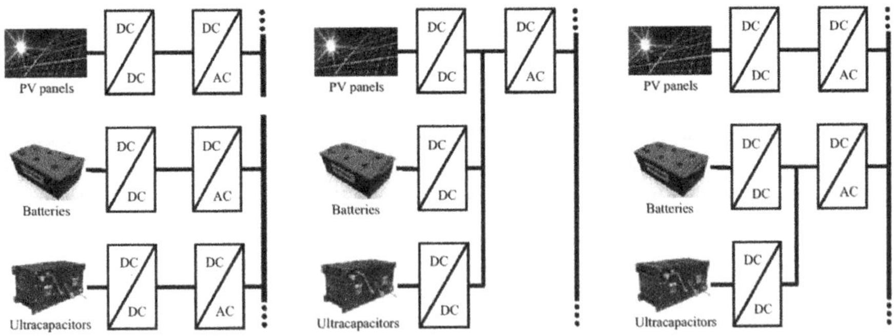

(a) AC-coupled structure (b) DC-coupled structure (c) Mixed structure

Figure 4.8. *Structure of a hybrid power generator [LU 10b]*

4.4. Photovoltaic generator with hybrid storage system

4.4.1. *Case study*

Figure 4.9 shows an outline of the photovoltaic generator considered in this chapter, highlighting the role of the management supervisor in the system. Its nominal photovoltaic power is 30 kW in standard conditions (1000 W/m^2, 25°C). The storage system combines two complementary technologies: a rapid power storage source (e.g. supercapacitors) and lithium batteries with high specific energy capacity (energy source). The main aim in adding a power source is to reduce the cost of the storage function across the whole lifecycle of the system [HOU 15].

The generator is intended to contribute to powering isolated homes. It is simply a case study and is not intended for any specific project.

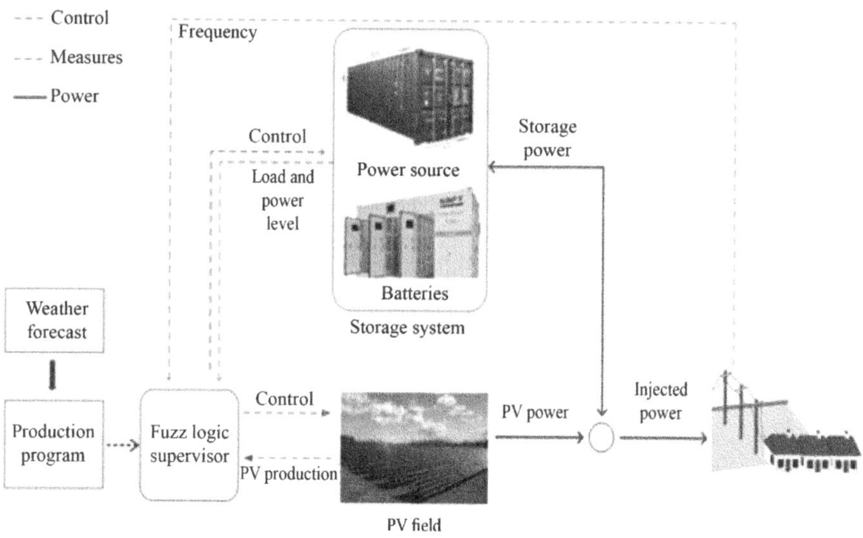

Figure 4.9. *Outline diagram for the PV generator linked to a hybrid storage system considered in this chapter*

We have selected a DC coupled structure, as proposed in thesis [LU 10b] (Figure 4.10).

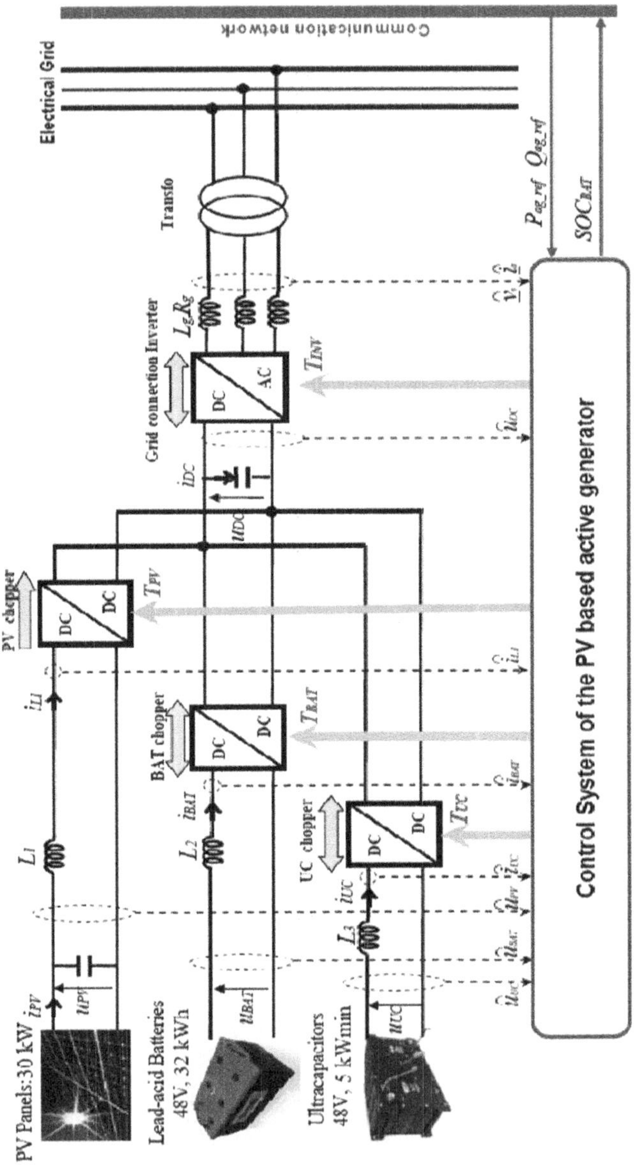

Figure 4.10. *Full photovoltaic system with hybrid storage (electrochemical battery and supercapacitors), connected to the grid [LU 10b]*

The energy management strategy applied to this hybrid storage system should enable it to:

– comply with an hourly production program;

– provide support to the grid in terms of frequency regulation;

– reduce the cost and increase the life duration of storage elements.

Furthermore, the management algorithm should enable the system to attain the following goals:

– life expectancy of 15–20 years;

– energy cost of less than €200/MWh;

– overall electrical yield of more than 90%.

Before discussing the stages involved in creating a fuzzy logic supervisor, it is helpful to understand the methods and standards used in this study for frequency supervision and for calculating battery degradation. These are presented below.

4.4.2. *Principles and standards for frequency support*

4.4.2.1. *The problem of frequency variation*

An imbalance between electrical power consumption and generation will have a direct impact on the evolution of frequency: frequency increases when production exceeds consumption, and decreases when consumption exceeds production. For example, if an unexpected event occurs, such as the loss of a production plant, then grid managers are faced with an immediate drop in frequency [ROB 12c].

To maintain balancing between production and consumption, grid operators require production plants to participate in frequency regulation via the use of primary and secondary power reserves [ROB 15]. For conventional plants above a certain power level, this participation is compulsory, at least in part (and is regulated by decree). In liberalized electricity markets, operators may offer compensation for this service through private agreements or using a dedicated capacity market.

For our case study, we will presume that the specification stipulates that the production plant must be able to provide frequency support to the grid. Primary and secondary support must be possible up to a maximum of 10% of total storage power. Furthermore, the system must respond quasi-instantaneously within 300 ms over a period of 15 min for primary support, and within 15 min over a period of 30 min for secondary support (Figure 4.11). Half of the primary reserve must be able to be provided in less than 15 s and all in less than 30 s.

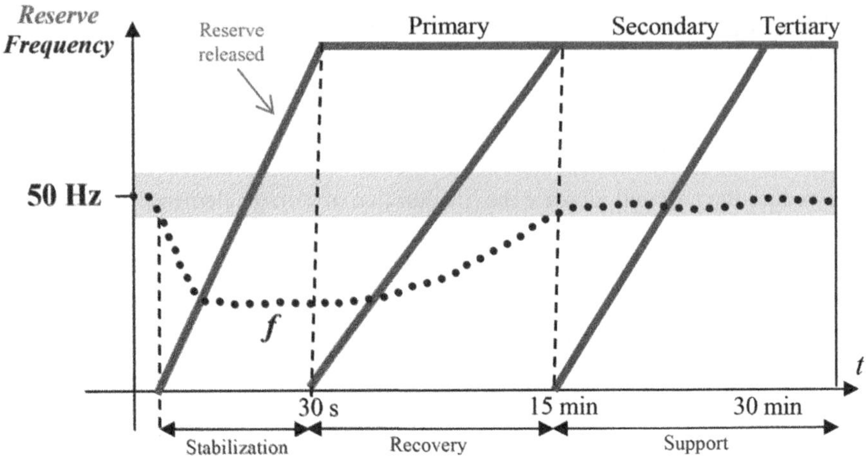

Figure 4.11. *Frequency regulation principle [ROB 15]*

In terms of regulation for hybrid storage (Figure 4.8), this mechanism acts on the active power reference for injection into the grid via the power electronic converters connected to storage elements in order to provide both primary and secondary frequency regulation.

4.4.2.2. *Power–frequency relationship*

A. Calculating frequency variation from power

The instantaneous power variation (increase or decrease in primary reserve) is implemented via a static linear characteristic, establishing the variation in primary power ΔP_{g_ref} in relation to the maximum power ($P_{\max\_st}$) as a function of the variation in the measured frequency Δf in relation to the normal frequency value (f_0 = 50 Hz) [LU 10, ROB 15]:

$$\frac{P_g(t)-P_{g\_ref}}{P_{max\_st}} = \frac{1}{s_f} \cdot \frac{(f_{mes}(t)-f_0)}{f_0} <=> \frac{\Delta P_{g_{ref}}(t)}{P_{max\_st}} = \frac{1}{s_f} \cdot \frac{\Delta f(t)}{f_0} \qquad [4.1]$$

where f_{mes} is the measured frequency and s_f is the slope. The droop parameter of the generator is defined as $k_f = \frac{P_{max\_st}}{s_f \cdot f_0}$, leading to the characteristic shown in Figure 4.12: $\Delta P_{g_{ref}}(\mathrm{t}) = k_f . \Delta \mathrm{f}(\mathrm{t})$.

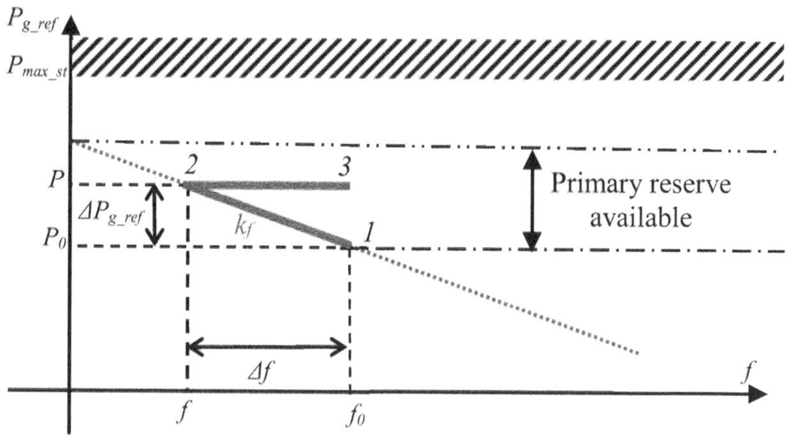

Figure 4.12. *Ideal static characteristic between active power and frequency [LU 10b, ROB 15]*

The instantaneous action of this primary regulation causes the generator to modify its output power (transition from operating point 1 to operating point 2 in Figure 4.12). Following action by all of the generators involved, and according to the inertia of the electric grid, the frequency will return to its normal value (transition from point 2 to point 3 in Figure 4.12). A static error thus appears in the output power of the generator. Thus, a second regulatory stage occurs following frequency stabilization, re-establishing the reference value for the power (i.e. the value it had before the initial regulation), modifying the program of planned power references for the generators which provide this service. This is carried out using the secondary power reserve, which takes over from those generators which supplied instantaneous power to the primary reserve.

To account for the delay between frequency changes and changes in generator power, the dynamics of the electrical system are modeled using a first-order transfer function [LEE 08]:

$$G_{sys} = \frac{\Delta f}{\Delta P_{g\_ref}} = \frac{1}{k_f\,(1+S\,T_{sys})} \qquad [4.2]$$

where T_{sys} is the time parameter which is identified to obtain the observed response time of plant storage reserves. Note that primary and secondary reserves are required to respond within 300 ms and 15 min, respectively.

B. Selected frequency support technique

The deviation in the short-term frequency Δf_{C_t} is directly calculated by rearranging relationship [4.3] with a time constant ($T_{primary}$) equal to 100 ms, giving a response time of 300 ms:

$$\Delta f_{C\_t} = \frac{\Delta P_{g\_ref}}{k_f\,(1+S\,T_{primary})} \qquad [4.3]$$

The calculation of long-term frequency variation is based on the conditions expressed in equation [4.4]. The activation condition is based on the mean value of short-term variations over a period T equal to 15 min.

$$\begin{cases} \Delta f_{L\_t} = 0 & If \;\; \frac{1}{T}\sum_{t=0}^{T}\left|\Delta f_{C\_t}\right| \leq 0.05 \\ \Delta f_{L\_t} = \frac{\Delta f_{C\_t}}{(1+s\,T_S)} & Else \end{cases} \qquad [4.4]$$

where T_s is the time constant, equal to **1s**, corresponding to a response time of 3s. The proposed slope of the reaction curve for primary regulation includes a dead zone (Figure 4.13). The selected parameters are shown in Table 4.2.

Parameter	Value
Percentage of maximum power assigned to primary support	8%
Percentage of maximum power assigned to secondary support	2%
Dead zone	100 mHz
Slope s_f	5%

Table 4.2. *Support curve parameters for primary and secondary reserves [ENT 13]*

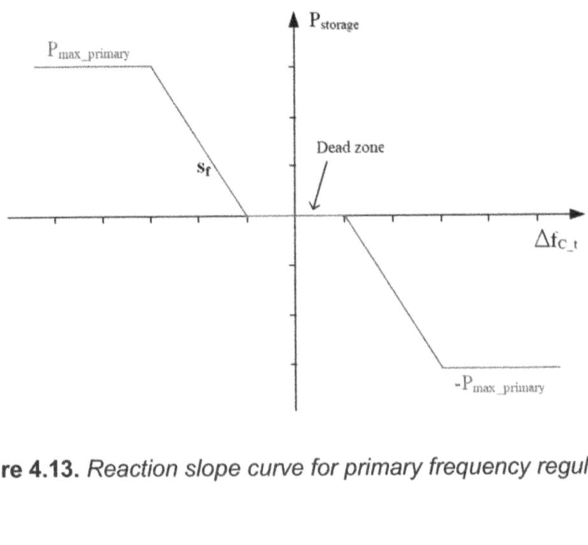

Figure 4.13. *Reaction slope curve for primary frequency regulation*

The reaction slope curve for secondary regulation is similar to that for primary regulation. Two possibilities should be considered:

– Centralized regulation: a central operator (the network operator) transmits instructions to production units remotely. In this case, secondary frequency support is a function of the indication or measurement which is transmitted. For any given frequency variation received by the system, the reaction slope implemented by the fuzzy logic supervisor is used to calculate the required power support.

– Decentralized regulation: the operator of the photovoltaic generator plant implements secondary frequency support directly via local measurements. In this case, the frequency variation is calculated by using the response slope implemented by the fuzzy logic supervisor in order to determine the required power support.

4.4.3. *Calculating battery wear*

To quantify the total damage to storage elements resulting from charge–discharge cycles, we will use a cumulative damage law based on Miner's linear cumulative rule [MAR 09]. This damage law is defined as follows:

$$D = \sum_{i=1}^{N} \frac{n_i}{N_i} \qquad [4.5]$$

where n_i is the number of cycles with depth of discharge DoD_i. N_i is the life expectancy corresponding to the depth of discharge DoD_i. This factor enables us to calculate the expected end of life of storage elements corresponding to degradation D.

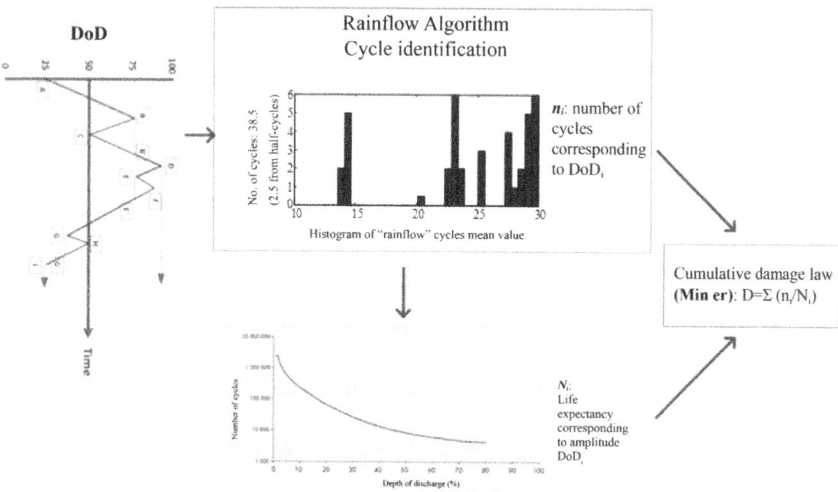

Figure 4.14. *Calculation principle for battery damage*

Figure 4.14 shows the calculation principle for cumulative battery damage. This method involves two steps [LAY 15]. The first step uses a cycle counting algorithm (rainflow) to precisely identify the parameters of a charge cycle (number of cycles, depth of cycles, types of cycles (full or half) and cycle period). For the second step, the aging curve for the storage unit in question is used to identify the life expectancies corresponding to the respective cycle depths identified in step 1. Figure 4.15 shows the aging curves for the different storage elements used in this study [ABB 15].

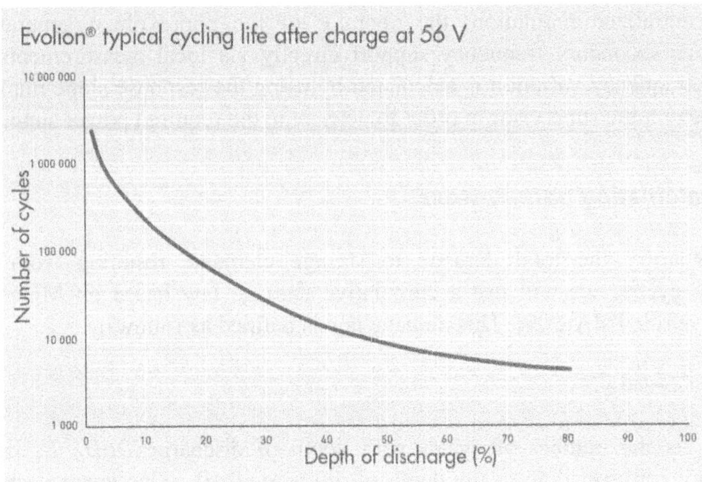

a) Lithium NCA battery (Saft) (Lithium Nickel Cobalt Aluminum Oxide)

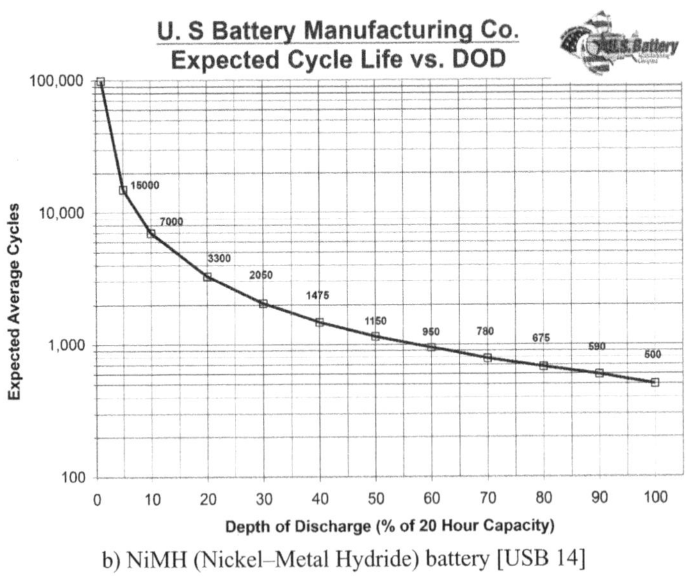

b) NiMH (Nickel–Metal Hydride) battery [USB 14]

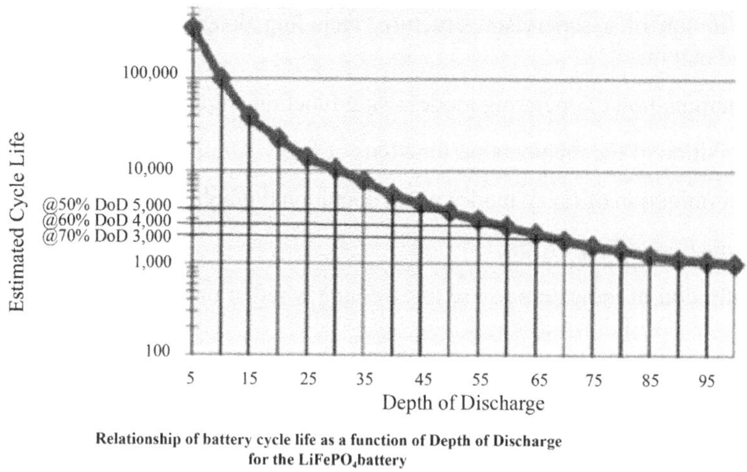

c) Lithium $LiFePO_4$ (Lithium Iron Phosphate) battery [SWI 11]

Figure 4.15. *Aging curves for the different storage technologies used in this study [ABB 15]*

4.5. Energy management

Optimal energy management occurs at the highest level of the control structure. It is central to the performance of the photovoltaic plant in terms of energy production. Power flows are managed by determining modes of operation which best respond to the main objectives of the application: satisfying a production program determined on D-1 as a function of production data and storage level, and participating in frequency support (primary and secondary). An additional, innovative objective is to improve the life expectancy of storage elements through optimized management.

The management methodology proposed below has been validated through simulation and testing using a laboratory test bed.

4.5.1. *Methodology*

Our supervisor is built using a multi-step methodology, determining the management rules for different parts of the system. The eight steps in our approach are [ROB 13a, ROB 15, ROB 16]:

– determination of a system specification, with objectives, constraints and resources clearly laid out;

– definition of a supervisor structure, including determination of the required input and output;

– determination of operating modes using functional graphs;

– definition of the membership functions of fuzzy variables;

– determination of fuzzy modes using operational graphs;

– extraction of fuzzy rules;

– definition of indicators to use in assessing whether or not objectives have been met;

– optimization of supervisor parameters.

4.5.2. *Operating specifications*

The operating specifications for the supervisor are determined by identifying objectives for the management of the photovoltaic plant and by characterizing the constraints and issues involved in operations, highlighting the means of action which may be used to attain these objectives. The operating specifications for the fuzzy logic supervisor are set out in Table 4.3.

Objectives	Constraints	Means of action
– Apply a production program, smoothing power injection – Participate in frequency support (primary and secondary) – Improve the life expectancy of storage units by optimizing the energy management	– Size of storage units – Primary support must respond within 300 ms over a period of 15 min. Secondary support must respond within 15 min over 30 min – Error margin for production program: less than 5% of mid-hour plant production in relation to the program. Outside of these margins, there is a risk that the photovoltaic producer may lose their production hour (no payment for the corresponding production) and may even face financial penalties	– Power instructions for storage: long-term (batteries) – Short-term power instructions (power source) – Downgrading factor for photovoltaic production

Table 4.3. *Operating specifications of the supervisor*

4.5.3. *Supervisor structure and determination of input/output*

In order to attain the management goals established above, each objective is associated with at least one input and each means of action is associated with at least one output.

Management inputs:

There must be at least one input per objective. These include:

1) The difference between planned power and instantaneous PV production, denoted by (ΔP). Two sub-inputs (ΔP_{bat} and ΔP_{SdP}) are specified in order to account for the two different storage technologies (electrochemical batteries, as an energy source, and power storage). The sub-inputs are obtained by decoupling dynamics, using a first-order filter following the internal specifications of the two storage units, denoted by (τ_{bat} and τ_{SdP}). The power expressions (ΔP, ΔP_{bat} and ΔP_{SdP}) are given by:

$$\Delta P = P_{st} = P_g - P_{pv} \qquad [4.6]$$

$$\Delta P_{Sdp} = \left(1 - {}^{1}/_{(\tau_{sc}\, s + 1)}\right) \times \Delta P \qquad [4.7]$$

$$\Delta P_{sdp} = \left({}^{1}/_{(\tau_{bat}\, s + 1)}\right) \times \Delta P = \Delta P - \Delta P_{bat} \qquad [4.8]$$

ΔP_{bat} and ΔP_{SdP} are required to define power references for long-term (battery) and short-term (power source) storage.

2) The state of charge (***SoC***). A distinction is made between the levels of charge for the two different storage systems: SOC_{bat} and SOC_{SdP}. These levels must satisfy the constraint imposed by the maximum capacity of the storage systems.

3) The frequency variation (Δf) is sub-divided into two sub-entries: Δf_{C_t} for short-term frequency variations (HF) and Δf_{L_t} for long-term frequency variations (BF). These two sub-inputs are required for frequency support (primary and secondary).

Management outputs:

1) A reference power for the energy source (batteries), denoted by P_{bat_ref}. This reference is the sum of three sub-outputs: the initial reference ($P_{bat_{ref0}}$), the secondary frequency support power (Δp_{bat_ref}) and an optional battery management power (P_{GES_bat}), for evening use.

2) A reference power for the power source (supercapacitor), denoted by P_{SdP_ref}. This reference is the sum of two sub-outputs: the initial reference (P_{SdP_ref0}) and a primary frequency support power $\left(\Delta p_{SdP\_ref}\right)$.

3) A downgrading factor for photovoltaic production, denoted by K_{pv}.

Each action has a corresponding output. A block diagram of the fuzzy supervisor is shown in Figure 4.16.

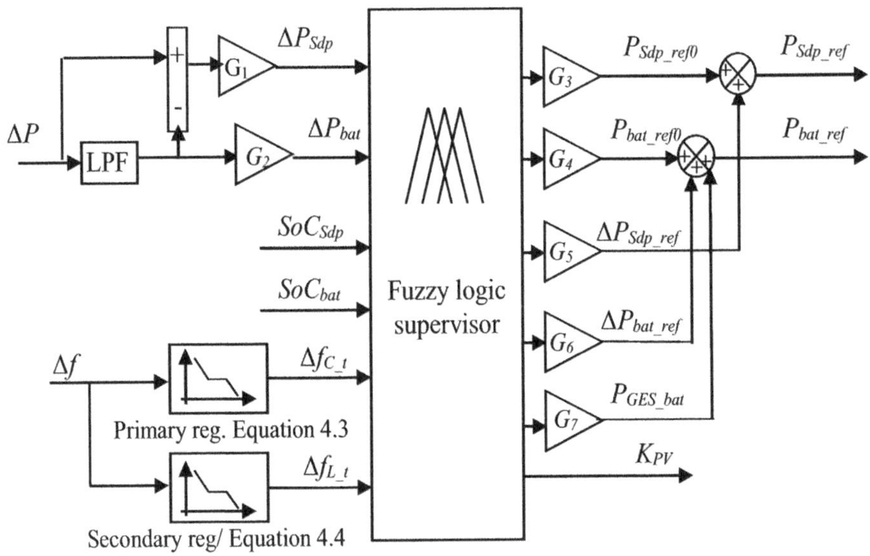

Figure 4.16. *Block diagram of the fuzzy supervisor*

LPF: low-pass filter (first-order filter). ($\boldsymbol{G_{i=1:7}}$) are normalization gains, determined using the methodology presented in Table 4.4:

Gains	G_1	G_2	G_3	G_4	G_5	G_6	G_7
Values	$\frac{1}{P_{\max\_Sdp}}$	$\frac{1}{P_{\max\_bat}}$	$P_{\max\_Sdp}$	$P_{\max\_bat}$	$0.08 \times P_{\max\_st}$	$0.02 \times P_{\max\_st}$	$0.05 \times P_{\max\_bat}$

Table 4.4. *Normalization gains*

The inputs devoted to the levels of charge of frequency and to short- and long-term frequency variations are bounded at $[0,1]$ and $[-1,1]$ respectively. Gains G_5 and G_6 are selected in accordance with the primary and secondary regulation parameters defined in Table 4.2.

Gain G_7 is selected in such a way that battery management (charge/discharge) does not affect the objectives of respecting the production program and of frequency support.

4.5.4. *Functional graphs*

Figure 4.17 shows the functional graph for the system, along with the different sub-graphs for the three main operating modes. These correspond to different states of charge in the storage elements.

Normal or main mode (N1): medium or nominal state of charge. The objective of this mode is to respond to the production program (established on D-1). The storage system compensates for any differences between instantaneous PV production and planned production, carrying out smoothing and frequency support.

Overcharged mode (N2): intended to protect the storage system from the damaging effects of overcharging (in excess of maximum capacity) on life expectancy. The aim is to minimize PV production in order to discharge the storage systems until the nominal value is reached.

Deep discharge mode (N3): intended to protect the storage system from the damaging effects of deep discharge on life expectancy. A few hours before the start of the production program, a reserve is established in storage in order to fulfill the next day's production requirements. Ideally, storage units should be charged up to their nominal value. This may be done using PV production before the program starts (e.g. on a day in summer) or using electricity from the grid.

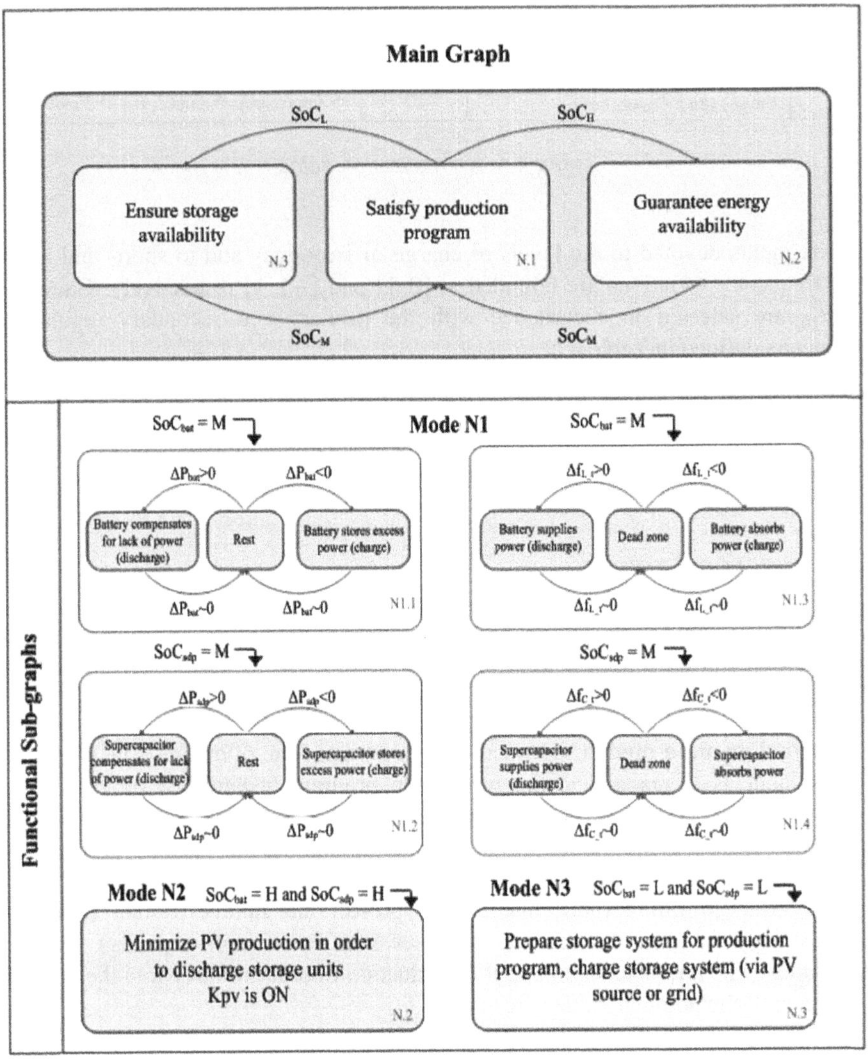

Figure 4.17. *Functional graphs of the system*

4.5.5. *Membership functions*

This step allows us to express the numerical values of input signals as fuzzy values (fuzzification). The notation used here is set out in Table 4.5.

L	*M*	*H*	*Z*	*P*	*N*
Low	Medium	High	Zero	Positive	Negative

Table 4.5. *Notation for fuzzy states*

The functions associated with the different direct supervisor inputs and outputs are shown in Figures 4.18 and 4.19. Each objective is associated with at least one input. A detailed description of inputs and outputs is shown below, with reasons for each choice and the ranges of variation.

Input choice methodology:

Input	**Description of range of membership function**
ΔP_{bat} and ΔP_{SdP} *(Figure 4.18a and 4.18b)*	The membership functions of these two sub-inputs are presented over seven levels in order to precisely characterize the difference between planned power and instantaneous PV production. The principal objective is to follow a production program while smoothing the injected power
SOC_{bat} *(Figure 4.18c)*	The limits between the different levels associated with this sub-input are selected in such a way as to prolong battery life. The state of charge of the batteries for the main operating mode (mode N1) is set between 30 and 90% of capacity. The upper and lower bounds form a buffer zone against potential overcharging and deep discharges
SOC_{SdP} *(Figure 4.18d)*	The operating range for this input is wide, responding to the significant irregularities in PV production
$\Delta f_{C\_t} = \Delta f_{SdP}$ and $\Delta f_{L\_t} = \Delta f_{bat}$ *(Figure 4.18e and 4.18f)*	Reference values for active power production by the photovoltaic plant, for primary and secondary power regulation respectively. Three levels are associated with long- and short-term frequency variations in order to satisfy the applicable norms and regulations for primary and secondary frequency support (see section 4.4.1)

Table 4.6. *Description and justification of input membership functions*

Secondly, we must establish the output membership functions (defuzzification functions) required to attain the desired objectives.

Output choice methodology:

Output	**Description of range of membership function**
$\boldsymbol{P_{bat\_ref0}}$ and $\boldsymbol{P_{SdP\_ref0}}$ *(Figure 4.19a and 4.19b)*	Seven fuzzy sets are associated with each of these two outputs in order to fully satisfy the production program, smoothing the injected power
$\boldsymbol{\Delta p_{bat\_ref}}$ and $\boldsymbol{\Delta p_{SdP\_ref}}$ *(Figure 4.19c and 4.19d)*	Three fuzzy sets are associated with each of these two outputs, corresponding to charge, discharge and rest. These two outputs are supplementary storage indications for primary and secondary frequency regulation
$\boldsymbol{P_{GES\_bat}}$ *(Figure 4.19e)*	Supplementary storage indication for night-time charging. Three fuzzy sets are used here for charge, discharge and rest states
$\boldsymbol{K_{pv}}$ *(Figure 4.19f)*	A security function used to protect the storage system against overcharging. This consists of a rapid decline in PV production, causing storage to be discharged. Other choices may be made for this function depending on the desired response. This specific option provides rapid protection for batteries. $\boldsymbol{K_{pv}}$ indicates activation ($\boldsymbol{K_{pv}}$ = 0) or non-activation of the decline function ($\boldsymbol{K_{pv}}$ = 1). This is a Boolean representation

Table 4.7. *Description and justification of output membership functions*

Note that:

– the power indication (reference) of the energy source (batteries), denoted by P_{bat_ref}, is the sum of three sub-outputs: the initial reference (P_{bat_ref0}), the secondary frequency support power (Δp_{bat_ref}) and a battery management power (P_{GES_bat});

– the power indication (reference) of the power source (supercapacitor), denoted by P_{SdP_ref}, is the sum of two sub-outputs: the initial reference (P_{SdP_ref0}) and a primary frequency support power (Δp_{SdP_ref}).

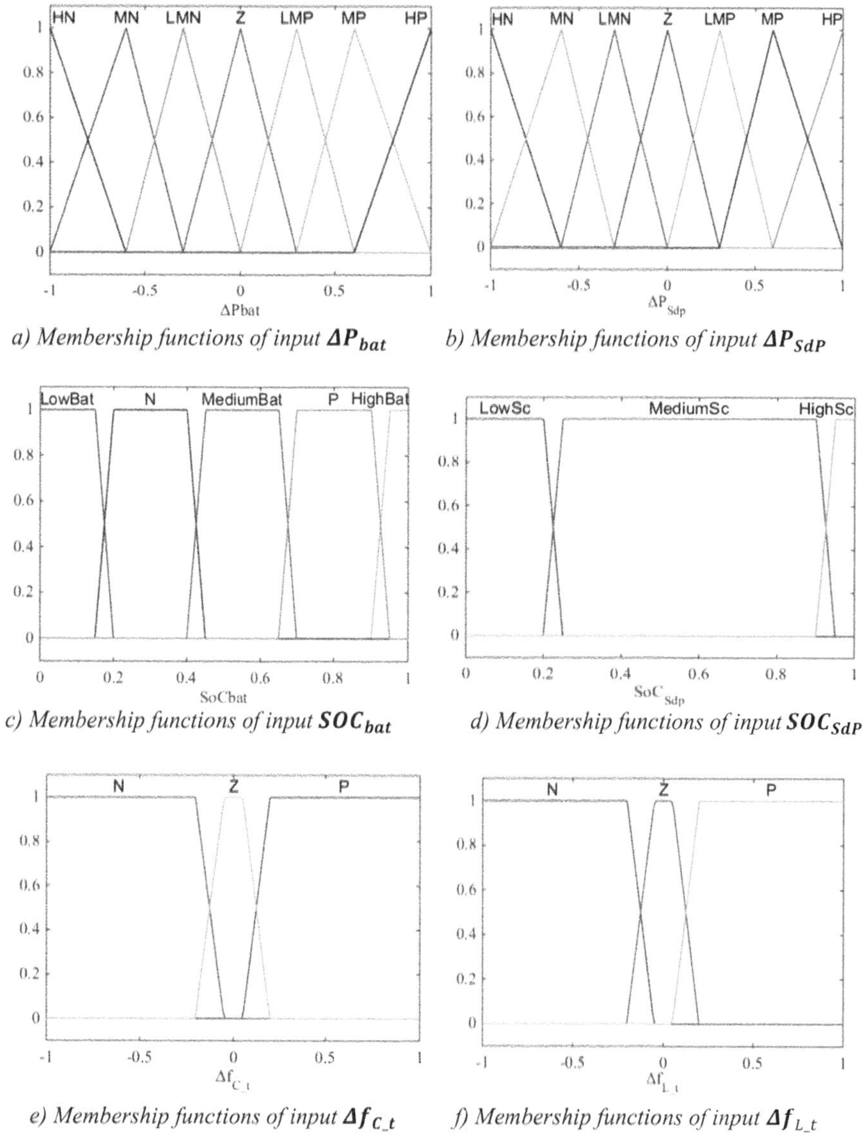

a) *Membership functions of input* $\boldsymbol{\Delta P_{bat}}$

b) *Membership functions of input* $\boldsymbol{\Delta P_{SdP}}$

c) *Membership functions of input* $\boldsymbol{SOC_{bat}}$

d) *Membership functions of input* $\boldsymbol{SOC_{SdP}}$

e) *Membership functions of input* $\boldsymbol{\Delta f_{C\_t}}$

f) *Membership functions of input* $\boldsymbol{\Delta f_{L\_t}}$

Figure 4.18. *Membership functions associated with supervisor inputs*

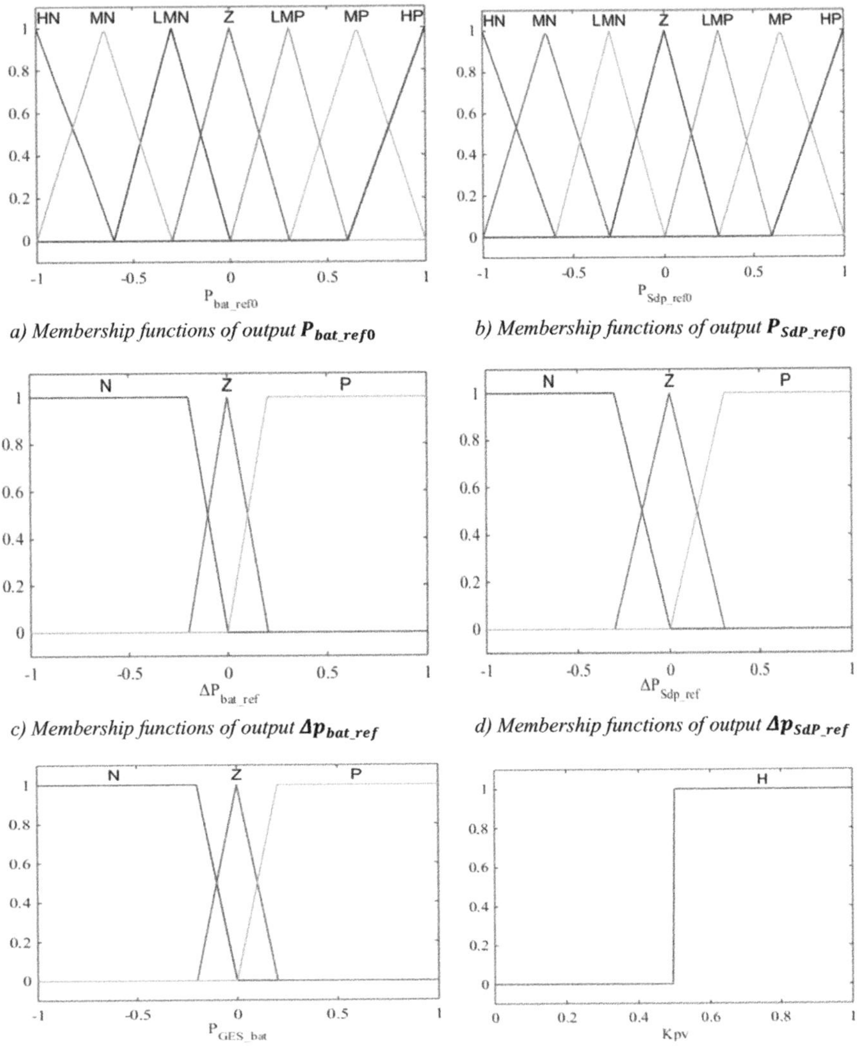

a) Membership functions of output $\boldsymbol{P_{bat\_ref0}}$

b) Membership functions of output $\boldsymbol{P_{SdP\_ref0}}$

c) Membership functions of output $\boldsymbol{\Delta p_{bat\_ref}}$

d) Membership functions of output $\boldsymbol{\Delta p_{SdP\_ref}}$

e) Membership functions of output $\boldsymbol{P_{GES\_bat}}$

f) Membership functions of output $\boldsymbol{K_{pv}}$

Figure 4.19. *Membership functions associated with supervisor outputs*

4.5.6. *Operating graphs*

This step consists of translating and developing our functional graphs using membership functions. Figure 4.20 shows the set of operating graphs corresponding to the three operating modes defined above.

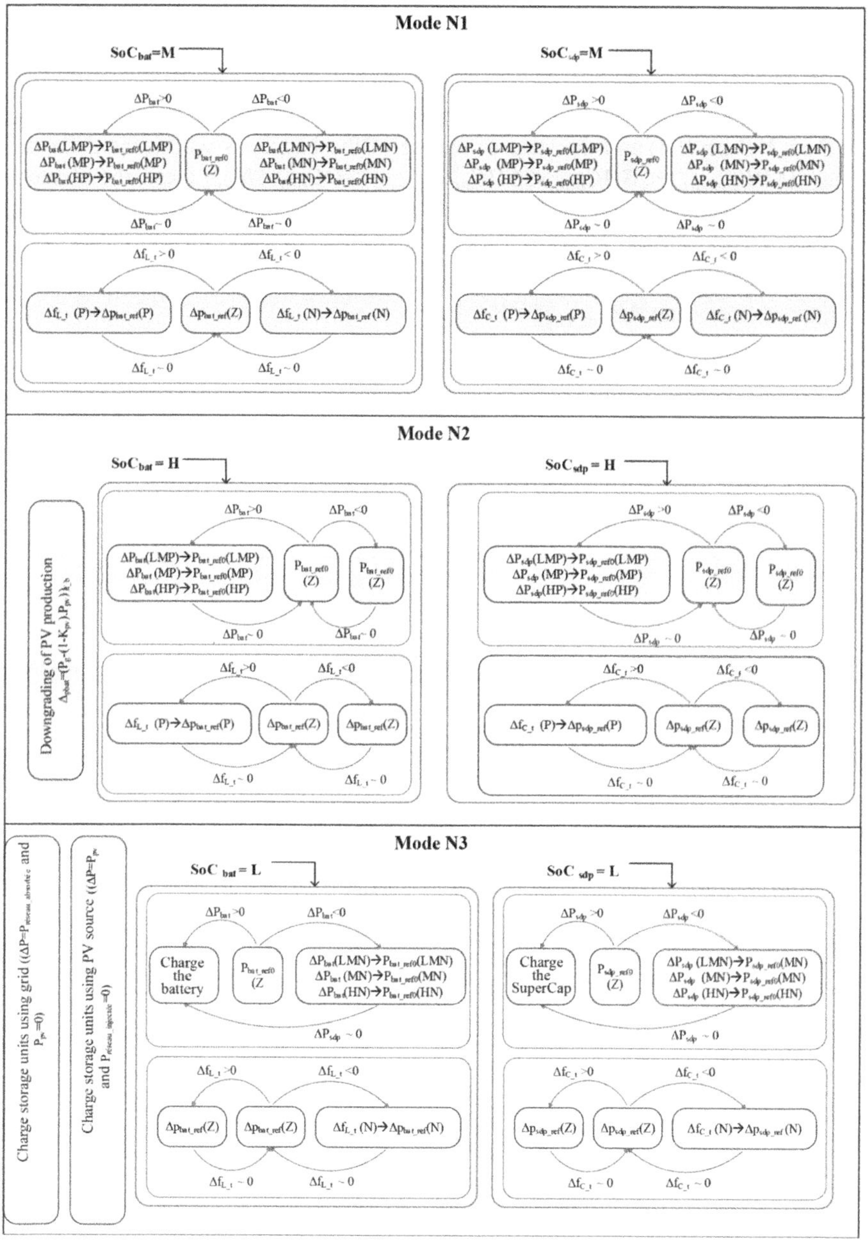

Figure 4.20. *Operating graphs*

These graphs show the state of inputs and the conditions in which decisions are made to determine outputs (actions). They provide a graphical representation of management rules, respecting the distribution and notation of membership functions. For example, for mode N1.1, if there is a shortfall in production $\boldsymbol{\Delta P_{bat}} > 0$ ($\boldsymbol{\Delta P_{bat}}$: $\boldsymbol{LMP\ or\ MP\ or\ HP}$) and the state of charge of the battery is within the authorized limits ($\boldsymbol{SOC_{bat_{min}}} < \boldsymbol{SOC_{bat}} < \boldsymbol{SOC_{bat_{max}}} \rightarrow \boldsymbol{SOC_{bat}} : \boldsymbol{M}$), then the batteries must be discharged ($\boldsymbol{\Delta P_{bat\_ref0}}$: $\boldsymbol{LMP\ or\ MP\ or\ HP}$); if the opposite is true, then they must be charged ($\boldsymbol{\Delta P_{bat\_ref0}}$: $\boldsymbol{LMN\ or\ MN\ or\ HN}$). In the case where $\boldsymbol{\Delta P_{bat}} = \boldsymbol{0}$ ($\boldsymbol{\Delta P_{bat}} : Z$), the current state of charge of the battery should be maintained (rest state: $\boldsymbol{\Delta P_{bat\_ref0}} : \boldsymbol{Z}$).

4.5.7. *Fuzzy rules*

The fuzzy rules connecting the different inputs and outputs of the supervisor are shown in Table 4.8. They may be deduced from the operating graphs in Figure 4.17.

<table>
<tr><td rowspan="14">N1</td><td rowspan="14">Respect a production program while smoothing injected power</td><td rowspan="7">SoC_{sc} is M
[0.3 0.9]</td><td>ΔP_{sc} is HN then P_{sc_ref0} is HN</td></tr>
<tr><td>ΔP_{sc} is MN then P_{sc_ref0} is MN</td></tr>
<tr><td>ΔP_{sc} is LMN then P_{sc_ref0} is LMN</td></tr>
<tr><td>ΔP_{sc} is Z then P_{sc_ref0} is Z</td></tr>
<tr><td>ΔP_{sc} is LMP then P_{sc_ref0} is LMP</td></tr>
<tr><td>ΔP_{sc} is MP then P_{sc_ref0} is MP</td></tr>
<tr><td>ΔP_{sc} is HP then P_{sc_ref0} is HP</td></tr>
<tr><td rowspan="7">SOC_{bat} is
N[0.2 0.4] or M [0.45 0.65] or P[0.7 0.9]</td><td>ΔP_{bat} is HN then P_{bat_ref0} is HN</td></tr>
<tr><td>ΔP_{bat} is MN then P_{bat_ref0} is MN</td></tr>
<tr><td>ΔP_{bat} is LMN then P_{bat_ref0} is LMN</td></tr>
<tr><td>ΔP_{bat} is Z then P_{bat_ref0} is Z</td></tr>
<tr><td>ΔP_{bat} is LMP then P_{bat_ref0} is LMP</td></tr>
<tr><td>ΔP_{bat} is MP then P_{bat_ref0} is MP</td></tr>
<tr><td>ΔP_{bat} is HP then P_{bat_ref0} is HP</td></tr>
</table>

	Primary frequency support (8% × $P_{\max\_st}$)	SoC_{sc} is M [0.25 0.9]	Δf_{C_t} is N then Δp_{sc_ref} is N
			Δf_{C_t} is Z then Δp_{sc_ref} is Z
			Δf_{C_t} is P then Δp_{sc_ref} is P
	Secondary frequency support (2% × $P_{\max\_st}$)	SOC_{bat} is N [0.2 0.4] or M [0.45 0.65] or P [0.7 0.9]	Δf_{L_t} is N then Δp_{bat_ref} is N
			Δf_{L_t} is Z then Δp_{bat_ref} is Z
			Δf_{L_t} is P then Δp_{bat_ref} ist P
	Minimize PV production in order to discharge storage units	SoC_{sc} is H [0.95 1]	ΔP_{sc} is HN-MN-LMN then P_{sc_ref0} is Z and $\boldsymbol{K_{pv}}$ is H
			ΔP_{sc} is Z then P_{sc_ref0} is Z
			ΔP_{sc} is LMP then P_{sc_ref0} is LMP
			ΔP_{sc} is MP then P_{sc_ref0} is MP
			ΔP_{sc} is HP then P_{sc_ref0} is HP
		SOC_{bat} is H [0.95 1]	ΔP_{bat} is HN-MN-LMN then P_{bat_ref0} is Z and $\boldsymbol{K_{pv}}$ is H
			ΔP_{bat} is Z then P_{bat_ref} is Z
N2			ΔP_{bat} is LMP then P_{bat_ref0} is LMP
			ΔP_{bat} is MP then P_{bat_ref0} is MP
			ΔP_{bat} is HP then P_{bat_ref0} is HP
	Primary frequency support (8% × $P_{\max\_st}$)	SoC_{sc} is H [0.9 1]	Δf_{C_t} is N then Δp_{sc_ref} is Z
			Δf_{C_t} is Z then Δp_{sc_ref} is Z
			Δf_{C_t} is P then Δp_{sc_ref} is P
	Secondary frequency support (2% × $P_{\max\_st}$)	SOC_{bat} is H [0.9 1]	Δf_{L_t} is N then Δp_{bat_ref} is Z
			Δf_{L_t} is Z then Δp_{bat_ref} is Z
			Δf_{L_t} is P then Δp_{bat_ref} is P

<table>
<tr><td rowspan="13">N3</td><td rowspan="13">Prepare storage system for production program
–
Charge storage system via grid or PV system</td><td rowspan="7">SoC_{sc} is L
[0 0.2]
Power smoothing and frequency support</td><td>ΔP_{sc} is HN then P_{sc_ref0} is HN</td></tr>
<tr><td>ΔP_{sc} is MN then P_{sc_ref0} is MN</td></tr>
<tr><td>ΔP_{sc} is LMN then P_{sc_ref0} is LMN</td></tr>
<tr><td>ΔP_{sc} is Z-LMP-MP-HP then P_{sc_ref0} is Z</td></tr>
<tr><td>Δf_{C_t} is P then Δp_{sc_ref} is Z</td></tr>
<tr><td>Δf_{C_t} is Z then Δp_{sc_ref} is Z</td></tr>
<tr><td>Δf_{C_t} is N then Δp_{sc_ref} is N</td></tr>
<tr><td rowspan="6">SOC_{bat} is L
[0 0.15]
Power smoothing and frequency support</td><td>ΔP_{bat} is HN then P_{bat_ref0} is HN</td></tr>
<tr><td>ΔP_{bat} is MN then P_{bat_ref0} is MN</td></tr>
<tr><td>ΔP_{bat} is LMN then P_{bat_ref0} is LMN</td></tr>
<tr><td>ΔP_{bat} is Z-LMP-MP-HP then P_{GES_bat} is N (**Charge from grid**)</td></tr>
<tr><td>Δf_{L_t} is N then Δp_{bat_ref} is N</td></tr>
<tr><td>Δf_{L_t} is Z then Δp_{sc_ref} is Z</td></tr>
<tr><td></td><td></td><td></td><td>Δf_{L_t} is P then Δp_{sc_ref} is Z</td></tr>
<tr><td colspan="4">Note: additional rules in case of evening storage use:
ΔP_{bat} is Z, SOC_{bat} is N and Δf_{L_t} is Z then P_{GES_bat} is N.
ΔP_{bat} is Z, SOC_{bat} is P and Δf_{L_t} is Z then P_{GES_bat} is P.</td></tr>
<tr><td colspan="4">Color code: Charge: ▬ Discharge: ▬ Rest: ▬</td></tr>
</table>

Table 4.8. *Fuzzy rules. For a color version of this table see www.iste.co.uk/robyns/buildings.zip*

Note that the number of mathematical combinations of different inputs and outputs gives a total of 6615 (7×7×5×3×3×3) rules. However, taking account of the fact that the variables are independent, respecting the physical limits of the system and following the methodology set out in section 4.5.1, we obtain only 52 rules [ROB 15].

4.5.8. *Evaluation indicators*

4.5.8.1. *Performance indicators*

The purpose of these performance indicators is to measure the efficiency of the supervisor. In accordance with our management objectives, we have selected two indicators, intended to measure fulfillment of planned power requirements and frequency support, respectively:

– Production program fulfillment score: this indicator is based on the MAPE (Mean Absolute Power Error) score, representing the mean error between the energy injected into the grid E_{inj} and the programmed energy production determined by the producer E_{pg} over a period of 30 min. It is evaluated each 30 min step using the following equation:

$$M = \frac{100}{n} \sum_{t=0}^{n} \frac{|E_{inj} - E_{pg}|}{E_{pg}} \quad [4.9]$$

– Frequency variation indicator: this indicator is based on the mean frequency variation over 15 min:

$$F = \frac{1}{T} \sum_{t=0}^{T} |\Delta f_{C\_t}| \quad [4.10]$$

4.5.8.2. *Financial indicator*

The financial indicator proposed in this study is based on Appendix 10 of the CRE 2015 bid for Non-Interconnected Zones [ZNI 15]. For each 1 min time step, the compensation for the energy produced is:

$$Rem = P_{rod} \times P_{rix}/60 \quad [4.11]$$

For each 1 min step, if the producer fails to respect his engagement within a margin of ±5% of installed power, penalties will apply. The calculation of these penalties per minute is described in Table 4.9.

Penalty	Circumstance
0	Program is followed (if $P_{rod} \leq P_{pg} + 5\% \times P_{inst}$ and $P_{rod} \geq P_{pg} - 5\% \times P_{inst}$)
$Prod \times Prix/60$	Overproduction (if $P_{rod} > P_{pg} + 5\% \times P_{inst}$)
$Prix/60 \times Eq[4.12]$	Underproduction (if $P_{rod} < P_{pg} - 5\% \times P_{inst}$)

Table 4.9. *Penalty calculation method*

$$P_{rod} * \frac{P_{rod}}{P_{inst}} - \left(0.1 + 2 \times \frac{P_{rev}}{P_{inst}}\right) * P_{rod} + (P_{rev} - 0.05 \times P_{inst}) * (0.15 + \frac{P_{rev}}{P_{inst}}) \qquad [4.12]$$

where:

P_{rod} is the mean power produced during the minute in question at the grid injection point (including PV panels, storage and auxiliary consumption) (in kW, positive in case of injection into the grid, negative when energy is drawn from the grid);

P_{rix} is the price communicated by the producer in his bid (in €/kWh);

P_{pg} is the production which the producer has committed to deliver per minute (in kW);

P_{inst} is the installation power (in kW).

4.5.8.3. *Life expectancy of storage components*

The life expectancy of storage components, particularly electrochemical batteries, expressed in years, is considered as a new evaluation criterion. The objective for the management strategy is to maximize the life expectancy of these components, calculated using the method described in section 4.4.2, with the addition of calendar aging (5% degradation per year (Saft)). To account for the uncertain elements of calculation and for chemical phenomena which are not linked to charge and discharge cycles, we will consider that batteries have reached end of life at 80% total degradation.

4.6. Simulation results

The purpose of this section is:

– to validate our supervision strategy through simulation;

– to demonstrate the technical and/or economic advantages of a hybrid system combining two storage technologies.

Powers of hardware are:

– peak power for the photovoltaic source is 30 kW under standard production conditions (radiation 1000 kW/m², temperature 25°C);

– nominal power of lithium NCA batteries (energy source): 15 kW/39 kWh: available when charging and discharging;

– nominal power of supercapacitors (power storage source): 5 kW (Psc_max = 15 kW, 165 F–48 V, 52.8 Wh). This choice is justified by the fact that transient peaks do not exceed 15 kW and by financial considerations.

The photovoltaic production scenarios and programmed power (contracted with the grid operator on D-1) correspond to a real site in an isolated zone, using weather data supplied by the NREL in the USA in 2007 [NRE 16]. This data is collected on a per-minute basis.

4.6.1. *Supervisor validation*

4.6.1.1. *Operating mode N1*

Figure 4.21a shows the evolution of storage power (in yellow) needed to satisfy the production program (in blue) described in Figure 4.21b. Figure 4.22 shows the elements of response provided by our two different storage technologies. According to our plan, the lithium NCA batteries (energy source) provide long-term storage (red curve), while the power source responds to rapid variations in photovoltaic production (blue curve). Figure 4.23 shows the evolution of the state of charge of the lithium NCA batteries (SoC_{bat}) in normal mode (N1) (section 4.5.4). The maximum charge and minimum discharge levels of the storage components are not exceeded (SoCbat_max = SoCsc_max = 90% of charge in Ah, SoCbat_min = 20%, SoCsc_min = 30% of charge in Ah).

Figure 4.24 shows the percentage of error in respecting the production program (MAPE score, evaluated every 30 min). This error remains well within the 5% limit for the case in question.

Figure 4.25 shows the evolution of frequency in 15 min steps. The grid support function successfully reduces the amplitude of frequency variations.

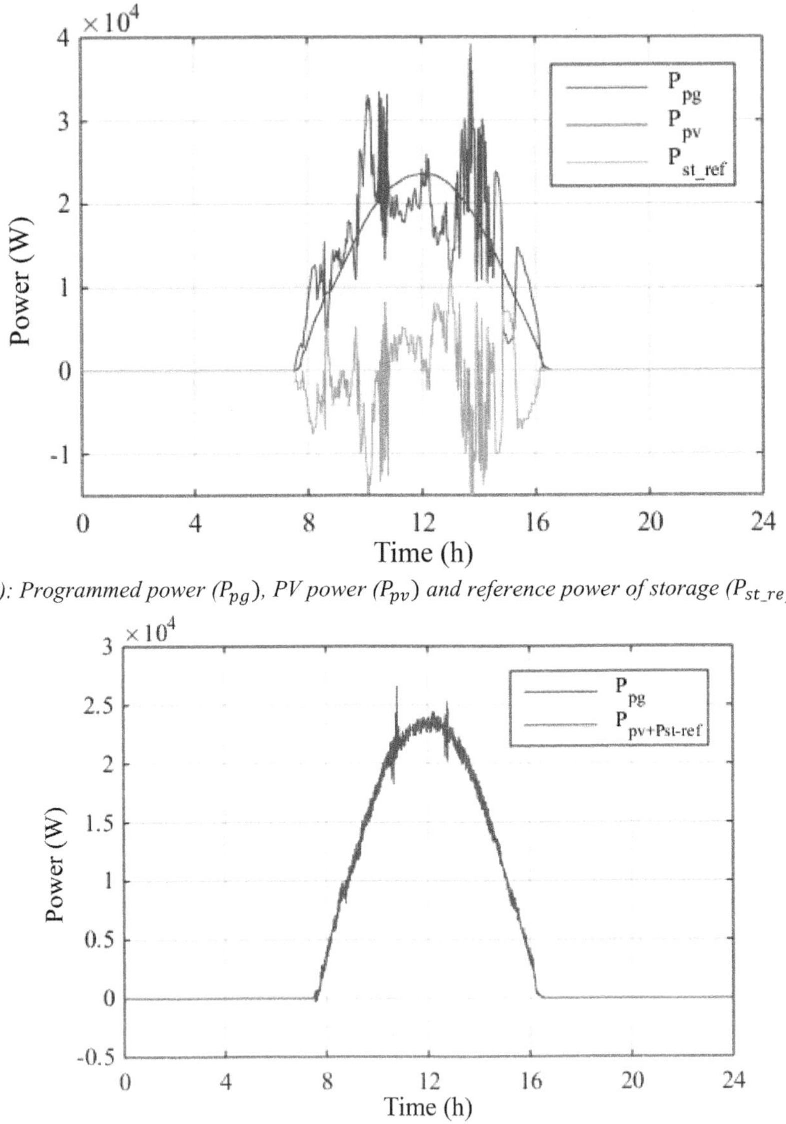

(a): Programmed power (P_{pg}), PV power (P_{pv}) and reference power of storage (P_{st_ref})

(b): Programmed power (P_{pg}) and the sum of PV power and storage power (P_{pv} + (P_{st_ref}) (Scenario N1)

Figure 4.21. *Evolution of powers for the PV plant over the course of a day*

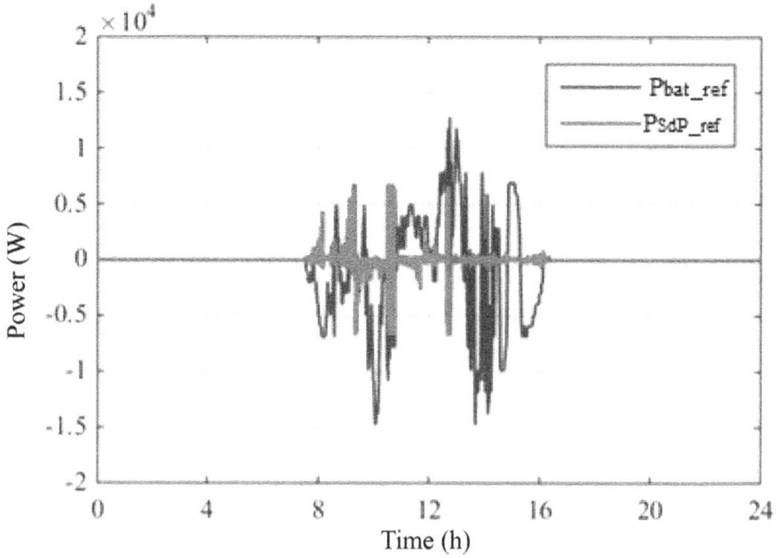

Figure 4.22. *Complementary operation of lithium NCA batteries (P_{bat_ref}) and the power source (P_{Sdp_ref})*

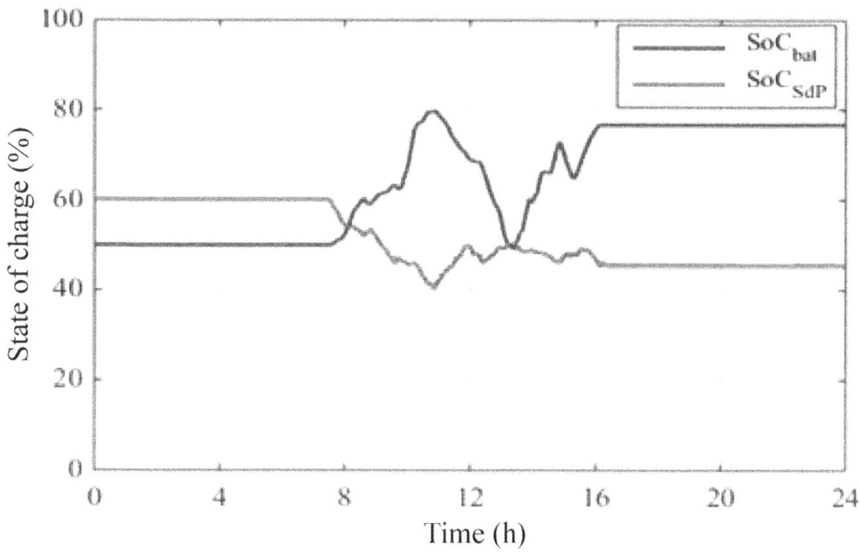

Figure 4.23. *State of charge of lithium NCA batteries (SoC_{bat}) and supercapacitor (SoC_{SdP})*

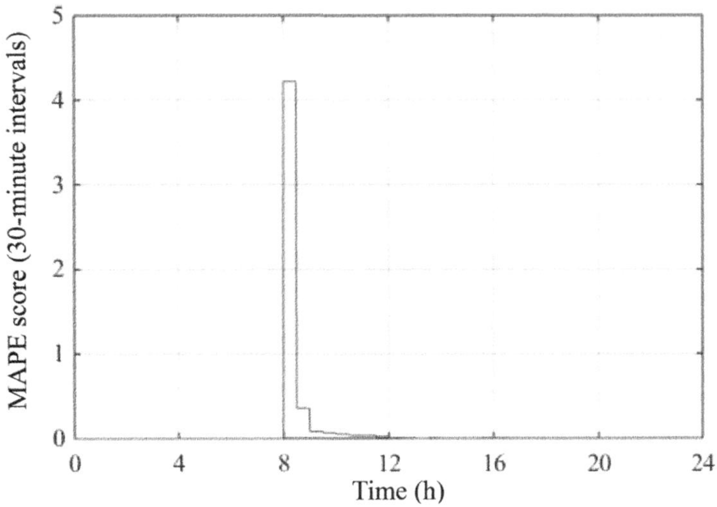

Figure 4.24. *Percentage of error in respecting the production program (MAPE score, evaluated every 30 min)*

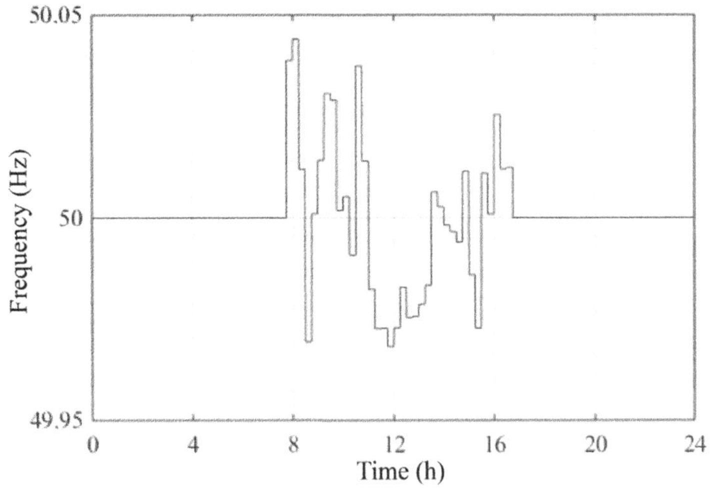

Figure 4.25. *Evolution of measured grid frequency at 15 min intervals*

4.6.1.2. *Operating mode N2*

To evaluate supervisor behavior in mode N2 (overcharge of the storage system), we chose to test the PV production scenario shown in Figure 4.26. This scenario corresponds to a sunny day and poor predictions. Figure 4.27 shows that the PV downgrading factor is activated each time the desired SoC limit for the lithium battery (90%) is exceeded. Figure 4.28 shows that the percentage of error in respecting the production program remains low while the PV downgrading factor is active.

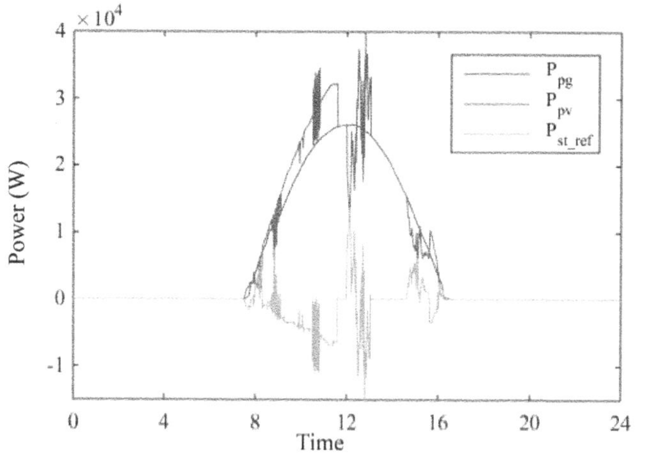

Figure 4.26. *Evolution of powers in the PV plant over the course of a day. Programmed power (P_{pg}), PV power (P_{pv}) and reference power of storage (P_{st_ref}) (Mode N2)*

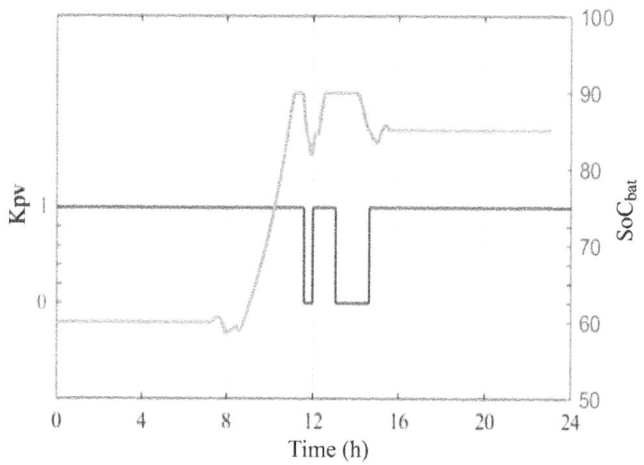

Figure 4.27. *State of charge of batteries (SoC_{bat}) and photovoltaic production downgrading factor (K_{pv})*

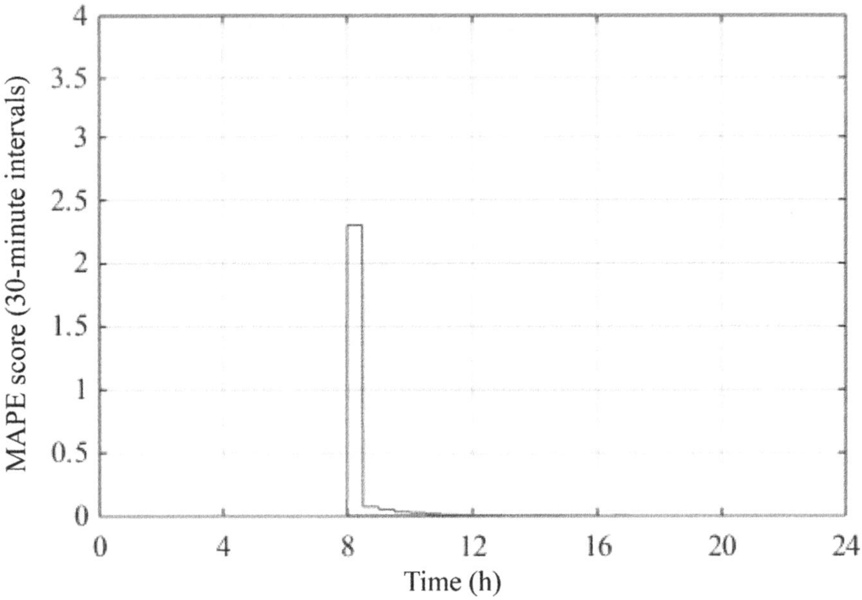

Figure 4.28. *MAPE (Mean Absolute Power Error)*

4.6.2. *Life expectancy of storage elements*

To analyze the contribution made by combining two storage technologies, we will estimate the life expectancy of the lithium battery using the cumulative damage rule (see section 4.4.2) in two different configurations (hybrid and non-hybrid):

A) Lithium-ion battery (NCA or NiMH or FePO4) **in combination with a supercapacitor (power storage source)**:

– electrochemical batteries (NCA, NiMH or FePO4) with a unitary power of 1.5 kW and a unitary capacity of 74 Ah under 48V and global powers from 9 kW to 31.5 kW with intervals of 4.5 kW, giving a total capacity of between 444 Ah and 1554 Ah;

– supercapacitors (Maxwell) with a nominal power of 5 kW and a capacity of 52.8 Wh;

– the peak power of the photovoltaic source is 30 kW.

B) Lithium-ion battery (NCA, NiMH and FePO4) **alone (not combined with a power source)**:

– electrochemical batteries (NCA, NiMH or FePO4) with a unitary power of 1.5 kW and a unitary capacity of 74 Ah under 48V and global powers from 9 kW to 31.5 kW with intervals of 4.5 kW, giving a total capacity of between 444 Ah and 1554 Ah;

– the peak power of the photovoltaic source is 30 kW.

For the purposes of this study, we will consider the scenario for a typical month (data from June 2007 for a non-interconnected zone), covering photovoltaic production and programmed power. The profiles in question are shown in Figure 4.29.

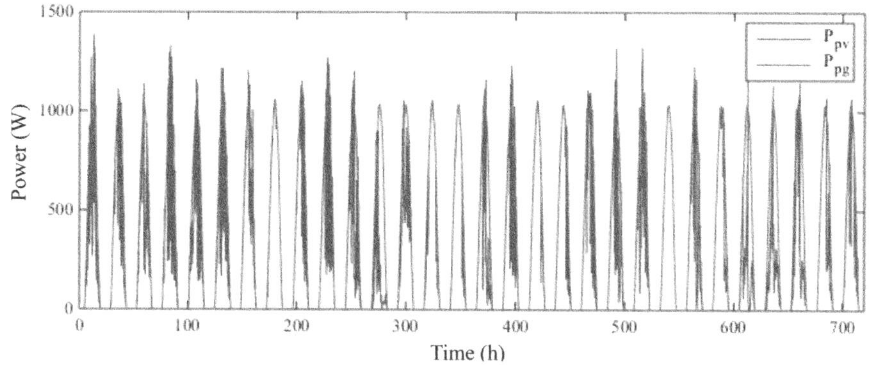

a) Programmed power and PV power production over a typical month

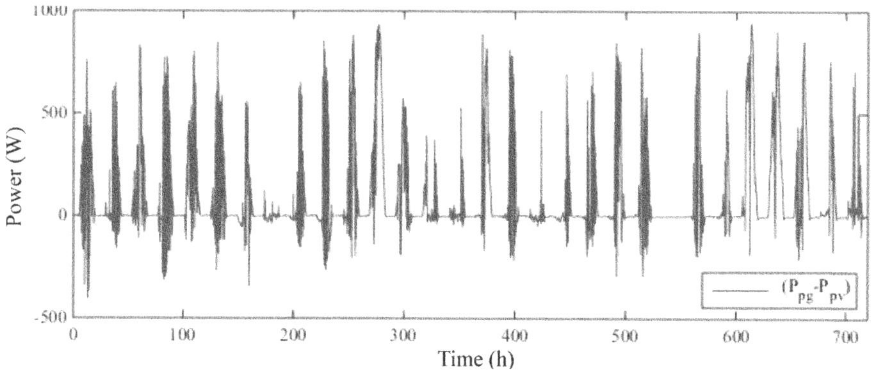

b) Difference in power $\Delta P = P_{st} = P_g - P_{pv}$ over a typical month

Figure 4.29. *Evolution of powers in a photovoltaic plant over a typical month*

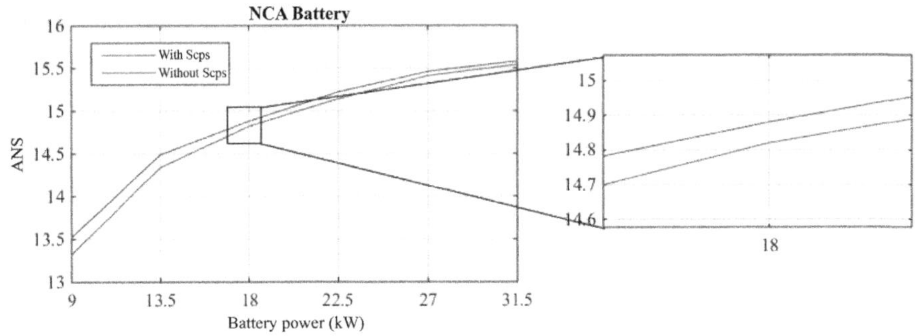

a) Results for lithium-ion NCA batteries

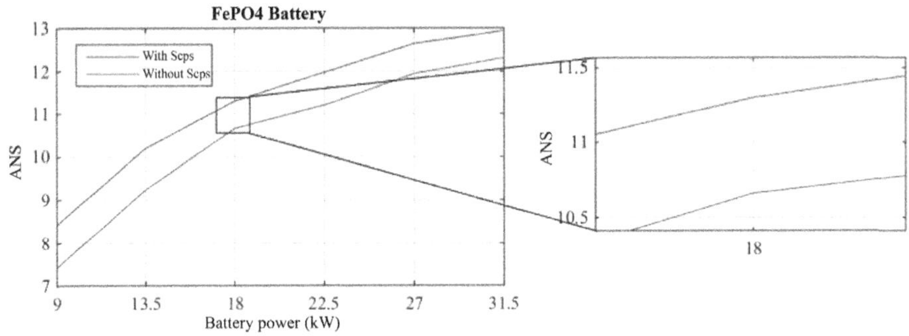

b) Results for lithium-ion FePO4 batteries

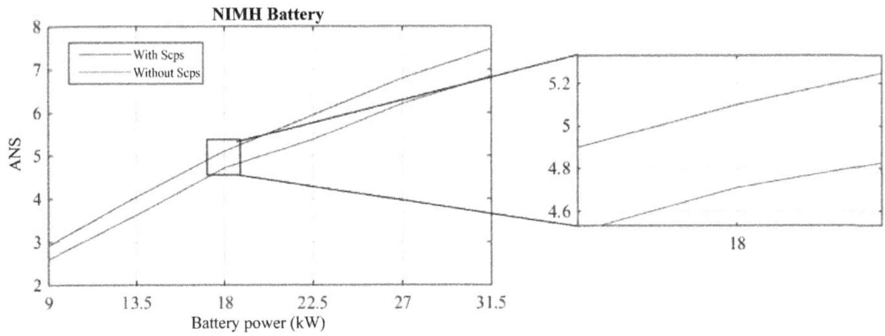

c) Results for NiMH batteries

Figure 4.30. *Life expectancy of three different electrochemical battery types with (blue) or without (red) hybridization with a 5 kW supercapacitor*

The results showing the life expectancy of three types of electrochemical battery, separately and in combination with supercapacitors, are shown in Figure 4.30. We see that:

– the addition of a power storage source results in an increase of a few months in the life expectancy of lithium batteries (energy source). This is dependent on the power profiles used and the energy capacity of the power source (in this case, the capacity is low, at 52.8 Wh);

– the life expectancy of NCA technology is significantly higher than that of other battery technologies, with a duration of around 15 years, compared to 13 years for FePO4 and around 8 years for NiMH batteries. The use of NCA batteries, alone or in conjunction with supercapacitors, appears to be the best choice in terms of life expectancy;

– the life expectancy of supercapacitors has not been precisely quantified for two reasons: 1) the aging profile of these units is not available, and 2) the number of cycles specified by the manufacturer is considerably higher than the number of cycles for the batteries in question ($> 10^6$ at 70% DoD) [MAX]. We assigned a life expectancy of 14 years;

– system dimensioning is crucial, affecting both the life expectancy of electrochemical batteries and frequency regulation. The choice of power separation filter between the two sources is also an important parameter. It should ideally be chosen with regard to the maximum charge and discharge time of the supercapacitors. From our tests, a charge or discharge time of 2 min appears to be a good choice.

4.6.3. *Efficiency*

The efficiency of the full storage system is obtained using a complex multi-component calculation. This calculation is particularly important as the parameter is crucial when determining system sizings and models.

In this context, efficiency is taken to be an indicator of the energy dissipated when the system is charging and discharging. A distinction is made between charge, discharge and round-trip efficiency. In a storage system, round-trip efficiency is defined as discharged energy divided by charged energy.

The global or "energy" efficiency of a battery is the product of two components:

– **A faradic component**, corresponding to a loss of electrical energy resulting from parasite electrochemical reactions. In the case of a lithium-ion battery, decomposition of the electrolyte leads directly to a loss of capacity (or battery health). Thankfully, faradic efficiency is very close to 1, meaning that batteries can withstand thousands of cycles.

– **An ohmic component**, corresponding to a loss of electrical energy in the form of heat, linked to the internal resistance of batteries. For a lithium-ion battery, this phenomenon is more significant when charging than discharging. We have thus used an efficiency of 0.9 when charging and 0.95 when discharging.

It is also essential to take account of the efficiency and self-discharge of the supercapacitor, figures which are significantly higher than for batteries.

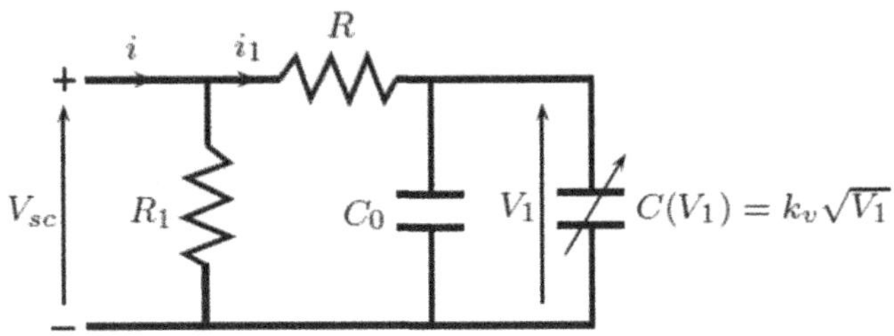

Figure 4.31. *Model of the supercapacitor*

Our model thus includes the following parameters:

– nominal capacity;

– internal resistance;

– parallel resistance (linked to self-discharge);

– variable capacity.

These are obtained from the data sheets provided by manufacturers, except for variable capacity, which is found in the bibliography [BEN 13].

The parameters used in our simulations are as follows:

Designation	Lithium-ion battery: Saft Evolion
Nominal capacity	74 Ah/3.55 kWh
Maximum/nominal/minimum voltage	56/48/42 V
Nominal charge/discharge current	32/44 A
Charge/discharge yield (round-trip)	>95%

Table 4.10. *Manufacturer data used to parameter Li-ion batteries [SAF 12]*

Designation	Supercapacitor: Maxwell BMOD0165 P048 BXX
Nominal capacity	165F
Nominal voltage	48V
Internal resistance ESR_{DC}	6.3mΩ
Absolute maximum current	1900A
Maximum continuous current	79A
Self-discharge current	5.2mA

Table 4.11. *Manufacturer data used to parameter supercapacitors [MAX 13]*

We used measured powers to calculate the mean global efficiency of the system (every 30 min) using the following relationship:

$$Mean\ efficiency\ in\ \% = \frac{E_{injected\ into\ grid}}{E_{pv}} * 100 = \frac{E_{pv_{inj}} + E_{sto\_dech}}{E_{pv}} * 100 \qquad [4.13]$$

where:

E_{sto}_dech is the stored energy (negative when discharging);

E_{pv}_inj is the photovoltaic energy actually injected into the network (taking account of downgraded production);

E_{pv} is the photovoltaic energy produced at the panel terminals.

Figure 4.32 shows an example of global efficiency calculated for a lithium-ion NCA 13.5 kW/32 kWh battery used in conjunction with a supercapacitor with a nominal power of 5kW. This is a mean energy yield evaluated in 30 min steps. It does not dip below 90%.

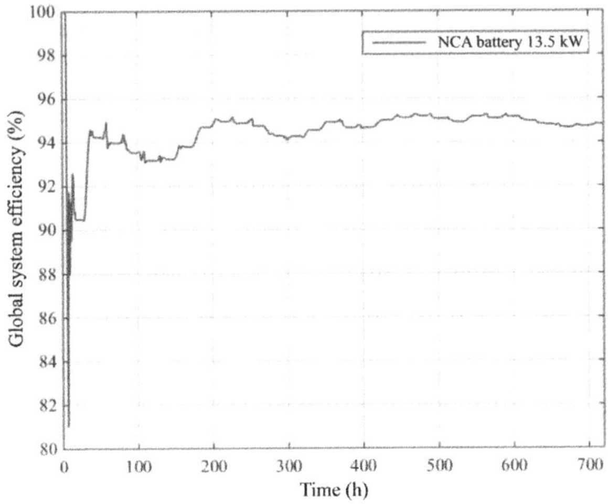

Figure 4.32. *Global efficiency of photovoltaic system with hybrid storage (mean value over 30 min)*

4.6.4. *Levelized cost of energy*

In this section, we will calculate the levelized cost of energy (***LCOE***). This is the full price of energy across the whole lifespan of the producing equipment. For this calculation, we used the following data [BEN 16]:

For the photovoltaic plant:

– life expectancy of 30 kW plant: 25 years;

– actualization rate (r): 2%;

– construction cost of PV plant: €1/Wc;

– operation and maintenance (O&M) costs per year (PV): 20 k€/MWc/year (i.e. 19 k€ over 25 years, taking account of the 2% actualization rate);

– quantity of energy producible per year: 1000 kWh/kWc;

– rate of degradation of PV productivity: 0.5%/year;

– price of energy specified by operator: €200/MWh.

For the storage system:

– price of lithium-ion NCA batteries including transportation, installation, power converter and transformation station: €550/kWh with a replacement cost of €350/kWh after 12–15 years;

– O&M for storage per year (estimated): 10 k€/MW/year (9.6 k€ over 25 years including the actualization rate of 2%);

– price of 5 kW, 52.8 Wh supercapacitor battery including transportation, installation, power converter and transformation station: €500/kW with a replacement cost of €200/kW after 14 years.

Table 4.12 shows the cost of storage systems with different configurations over a period of 25 years.

Power of NCA batteries (kW)	Batt. cap. (kWh)	Batt. price (€)	No. rep. batt.	Total batt. price (€)	5kW supercap. price (€)	Total storage cost (€)
Case 1: 9	21	11550	2	18900	3500	22400
Case 2: 13.5	**32**	**17600**	**1**	**17600**	**3500**	**21100**
Case 3: 18	43	23650	1	23650	3500	27150
Case 4: 22.5	53	29150	1	29150	3500	32650
Case 5: 27	64	35200	1	35200	3500	38700
Case 6: 31.5	75	41250	1	41250	3500	44750

Table 4.12. *Cost of storage for different powers over a period of 25 years*

The LCOE is calculated using the following equation, which expresses the net present value of all costs over the lifetime of the asset divided by the total electrical energy output of the asset. Index *t* represents the number of years *(t = 25 years)*:

$$\text{Price} = \frac{\text{Total cost}}{\sum \text{Energy}/\text{t}} \qquad [4.14]$$

where:

Total cost = (cost of PV + cost of storage + O&M for PV + O&M for storage + price of storage transformation storage + penalties) [4.15]

with: $O\&M_{total} = (O\&M)_{year} \times (1 + r)^{nb\_year}$ [4.16]

Penalties are calculated using the financial indicators described in section 4.5.8. Table 4.13 shows the penalties for grid operators (in case of failure to respect the production program) for different battery powers over 25 years. From this, we can deduce the corresponding LCOE.

Battery power kW	Penalties in € over 25 years	LCOE (€/MWh)
9 kW	927	185
13.5 kW	742	182
18 kW	657	195.5
22.5 kW	552	208
27 kW	541	221.5
31.5 kW	576	235

Table 4.13. *Penalties payable to the grid operator for different battery powers over 25 years*

Taking the results from the previous tables for case 2, the option with the lowest storage cost, we obtain a levelized energy cost of about €182/MWh.

4.7. Experimental validation of energy management

The aim of our experimental tests is:

– to demonstrate the usefulness of a hybrid storage system (batteries plus supercapacitor);

– to enable concrete testing of our multi-objective management strategy;

– to evaluate the energy management indicators in an experimental manner.

These tests were carried out at the Electrical Power Management Lab platform (EPM Lab) at the Electrical Engineering and Power Electronics Laboratory (L2EP) in HEI Lille, on a reduced-scale.

4.7.1. *Definition of tests*

Our test configuration included two hybridized storage elements: a supercapacitor and a Li-Ion battery. The dimensions, determined by the material available in the laboratory, were as follows:

PV generator	1800 W
Battery	1500 W/3700 Wh
Supercapacitor	3800W/53 Wh

Table 4.14. *Test system configuration*

An outline of the platform and the control/command interface developed using ControlDesk is shown in Figure 4.33. To demonstrate the reliability and robustness of our management method, testing was carried out on a day when PV production was highly variable. Our test period was limited to times with PV production; thus, each test ran for between 6 and 9 hours.

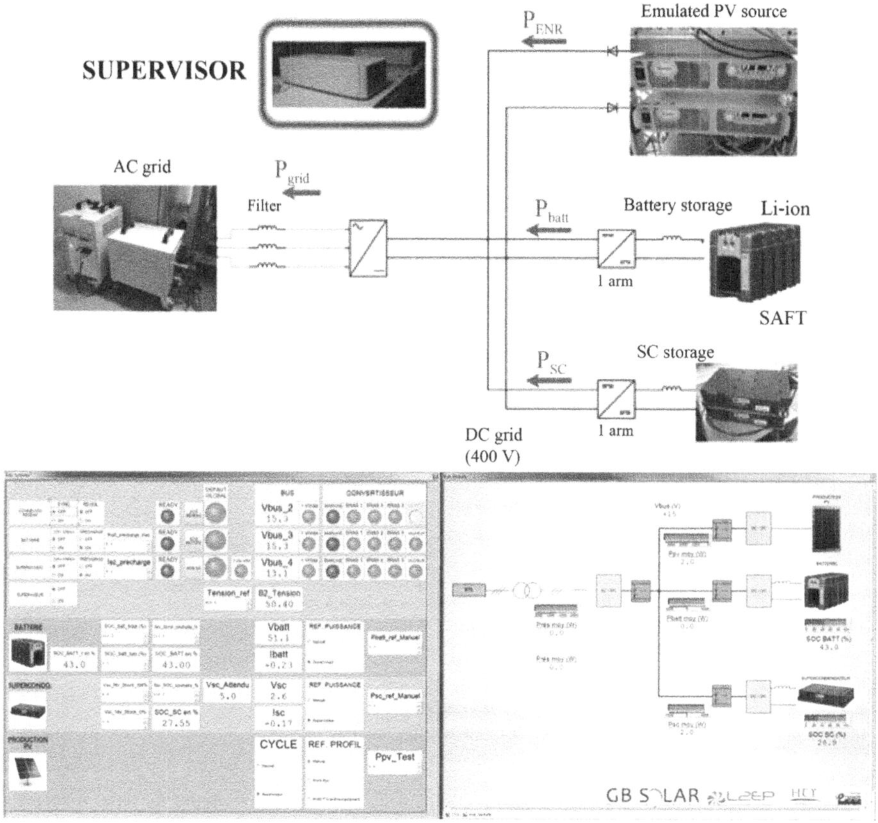

Figure 4.33. *Diagram of test platform and control/command interface (reduced-scale prototype, 1800 W)*

4.7.2. *Experimental results*

The analysis of our experimental results (Figures 4.34–4.37 and Table 4.15) shows that:

– the system is well-managed, ensuring, as far as possible, that the production program is respected (Figure 4.34) with a MAPE score of under 5% and with maintenance of frequency within the authorized range as far as possible (Figure 4.37);

– the two storage components are well-managed, as we planned during the design phase, the lithium NCA batteries (energy source) are used for long-term storage cycles, while the supercapacitors (power source) respond to rapid variations in PV production (Figure 4.34);

– the PV production downgrading factor (limitation of PV production via the inverter) is well-managed, and is activated each time the desired SoC limit for the lithium NCA batteries (90%) is reached (Figures 4.35 and 4.36);

– the indicators chosen for this study (Table 4.15) were successfully evaluated. The extrapolated battery degradation factor, estimated over the course of the year based on tests for a single day, gave a value of 1.93%.

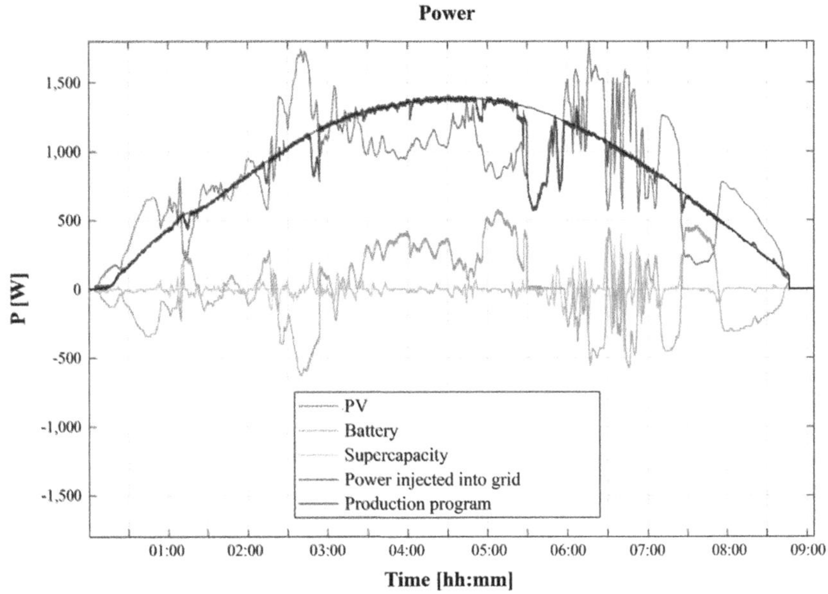

Figure 4.34. *Evolution of powers of PV plants over a day. PV power (red), electrochemical battery power (light green), supercapacitor power (light blue), actual power injected into grid (dark blue) and power programmed by the grid operator (black)*

While these results effectively validate the supervisor and testing systems, there are a few points for improvement before the operation of the supervisor may be validated in its entirety, particularly in terms of micro-management of the battery (notably avoiding deep discharge and maintaining the minimum SoC in case of low or zero PV production) and in improving life expectancy.

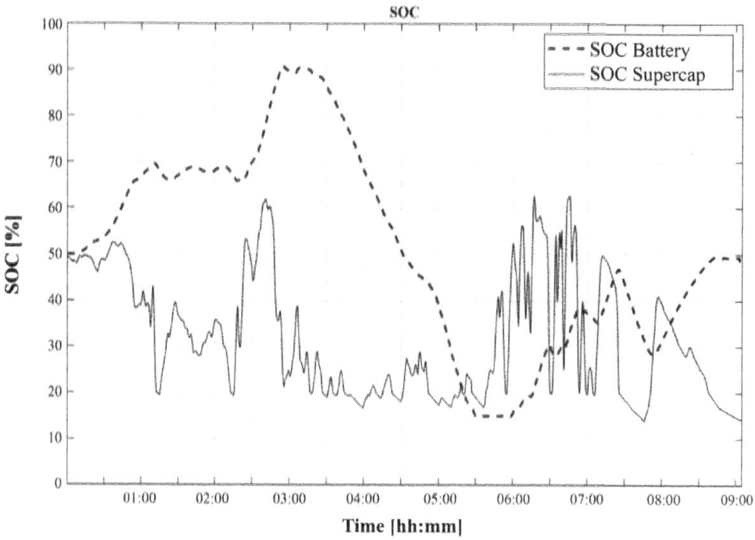

Figure 4.35. *Evolution of the state of charge (SOC) of the battery (blue) and supercapacitor (black)*

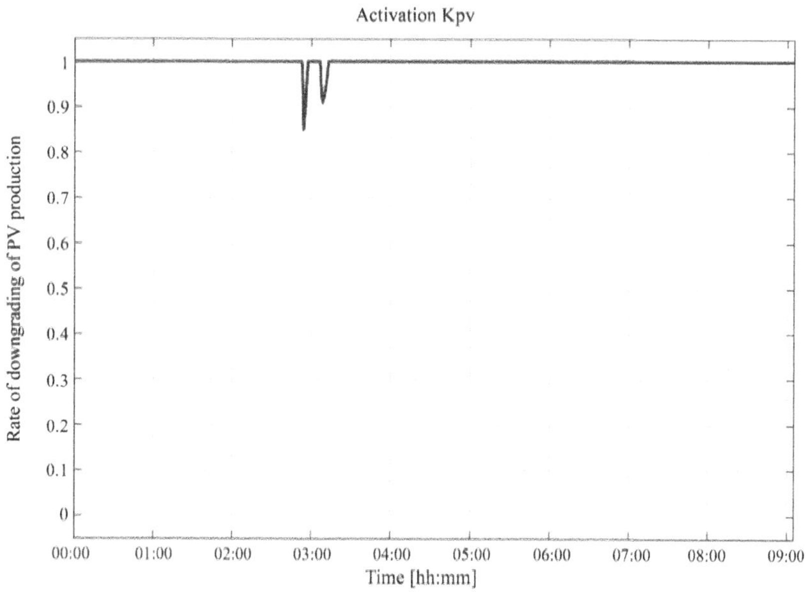

Figure 4.36. *Activation of the downgrading function Kpv*

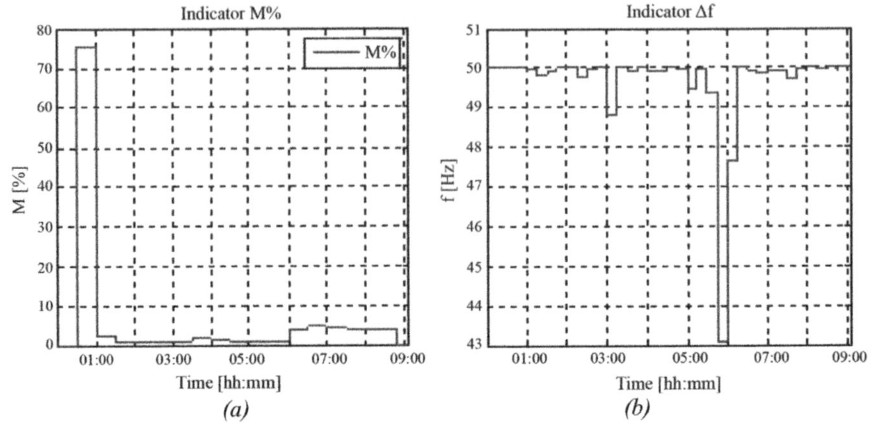

Figure 4.37. *(a) Percentage error in satisfying production program (MAPE score, evaluated at 30 min intervals); (b) Evolution of frequency at 15 min intervals (see section 4.5.8)*

Test with highly variable PV profile	
Remuneration (€) for test day	3.85
Penalties (€) for test day	0.056
Battery degradation for test day (%)	0.005
Annual battery degradation (%)	1.93

Table 4.15. *Results for remuneration, penalties and degradation of electrochemical battery*

4.8. Conclusion

In this chapter, we developed a photovoltaic system combining two different storage technologies (one for energy, using electrochemical batteries, and one for power, using Maxwell supercapacitors). A “smart” supervision algorithm based on fuzzy logic was implemented following a structured methodology. Simulations demonstrated that the system attained the desired objectives in terms of conformity to the production program, while respecting the different constraints imposed by the electric grid operator.

A comparative study of different storage configurations, particularly in terms of the life expectancy of storage components and of system energy costs (LCOE), was also implemented. The costs involved over the lifetime of the system were analyzed in order to assist designers. We deduced that:

– the use of lithium-ion batteries with NCA technology (Saft), alone or in conjunction with supercapacitors, appears to be one of the best choices in terms of life expectancy;

– the sizing of the storage system, in terms of energy and power, is crucial: it affects both the life expectancy of batteries and the capacity of the system for frequency regulation;

– in order to successfully adjust production in relation to the production program and to regulate frequency correctly while ensuring that costs remain reasonable, power storage should represent ≥50% of the installed PV power. A 1-to-1 ratio is recommended;

– in our case study, the proposed system gave an LCOE of less than €200/MWh.

Finally, we validated the fuzzy logic approach to energy management using a test bed with a realistic reproduction of the system in terms of power and emulation time. This demonstrated the effectiveness of our proposed method.

The proposed system offers a solution for ensuring a continuous service for buildings and neighborhoods, particularly in zones which are not connected to the grid. In this context, storage may contribute to:

– securing the energy supply, for example if a production turbine stops;

– compensating for irregularities in intermittent energy production;

– balancing demand and production, contributing to the stability of the electrical grid;

– improving the reliability of renewable energy sources operating off-grid;

– increasing the part played by renewable energy sources in the electrical mix;

– promoting self-consumption.

However, a solution using Li-ion batteries alone may be more economical and may be sufficient, given the high power density of these units and the recent developments which have been made in both technological and economic terms.

4.9. Acknowledgments

This study was carried out in partnership with GB Solar Company, was funded in part by the Hauts-de-France regional authorities, and received support from the MEDEE (*Maîtrise Energétique des Entraînements Electriques*, Energy Management of Electrical Programs) cluster.

5

Economic and Sociological Implications of Smart Grids

5.1. Introduction

The evolution of electrical grids toward a "smarter" model is reliant on the development of new information and communication technologies (notably including smart meters). The full innovative potential of these smart grids lies in their capacity to promote interactions between different actors in the electrical system, increasing the electrical "smartness" of the actors themselves. These actors are extremely diverse, from large-scale centralized producers to smaller, decentralized production units, residential, tertiary and industrial consumers, transportation systems, storage systems and the grid operators who manage distribution and transmission grids. Their consumption and/or production profiles are very different, and the economic and societal objectives or constraints involved are highly varied. New types of actors may also emerge, leading to the development of new economic models and providing responses to energy and climate issues, promoting the development of renewable energy sources. One of the challenges of this development is to ensure that all actors benefit, including those in a situation of energy poverty. A related development concerns the emergence of service aggregators, grouping actors with similar profiles (renewable energy producers, residential or industrial consumers, etc.) into units sufficiently large to provide economically viable services. Similar aggregations could also be developed between actors with different but complementary profiles.

One of the main elements of the evolution of smart grids is the shift from consumers to "prosumers". This evolution concerns consumers with their own local renewable production sources and/or storage systems of any size, or who, more generally, possess storage capacity (thermal, in the form of hot water tanks; electrical, in the form of electric vehicles, etc.). These consumers technically have the capacity to modulate their consumption profile or even drop out of the grid on occasion. This capacity is potentially useful for grid operators, for economic reasons and for environmental reasons. However, these capacities can only be exploited with the consent of the actors concerned. The motivations involved may vary according to socio-economic profiles, including economic, ecological and technological aspects; some actors remain indifferent, whereas others are strongly opposed to these developments.

Recent advances in self-production and self-consumption[1] have boosted the use of these load modulation and load cut off. Collective self-consumption of local renewable production has taken off in many countries. Self-consumption is said to be collective when not all of the consumers of local production also own production units; it is individual when the consumer owns the production unit. The notion of "local" may be defined in a number of ways, for example as a specific part of the distribution grid (e.g. downstream of a transformation station [CRE 18] serving part of a residential neighborhood) or as a distance, for example a radius of 1 km around the production unit [MIN 18], within which energy may be exchanged between large-scale tertiary buildings as well as residential neighborhoods.

Individual self-consumption tends to have a positive effect on energy mastery and enables a reduction in electricity costs (potential reduction in the cost of connection to the distribution grid, subscribed power and subscription, and taxes). Collective self-consumption has the additional effect of increasing self-consumption levels by grouping buildings with different consumption profiles.

Additional benefits may be obtained in terms of energy balancing by adjusting consumption or via self-consumption of local production, with or without the use of storage systems.

Communities may also benefit in cases where self-consumed energy comes from renewable sources (a major consideration in self-consumption), distributed across multiple production sites (particularly photovoltaic panels on roofs), reducing or even eliminating the energy losses resulting from long-distance transportation.

We will begin by considering the different rationalities of actors in smart grids, which may vary widely and affect whole groups. Next, we will discuss the economic

1 As defined in section 1.4.2.

and social implications of smart grids, including changes to the chain of value, contract models, the socio-economic profiles of consumers or prosumers and regulation. We will consider the social acceptability of participation in energy management, specifically in the context of energy balancing in a multi-actor commercial building (shopping mall) and in a domestic setting (households in residential buildings).

5.2. Actor diversity in smart grids

The development of smart grids opens up new perspectives, by giving all actors, including consumers, the capacity to play an active role; previously, network stability was essentially managed by large-scale centralized producers and by certain storage systems, such as hydraulic (pumped) storage [ROB 15].

Energy producers may be grouped into two categories: large-scale centralized producers (thermal or nuclear power plants, large-scale hydraulic dams) and decentralized producers, often using renewable energy sources dependent on intermittent sources (photovoltaic, wind power, small-scale hydraulics) [ROB 12c]. Private individuals producing electricity, for example via solar panels installed on the roof of a house, fall into this category.

Similarly, there are three groups of consumers: residential (houses, apartments), tertiary (office buildings, stores, etc.) and industrial. Finally, the transport systems which come into contact with the electrical grid – rail, metro, tram, trolley, electric vehicles, etc. – must also be taken into account [ROB 16].

Electrical grid operators are responsible for providing consumers with a reliable supply of high-quality energy. The integration of new groups of actors is particularly important in this context, as operators must ensure connections between these actors and must balance production and consumption. The development of smart grids has a significant impact on operators, both in technical terms, necessitating changes to the network, and in economic terms, requiring the development of new economic models. These changes may have both positive and negative implications for operators in different cases.

It is important for electrical grid operators to identify actors offering a certain level of flexibility within the electrical network, for example storage systems such as STEPs. Centralized producers must also provide flexible production. In the context of smart grids, decentralized producers and consumers may provide an additional element of flexibility. A single actor may simultaneously act as producer, consumer and storer of energy, as in the case studies presented in Chapters 2 and 3.

In sociological terms, it is important to understand the specific rationalities of different actors (objectives, hesitations, strategies, etc.). In economic terms, we must analyze the value of the "resources" which different actors provide or represent. This is dependent on the costs of implementation and on the importance accorded to resources by each actor in responding to their own imperatives. The combination of actors with heterogeneous rationalities presents a challenge in terms of finding a balance acceptable to all (see definition below).

Rational individuals and the limited rationality of individual choices

"The individual making a choice is a being possessing the faculty of reason, with at least a certain distance from and room to maneuver with regard to the situation, a being with the faculty to analyze a situation and to select his behaviors in accordance with what seems sensible to him, i.e. in correspondence with what he considers to be his own interests. (...) Beginning with a more or less precise idea of what might constitute an acceptable solution (...), he will examine a series of possible solutions with regard to their likely consequences (...). He will cease to look for possible solutions as soon as he has identified an option which would allow him to attain his objectives, which he therefore judges satisfactory, and thus rational for him".

Erhard FRIEDBERG, Emeritus Professor,

co-author of l'Acteur et le Système with Michel CROZIER, 1977

Citation taken from Rationalité et analyse des organisations, Cairn Info, 2011/3 No. 165

Service aggregators represent groups of actors with similar profiles (renewable energy producers, residential or industrial consumers, etc.) into units sufficiently large to provide economically viable services. Similar aggregations may also represent actors with different but complementary profiles.

5.3. Economic and sociological implications of smart grids

5.3.1. *Introduction*

From printing to steam power, electricity and "our" new technologies, each technical revolution has brought about its own economic and sociological upheavals, some predictable, other completely unexpected. Schumpeter [KAR 04] identified innovation as the driving force of economic dynamics, arguing that growth is a continuous process of creation as well as that of destruction of economic activities. This is known as "creative destruction". Schumpeter also accorded a central role to entrepreneurs who seize the opportunities offered by technological developments, as

Watt and Boulton did in their time[2], without being aware of the social implications. This unrelenting renewal, which comes in bursts, forms part of modern life, whether we like it or not. All of the actors involved in the chain of value, along with the supporting institutions, are well aware of the need to manage these innovations in such a way that the resulting changes are acceptable, or at least any drawbacks are attenuated, for those actors who are most vulnerable.

Note that energy self-consumption refers to the fact of consuming some or all of the energy produced on-site. Currently, this self-consumption does not imply total independence from the electric grid. For this to be possible, self-consumers would require greater production and storage capacities, and possess the ability to maintain system stability.

Self-consumers thus alternate between self-production and drawing power from the grid. As the majority of self-production is photovoltaic, we will use this example to illustrate the flows involved. Production occurs during the daytime, and its intensity is directly correlated to the level of sunlight. For many self-consumers, production is not always simultaneous with consumption. Short of investing in a storage system or of altering consumption habits accordingly, these individuals thus retain the status of ordinary network users, and their photovoltaic production is injected into the network in its entirety.

Until recently, with the exception of countries such as Germany where electricity is twice as expensive as in France, there was no economic interest in self-consumption. It was better to sell all of the electricity produced to the grid operator and benefit from attractive purchase rate. However, reductions in the cost of producing photovoltaic electricity have resulted in grid parity, i.e. the cost of self-produced electricity is lower than the cost of grid electricity.

Technical and legal developments mean that the shift from consumers to prosumers is likely to speed up. In France, for the moment, self-consumption remains marginal. At the end of 2017, Enedis (the French distribution grid manager) reported a total of 20,000 self-consumers. Reductions in the price of production systems, the development of decentralized energy management, the motivation of consumers and increased regulatory flexibility are going to change this established fact. In 2017, almost half of all requests to connect new production units were for self-consumption purposes. Prosumers are likely to play an important role in future systems, increasing the horizontal trend in the electric industry and raising new

2 The Boulton & Watt company was founded by Matthew Boulton and James Watt in 1775 to construct steam machinery in their factory in Soho, Smethwick, near Birmingham, England. The company continued to exist for over 120 years, and was still producing steam engines in 1895.

questions, for example in terms of territorial equity[3] and on relative control over a rapidly changing sector.

The first interest of self-consumption is, evidently, to create value. Self-consumed energy is generally produced from renewable sources. Communities thus benefit from carbon-free production (photovoltaic production emits low levels of greenhouse gases over the full lifespan of the system, particularly if the energy used to produce the system itself is low-carbon).

More than anything, self-consumption provides value to the grid. The fact of consuming electricity produced on-site reduces the quantity of electricity to transport and thus eases pressure on the grid. This may mean that certain investments can be avoided or pushed back to a later date. In this context, there are two types of value to consider: "drawn" power and "injected" power. The "drawn" value contributes to reducing consumption peaks, if self-consumption takes place during this period. This has an impact in terms of grid investment: equipment (such as power lines, transformers, point production, etc.) can be smaller. The "injected" value relates to self-consumption in a consumption zone. In this case, self-consumption reduces the need for support from the grid in response to injection peaks. Self-consumption can thus reduce grid reinforcement costs, which are partly covered by the TURPE (*Tarif d'Utilisation des Réseaux Publics d'Electricité*, Public Electricity Grid Usage Tariff), paid by all users.

Self-consumption also has risks. One of these is increased consumption. This may be accepted if the additional consumption does not result in added cost or loss of earnings for the self-consumer, which occurs, for example, if injected electricity is not taken into account.

The main risk, however, relates to the breakdown of energy solidarity at the national level. Self-consumption challenges the French energy model. As we will see in the following sections, the current model is based on centralization (large-scale power plants) and prioritizes territorial equality[4] (tariff equalization principles). At national level, pricing is based on the average cost throughout the country. In some ways, energy costs could be said to be "socialized": the additional

---

3 Self-consumption challenges the current balance based on the principle of tariff equalization (customers pay the same price wherever consumption occurs, so wherever they use the grid) which forms the basis for territorial equality in this area.

4 Grid access pricing is independent of the distance between the injection and usage site. This foundational principle of the French electrical system is also enshrined in European law (provisions in article 14 of regulation (CE) no. 714/2009). The principle of tariff equalization consists of applying the same prices across the whole national territory for the same usage. This equalization is set out in clauses of L. 121-1 and L. 121-2 of the French energy code.

costs of supply to isolated or low-density zones are spread across all users, so those in less costly zones pay a price higher than the actual cost of their own supply.

Self-consumption, however, is a localized phenomenon to which consumers do not have equal access. Self-consumption thus constitutes a fundamental challenge to the principles laid out above. As in the case of collective self-consumption, a framework promoting individual self-consumption might allow consumers with the capacity to inject energy into the grid (net contributors to the system) to break away, i.e. to cut themselves off from the main grid, at least at certain times. If this is left unchecked, a new energy model could emerge based on energy communitarianism, whereby sub-groups of consumers ensure they profit from the best tariff via a system of local contracts with producers, making the electricity system harder to understand and increasingly complex.

Equity is a major challenge for public authorities and is a significant issue for the public economy. The consequences of "individual" self-consumption and of the development of smart grids are the same in this context: in both cases, the client only utilizes a small part of the grid.

If self-consumers became exempt from the public electricity grid usage tariff (TURPE), grid operators would be allowed to increase prices in order to cover costs. Some users would thus be called upon to subsidize the costs of others. Furthermore, it seems likely that self-consumption would enrich consumers who are already in a privileged situation (richer households with enough money to install solar panels on the roof of their home). Bluntly speaking, the major issue with individual or collective self-consumption (within a smart grid) lies in preventing the emergence of energy communitarianism if the technical and economic conditions for such a development become reality.

We thus need to identify the most likely implications of smart grids, along with the degree of flexibility available to actors in order to construct sustainable economic models. We aim to provide a partial response to these questions below, combining contributions from theoretical and empirical literature with our own observations and conclusions drawn from a real-world project. The purpose of this exercise is partly to provide a socio-economic perspective on the questions involved, as well as, and especially, to identify areas for future work and to promote collaboration between the humanities and the engineering sciences.

5.3.2. *Implications of smart grids for the value chain*

A value chain consists of all of the companies and actors involved in a production process, from raw materials to a final product. To understand the chain, we need to analyze relationships between these units in the context of production, research and commercial exchanges until final consumption.

Given the scale of future changes in the electricity sector, we will begin by outlining the general conditions for these changes before considering the major shifts which are in progress.

Issues relating to CO_2 emissions and the depletion of fossil fuel resources have led actors in the energy sector to focus more and more on renewable sources. However, the introduction of "green" energy into the production mix, particularly highly variable solar and wind power, has triggered a total modification of the value production chain in this sector. There are two types of reasons for this, which may be classified as technical and systemic:

– the technical specificities of variable green sources and the need for a means of storing this energy;

– the shift from a highly centralized approach to a demand-control model, with the emergence of prosumers.

Given the intermittent nature of variable renewable sources and the difficulty of controlling output, there may be dips and peaks in production which can saturate the grid. Storage appears to be an obvious solution, but there are various technical and economic uncertainties associated with storing electricity [ROB 15, ROB 16]. A complementary solution involves integrating demand into the overall management model for energy distribution.

Any new model also needs to account for the possibility of self-consumption. In France, self-consumption has been on the rise since 2017. It is now possible, from both technical and legal perspectives, to produce part of the electricity consumed by an installation on-site. Consumers thus become both producers and actors in a system, but the distribution grid remains a shared resource.

Management of demand and knowledge of users, or producers, has become more important than ever before. Unfortunately, very little data is available. We wish to understand the behaviors, acceptance of change, reactions to behaviors of other actors, etc. of both domestic and professional users. These questions will be discussed in greater detail later in this chapter.

Green energy and new technologies have thus resulted in a huge upheaval in the energy production and distribution system. As shown in Table 5.1, we are moving from centralized and mono-directional production, with a structured and structuring transport grid, to a decentralized mode of production including storage capabilities.

	Current organization	**Smart grids**
Production	Centralized	Less centralized, with storage capabilities and increasingly complex flow management (ICT + storage technologies)
Circulation of energy and information	Mono-directional	Bi-directional (between consumers and producers, between self-consumers and network operators) or even multi-dimensional (between several self-consumers and a grid operator, for example)
Management	Balance based on control of production	Balance based on managing consumption, classic production and storage systems
Actors	Pyramid structure	Vast increase in actor numbers, now including consumers

Table 5.1. *Major current and forthcoming changes in electrical grids*

5.3.2.1. *Main conditions for change*

Leaving aside the environmental issues (notably greenhouse gas emissions) and the likelihood of increasing fossil fuel prices [ROB 12c], we will now discuss the impact of technical, legal and economic conditions on actor perceptions.

There are a number of areas of uncertainty for actors in this industry:

– Upstream: concerning technical developments and the approaches taken by regulators to limit uncertainties, i.e. regulatory developments on various levels (global, European and local).

– Downstream: acceptability to consumers and changes in consumer behavior. At first glance, it seems that final consumers may not have a real choice in the matter. However, it is already apparent that the systemic modification of the value chain will give consumers a greater and more complex role to play, both in terms of consumption (or non-consumption) and of the potential for self-production:

- User behavior is also an unknown variable. Will users adopt voluntary load reduction practices, will they support these practices, and through what process?

- How willing are people to pay for new equipment and new services? This is a major factor when considering economic models.

The main conditions for change are:

– Technological and socio-economic (upstream):

- the development of renewable energy sources, highlighting the environmental benefits of reducing fossil fuel usage;

- energy savings and decentralized production (avoiding transport losses);

- energy storage technologies (to ensure a reliable and continuous supply);

- the development of new technologies plays an inter-operational but crucial role (integrating parameters defining the constraints of consumers and prosumers, learning capacity of management systems, etc.);

- the capacity to take account of consumer and prosumer behavior in developing management processes, in a context in which the roles of grid operators and public authorities themselves are not yet fully defined.

– Legal and regulatory (upstream):

- the scenarios which may be possible depend on the willingness of public actors to engage with. The question of public debt has also been discussed in this context, as it may affect the investment capacities of public and local authorities;

- the financial risk involved in these major investments needs to be limited, via:

- favorable legislative and institutional conditions,

- guarantees on investment (Third Industrial Revolution European Society[5]) and adjusted tariffs, defined in France by the CRE (*Commission de Régulation de l'Energie*, Energy Regulation Commission),

- regulations regarding the standardization of techniques.

- Environmental stabilization is the subject of regulations at a number of different levels:

- international (CO_2 market, European initiatives to create a framework for industry);

- national (guarantee investment, *Programmation Pluriannuelles de l'Energie* – multi-annual energy program);

- local (support from local and regulatory authorities);

- the role of regulatory bodies and public authorities on different levels is crucial. Regulators ensure the balance of forces involved, which determines the way in which added value, benefits and externalities are shared. Regulators are responsible for raising awareness, for security and for communications. In economic terms, the energy question therefore has an impact on public economics. The authorities responsible for territorial developments, alongside municipal authorities, have clearly understood the challenges presented by this reconfiguration; local

5 The Third Industrial Revolution European Society is an association of European professional experts in the domains of economics and the technical and social sciences.

authorities will doubtless take on the role of regulators in coming years. Our experiment is designed to correspond to this scale in order to identify future economic models. Note that these entities are already responsible for the grid.

– In order to go further and to take account of uncertainties relating to consumer behavior, a number of studies have been carried out in the context of experimental projects. The findings from California are particularly interesting in this regard, including tariff modifications [HER 10]:

- demand-based "Time of Use" pricing;

- a tariff modified once a year, "Critical Peak Pricing", where consumers may engage in load reduction to a point established in advance;

- a more variable tariff, "Peak Day Pricing" or "Critical Peak Pricing" for industrial users.

A summary of the conditions of change involved in the emergence of smart grids is shown in Figure 5.1. This figure shows the complexity of the system and enables us to identify the position of each user in terms of both opportunities and constraints. As sustainable resources are required for the management of water, energy and transport networks, the way in which pricing affects use and location or promotes innovation has largely been ignored. The diagram in Figure 5.1 shows that new pricing models, efficiently reconciling financial constraints with behavioral orientation, are required. This is all the more necessary as resources from public contributions become increasingly rare, or dependent on a globalized economic environment.

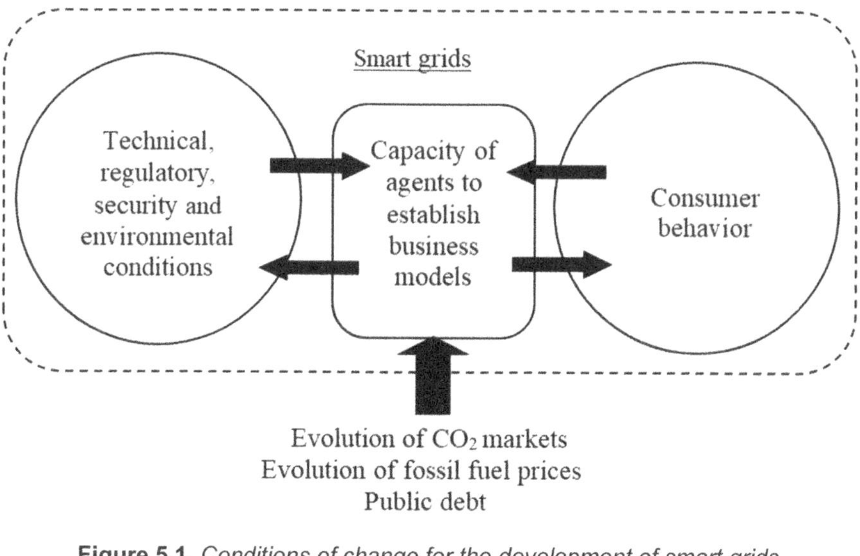

Figure 5.1. *Conditions of change for the development of smart grids*

5.3.2.2. *Current understanding of economic models*

A value chain is a set of interdependent activities, the implementation of which results in the creation of identifiable and, where possible, measurable value. It thus covers all steps of a production process, from the supply of raw materials to final consumption.

The classic definition of a value chain is as a precise representation of the activities of a company, highlighting key activities which have a real impact. The aim is to evaluate cost or quality effects which may give a company a commercial advantage. In more concrete terms, Michael Porter [POR 82, MAG 11] draws a distinction between main activities and support activities. Main activities contribute directly to the material creation and sale of a product (supply, manufacture, logistics and bringing to market, marketing and sales, services). Support services assist in the accomplishment of main activities, and constitute the company infrastructure (human resources, R&D, purchasing, etc.). However, this definition is overly restrictive in the context of energy production.

First, the energy question poses social and environmental challenges which, as ever, mean that actors are subject to political decisions, expressed in the form of regulations. Second, as we see from the conditions of change set out in Figure 5.1, the transformation of the industry and of the roles of actors has modified, and will modify, the value chain. In order to set out the principles for a business model, we need to identify and analyze two spheres, defining and helping to understand the challenges involved, areas of flexibility and the risks resulting from the reconstruction of the value chain:

– the macro dimension, i.e. a favorable top-down framework, related to environmental security (regulators);

– the micro (bottom-up) dimension, linked to the capacity of agents to establish business models which, as far as possible, successfully integrate consumer availability.

5.3.2.2.1. The major change: actor coordination

In building these models, we need to consider the way in which smart grids modify the value chain. It is also helpful to consider the way in which value will be distributed between the electricity sector and the ICT sector. On this level, the question of regulation between actors and of value sharing is crucial. The challenge is to identify sustainable models which also possess the capacity to adapt and to integrate technological innovations in an optimal manner.

Regulation should result in the sharing of added value. For this to be possible, it is necessary to coordinate between actors with potentially divergent interests. Regulation should ensure:

– effective competition within the energy market;

– security of investment;

– communications with customers.

The major change in this new value chain lies in the fact that final users play an active role in the energy system, moving from the periphery into the center. For users to play an active role, they need improved access to information on the cost/profit balance and better information regarding available services. This would solve the uncertainty regarding the capacity of users to become system actors, while protecting their data in accordance with existing legal standards.

The efficacy of the model is thus essentially dependent on the coordination of relations between the different actors involved, and on their capacity to form a coherent, collaborative network. Thus, as Patrice Goeffron states [GEO 14], the six "levers" for the formalization of models essentially center on the need for cooperation:

1) create spaces for cooperation;

2) share common visions via roadmaps (such as the ADEME roadmap);

3) share evaluations of dimensioning or evaluations of investment requirements;

4) spread risk across consortiums and take a step-by-step approach;

5) include the actors involved in the definition of local economic models;

6) ensure regulatory visibility.

New actors emerge with the role of promoting cooperation, and the roles of other actors are modified. The user becomes the center of a system featuring:

– the emergence and interaction of new actors, such as aggregators, load reduction managers, providers of post-meter services, electric vehicle charging managers, etc.;

– new services such as voluntary load reduction and demand management, complementing the existing energy offer.

5.3.2.2.2. Elements required to define business models

Within a defined environment and a specified perimeter, the following elements are required:

– identification of the actors involved and understanding of their interests and risks, objectives and constraints in relation to smart grids;

– a cost–benefit calculation within a coherent area. For example, the conclusions of cost–benefit analysis for smart meters will be different for the distributer alone than for all of those involved, from producers to final consumers;

– identification of costs and potential gains throughout the new value chain, once the calculation perimeter has been defined;

– an estimation of the sums associated with these costs and gains;

– identification of potential sources of funding in order to ensure that the minimum conditions for project success are met.

Demonstration projects (such as those funded as part of the ADEME calls to tender) form part of this approach, establishing a framework in which actors are relatively clearly identified. However, essential input data still needs to be defined:

– actor strategies and constraints, whether or not these are compatible with the creation of a collaborative workspace;

– the willingness of agents to modify their behaviors, costs (including opportunity costs) and potential gains.

Using these cost–benefit methods, it is possible to establish business models in relation to a relatively limited group of actors and behaviors. However, the problem is that these elements are not really stable; consumer behaviors have yet to be identified and are subject to change. Further research in this area is essential.

Finally, the vast increase in the number of actors and in technical complexity, alongside the chain reaction resulting from modifications to production margins, distribution or consumption, mean that a regulatory entity is required. The shift from a competitive to a cooperative approach is thus a natural one. Hence, in terms of methodology, two approaches may be considered when attempting to improve understanding of these systems:

– game theory [DEQ 11] (collaborative games) and experimental economics (feedback relating to consumer reactions);

– contract theory [SAL 12] as well as public economics, including the notion of governance and work on regulation.

Game theory may be applied to the electrical energy sector, enabling situations to be modeled in which the choice of actions and individual gains are dependent on the capacity to coordinate with, or anticipate, the decisions of others, in a way which may benefit all involved (win-win game).

This theory has already been applied in many fields, notably in the context of freight and public transportation, and, in a more widely known example, in economics (Jean Tirole was awarded the 2014 Nobel Prize for economics for applications of this type).

Strange though it may seem, this approach has received little attention in the context of energy. There are several situations in this sector (such as consumption management in an electrical distribution network, for example), in which more or less sophisticated incentive mechanisms may be used to create interactions between individual decisions which feed into optimal collective behaviors. Once again, this is in the best interests of the majority of actors.

The subject is becoming increasingly relevant; game theory is no longer the quasi-exclusive preserve of economists, and is making inroads in the energy sector. Note, for example, Olivier Beaude's thesis on game theory in this context: "*Modélisation et optimisation de l'interaction entre véhicules électriques et réseaux d'électricité: apport de la théorie des jeux*" [BEA 15] alongside the work of Benoît Durillon [DUR 18].

Contract theory concerns the way in which actors engage in contractual agreements. In this context, the notion of informational asymmetry is important. The theory may be applied to a variety of domains, including the labor market, banking, insurance and more. The aim is to increase understanding of the way in which contracts should be designed in order to profit all of the parties involved, starting from the principle that the fundamental interests of actors are divergent, and to identify ways of structuring contracts in order to create convergence between these interests in a mutually beneficial manner.

Despite the ongoing metamorphoses of the energy sector and the presence of competition, questions of contracts and of regulation are of crucial importance in the context of smart grids. Several theoretical approaches have been developed to tackle the question of market regulation in this area. Our purpose here is not to provide a precise description of these models; we will simply give a number of elements for reflection, based on the contracts established between the different actors in the newly transformed energy value chain.

The place of smart grids within a system and their ongoing evolution need to be analyzed in order to select the most suitable theoretical frameworks. In the context of this transitional period, the situation may be simplified in the form shown in

Figure 5.2, featuring "upstream" positions, i.e. energy suppliers and "downstream" positions, i.e. consumers.

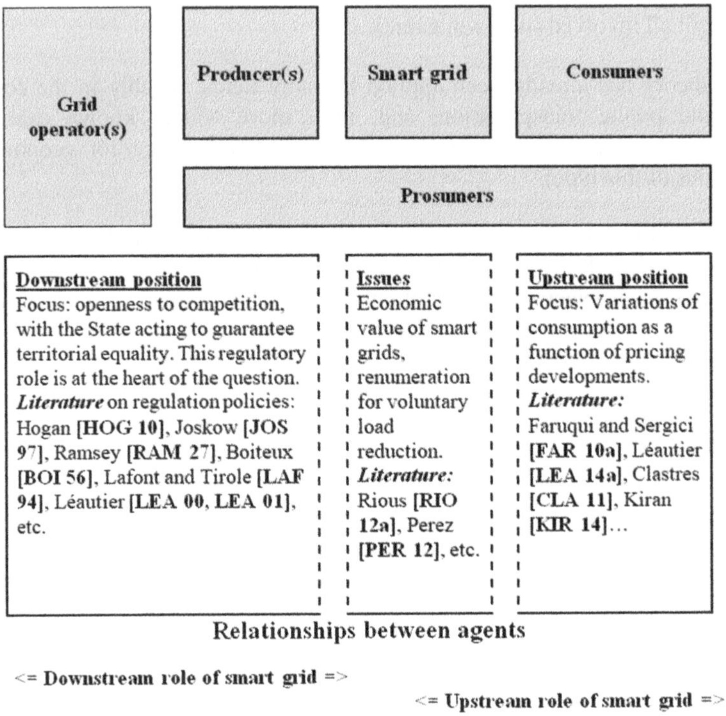

Figure 5.2. *Bibliographical references on the economic aspect of smart grids*

These perspectives will be investigated further in the following two sections. However, we will not specifically analyze the relationships between smart grids and other actors on the basis of different theoretical corpora (contract theory, public economics, game theory, etc.); we will use the term "governance" to denote the relationships between different actors.

5.3.3. *The "downstream" role of smart grids*

The elements selected and presented above give a clear illustration of the role of smart grids for energy producers within the context of evolving governance. Before going into detail concerning the situation of producers as presented in current energy purchasing contracts, we will examine a number of theoretical approaches to regulation, with reference to the structure of energy purchasing contracts and to the key role of grid operators.

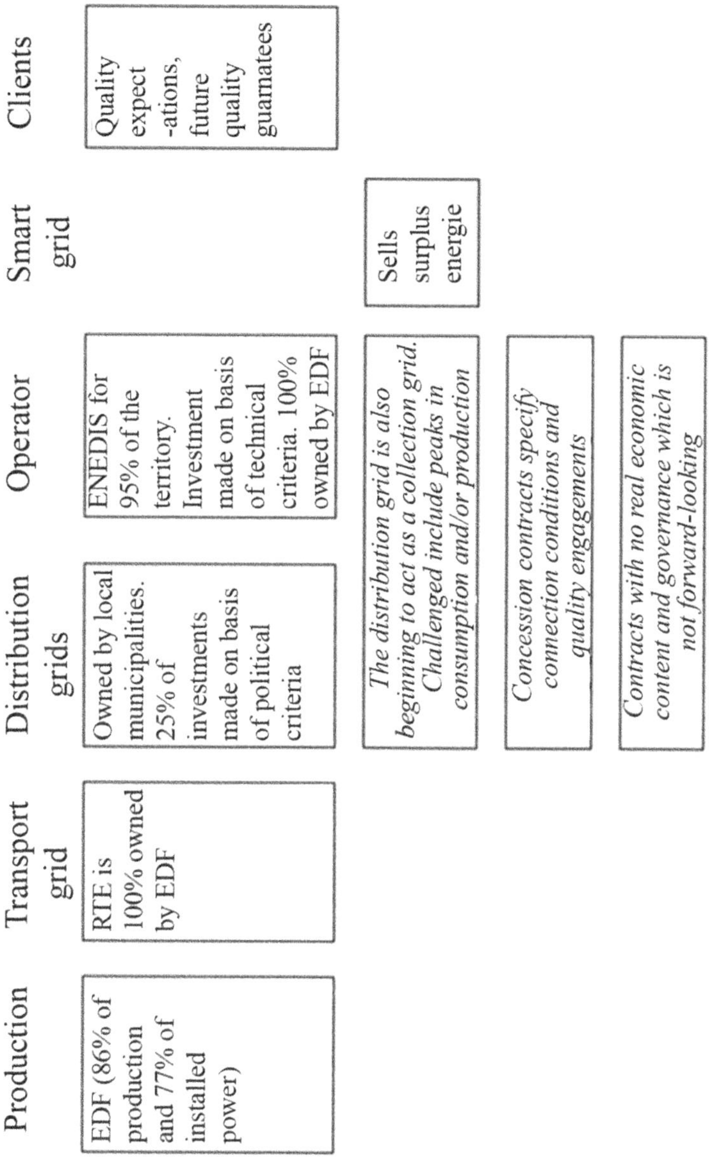

Figure 5.3. *Current governance of the electrical system for mainland France*

5.3.3.1. *Governance*

Figure 5.3 shows a summary of current governance of the sale and distribution of electricity in France.

5.3.3.2. *Outline analysis*

Economic analysis of these systems, considering the management of grid infrastructures, energy production and distribution, is currently directed by regulatory policies which take account of market imperfections and propose the means to control or direct the actions of large companies, in order to benefit consumers or support national strategies.

5.3.3.2.1. The regulatory role of the State in an open market

In France, the opening of the energy market to competition resulted in a re-definition of the role of the State, which now acts as a regulator via independent regulatory authorities (the CRE, *Commission de Régulation de l'Energie* – Energy Regulation Commission).

The State acts as the guarantor of territorial equalization. This principle has underpinned all territorial development policy in France, from the "golden age of DATAR" (*Délégation interministérielle à l'Aménagement du Territoire et à l'Attractivité Régionale* – Interministerial Delegation for Territorial Development and Regional Attractiveness), 1963–1973, to more recent times, as in the case of the Loi Voynet [LAN 1]. This concept is centered on the notion of solidarity between territories, equal opportunities for development, and on supporting those inhabitants and territories with the fewest resources.

In the energy sector, this principle is seen in the price equalization approach, whereby some consumers contribute to the cost of supplying other consumers. However, the introduction of competition has enabled a certain degree of flexibility for actors who are not obliged to supply non-profitable zones. In deregulated sectors, a compensation fund for public service missions is required in order to ensure that State obligations remain compatible with competition between operators.

Redistribution may also be carried out by other means, for example via taxation or direct revenue transfers. Atkinson and Stiglitz [ATK 76] demonstrated that any redistribution between individuals should be implemented via income tax to avoid skewing the consumption choices of economic agents.

5.3.3.2.2. Elements of contract theory

The perceived link between public services and public companies in France remains strong within both corporate and government spheres. The obligation to allow competition, imposed by European authorities, was somewhat unwelcome and has led many to take a defensive approach. The economic framework for infrastructure management needs to be updated, and quickly.

Jean Tirole and Jean-Jacques Laffont's work [LAF 94] on sector regulation established a coherent conceptual framework for:

– defining incentivized contracts between the State and regulated companies;

– establishing the way in which competition should be introduced into grid-based industries;

– setting prices for access to essential infrastructure;

– in more general terms, organizing the regulation of these economically sensitive sectors and defining the role of the State as regulator.

These works on competition law, the abuse of dominant positions, intellectual property and multi-face platforms have had an influence on the way this law is practiced and on the approach taken by the European Commission, alongside other competition authorities.

In low-incentive contracts, producers are sure that their costs – or, more generally, the majority of their costs – will be covered, either by increased subventions or by an increase in the price paid by users (cost-plus contracts for non-sales services, and cost of service regulations whereby consumption prices are indexed according to the level of cost involved).

High-incentive contracts take the form of maximum price ceilings which are not indexed according to effective production costs. In this case, it is more expensive for companies to provide high-quality services, as they are responsible for a greater proportion of the costs involved. Companies thus have a non-negligible incentive to reduce service quality. In order to respond to this problem, it must be possible to control service quality directly.

The temporal aspect of investments and the uncertainty involved make it harder to reach a decision and involve a risk in terms of willingness to invest in mutual relationships. Inversely, producers may spread losses if the service they provide is essential to the collectivity. To avoid this issue, contracts must be "complete" or include provisions for re-negotiation.

5.3.3.2.3. Grid access

Sale of access to the transport grid

The most evident solution is to consider the bilateral exchanges of physical flows of electricity and to define and exchange physical transport rights. For example, a French producer wishing to export production to Belgium must obtain and pay for the necessary capacity in the interconnection system.

Another solution has been put forward by William Hogan [HOG 10]. This consists of associating financial rights with a bid system. In this context, exchanges are no longer bilateral, and each agent (producers, consumers) indicates their willingness to pay via a supply or demand curve. The system, taking account of grid constraints, then allocates resources (e.g. given the demand details for different network nodes, the system will minimize costs according to the least-cost dispatch approach).

If financial rights are used, then producers no longer need to purchase physical capacity on the line; instead, they purchase a financial instrument that compensates for the cost. They may then produce and sell electricity on the French wholesale market and purchase electricity on the Belgian wholesale market to deliver to their client. Financial law provides protection against price variations, due to line congestion, for example. The financial right system is thus essentially a risk elimination strategy.

Paul Joskow [JOS 97] showed that, in imperfect markets where a producer holds a local monopoly or a buyer holds a local monopsony[6], these agents may increase their market power through clever use of physical or financial transport rights.

Monopolies have certain consequences for the establishment of grid access prices. The Ramsey–Boiteux theory [BOI 56] suggests that a natural monopoly will result in losses if obliged to set prices at the marginal cost level (as there is a significant initial cost to cover). To balance budgets, the operators need to apply prices higher than the marginal cost and inversely proportional to demand elasticity (see the box given below). The implementation of Marcel Boiteux's seminal article was hampered by a lack of regulator information concerning demand elasticity. Critics of this economic approach, and supporters of the status quo, have rightly highlighted this informational asymmetry.

6 A **monopsony** is a market in which there are multiple offers but only one demand. This is the opposite of a monopoly, in which there is only one offer for several demands.

In economics, elasticity measures the sensitivity of one value to the variation of another value. A distinction may be made between price elasticity, income elasticity and offer elasticity.

– *Price elasticity is the relationship between the relative variation of demand for a good and the relative variation of the price of this good. The relationship is generally negative, as when price rises, the quantity of demand falls, and vice versa. However, other cases are possible:*

- Perfectly inelastic demand: this does not change in either direction with price changes. This is notably true of essential goods. It is also the case, in the short term, for "pre-engaged" goods: rent, insurance contracts, telephone subscriptions, television, Internet, electricity supply, etc.

- Positive elasticity: this signifies that an increase in price will result in an increase in demand. This occurs for two types of goods:

- Giffen goods (named for Robert Giffen) [PIN 12], a vitally essential good (such as bread): price rises result in a significant reduction in customer purchase power. To balance their budgets, customers are forced to cut back on more costly alternatives in order to maintain their supply of the first product.

- Veblen goods (named for Thorstein Veblen) [PIN 12], a type of luxury good. Demand remains low for as long as the good is not expensive enough.

– *Demand elasticity in relation to income is the relationship between the percentage variation in demand for a good and the percentage variation in income. It measures the impact of a variation in consumer revenue on demand for a particular good. The classification defined by Ernst Engel [PIN 12] features three types of goods:*

- Inferior goods. Income elasticity is negative. These are the goods for which substitutions will be made as income permits.

- Normal goods. The budgetary coefficient for goods of this type is stagnant or varies little as income increases. Elasticity is between 0 and 1.

- Luxury goods. Income elasticity is strictly greater than 1. This is the case for many areas of leisure spending, transport, culture and health, for example.

– *Offer elasticity is the capacity of production to increase or decrease in volume in relation to price variations:*

- Offer is inelastic (elasticity = 0) when an increase in price or demand for a good does not increase the offer of the good.

- Offer is elastic, or highly sensitive, when a variation in the price of a given good results in a variation in the volume of the good being produced.

Setting prices

Access regulation is necessary in this case, as a company has monopoly over access (i.e. may demand excessive access costs) and has the means of shutting out certain competitors. This raises questions regarding access costs, i.e. the price charged by the operator to grant access to their infrastructure. The way in which access prices are established must strike a balance between enabling competition and preserving incentives for the former monopoly holder, ensuring they still have the capacity to maintain and/or develop their network.

Jean-Jacques Lafont and Jean Tirole [LAF 94] highlighted the similarities between this problem and that of a multi-product monopoly. They suggested the use of incentivized regulation with a global price-cap on a "basket" including goods sold per-unit and the standard cost of service access.

The development of green energy raises questions concerning the usage and evolution of the network, and thus regarding the role of pricing (price signals). Green [GRE 97] identified six objectives to consider when establishing pricing for electrical grids:

– promote the efficient day-to-day operation of the bulk power market (short-term price signal);

– signal locational advantages for investment in production and demand (long-term price signal);

– signal the need for investment in the transmission system (long-term price signal);

– compensate the owners of existing transmission assets (cover costs);

– be simple and transparent (to assist choice);

– be politically implementable.

The first formal analysis published in this domain was that of Marcel Boiteux [BOI 56] concerning peak pricing: electricity prices are established on the basis of marginal cost[7]. The second essential contribution was made by Fred Schweppe *et al.* [SCH 88], extending Boiteux's findings to an interconnecting network, reflecting the spatial dimensions of an electrical system. However, economic analysis does not provide solutions to the accompanying political problems. It does provide a

7 At off-peak times, marginal cost essentially covers fuel costs. At peak times, i.e. when demand is equal to capacity, the marginal cost also takes account of capacity cost. The price is determined by the intersection of offer (set at capacity) and demand.

measurement of the gains (or costs) associated with a decision – in this case, the cost which would result from "untrue" pricing, i.e. the costs resulting from controlled pricing.

The question, then, is to determine whether any form of pricing is more incentivizing in the absence of coordination. A distinction must be made between commercial interconnections and regulated interconnections or extensions. Joskow and Tirole [JOS 05, JOS 07] demonstrate that it is generally in the best interests of developers of commercial lines to under dimension these lines in order to maximize position rent.

We also need to know whether the overall regulatory framework (i.e. not simply that relating to the price of injected or drawn energy) provides sufficient incentive for the grid owner to make optimal investments. Léautier [LEA 00], Vogelsang [VOG 01, VOG 06] and Hogan *et al.* [HOG 10] examined the question from a theoretical perspective, while Léautier and Thélem [LEA 09] took an empirical approach.

Of course, the problem of interactions with the way in which producers exercise their market power goes above and beyond grid pricing. It is a well-known fact that the presence of constraints in the network favors the exercise of local market power (e.g. Cardell *et al.* [CAR 97], Borenstein *et al.* [BOR 00], Léautier [LEA 01, LEA 14b]). In this case, we must consider the more specific question, in the context of the interaction between grid pricing and the exercise of market power, of whether or not zonal prices reinforce market power.

In any case, the regulator (the State or the appointed authority) contributes to the way in which market imperfections are taken into account within pricing mechanisms, without challenging the idea of spatial and temporal differentiation in the costs of use of electric distribution networks, reflecting the cost of service provision.

<u>The position of the producer in the current legal situation</u>

With regard to legal and contractual considerations, it is important that we begin by describing the relationships between green energy producers and other system actors.

The contractual system may be said to include three parties: the producer, the buyer and the distributor. Two additional, secondary parties are also mentioned in the contract: the entity responsible for programming, named by the producer at the request of the distributor, and the entity responsible for load balancing (an operator working under contract to the electricity transport grid operator, such as RTE in

France; they are expected to bear the cost of any differences between injected and consumed electricity noted after the event). The balancing entity is nominated by the purchaser.

The purchase contract between buyer and producer is conditional on two other contracts between the producer and distributor. The major challenge with these contracts lies in the fact that any developments, however marginal, involve a new contracting process. In this context, the predominant role played by buyers in matters of governance is particularly evident in the way in which pricing is calculated and in the management of information. Pricing is closely regulated via a complex administrative process (production certificates, auxiliary diagrams, certification from the relevant authorities, etc.), through which the CRE (French energy regulation commission) and administrative authorities may verify conformity, where the costs of inspection are paid by the producer (in cases of non-conformity – otherwise the entity requesting the inspection pays), and where standards are used to guide investment.

In terms of self-consumption, the CRE [CRE 18] has highlighted the challenges of creating a legal and contractual environment to enable the development of self-consumption along lines beneficial to all of the actors involved. Following a consultation process in 2017, the CRE published a number of orientations and recommendations concerning the technical, contractual and economic framework for self-consumption in February 2018. These include:

– In the field of technical recommendations, grid operators are advised to:

- update their technical documentation;

- create a simplified virtual platform for new producers;

- prioritize the installation of smart meters suited to the needs of self-consumers.

– Changes to the contractual framework are designed to facilitate access to self-consumption, while implementing rules to guarantee the correct operation of the whole system. At present, the CRE:

- stipulates that grid operators should offer simple, flexible contracts to low-power individual self-consumers (< 36 kVA), guaranteeing their right to change supplier. As self-consumers do not sign contracts with the grid operator directly, contract models will change;

- recommends that changes be made to the regulatory framework in order to avoid deadweight effects, ensuring that collective self-consumption operations choosing to use dynamic coefficients maximize their self-consumption before injecting surplus.

Points to note

In this context, we speak of purchase contracts rather than sales contracts. This terminology highlights the connections established by the contract.

Currently, in France, while buyers are subject to purchase obligations, contracts may include a considerable degree of flexibility. This flexibility, provided by certain contractual clauses, can take a variety of forms. These contracts also demonstrate the importance of the role played by the grid operator and/or the State on several different levels:

– Technical:

- the imposed standards help to direct investment;

- control of the price calculation formula and identification of peak power should permit adjustment in accordance with the capacity of the distribution grid.

– Organizational:

- the contract effectively introduces a third actor, the entity responsible for balance in the grid;

- the grid operator or their appointed load-balancing manager is involved in defining the central elements of the contract.

– Contractual:

- these are incentive contracts, defined with reference to maximum price limits (CRE);

- in order to prevent producers from downgrading their quality of service, buyers have the option to monitor service quality, directly or indirectly.

In the following section, we will consider the "upstream" role of smart grids. Our aim in this case is to understand the role of energy producers in selling, storing, aggregating or organizing the distribution of energy to a network of consumers.

In addition to the questions discussed in the introduction – relating to actor identification, perimeter definition, cost–benefit calculations along the value chain, etc. – we must also consider the sustainability of the model with regard to client behavior and to the capacity of the regulator to organize governance in a way which optimizes gains as a function of the investments made by each party.

Very little work has been done in this area, and a basis for generic reflection is lacking. At best, certain difficulties have been observed during implementation. In the following section, we will propose a number of theoretical elements which may be used to direct reflection, and a number of experimental elements which demonstrate these difficulties.

5.3.4. *The "upstream" role of smart grids*

The main unknown variable in this area relates to consumer behaviors, and thus to the definition of the utility function[8] of final buyers in energy. This information is particularly essential as, for technical (variability of green energy production) and contractual reasons relating to the grid (peak power), grid operators or aggregators require new tools in order to adjust consumption levels, i.e. to trigger consumption behaviors which permit optimized distribution of produced energy.

The position is a complex one, requiring fast-acting and complex regulatory tools with the capacity to account for users' willingness to modify their consumption behaviors. In this respect, an examination of the price elasticity of energy alone is not sufficient. Utility may depend on other parameters, such as support for a system which promotes green energy sources, or appreciation for new services rendered possible by the existence of smart grids.

5.3.4.1. *Analytical elements: price elasticity and more*

The first question to ask relates to the acceptability of variable pricing. Consumer reactions need to be studied in order to determine the potential success of an energy management system: specifically, we wish to assess the scale of consumer reaction to price changes.

5.3.4.1.1. Price elasticity of demand

Faruqui and Sergici [FAR 10b] analyzed reductions in electricity consumption at peak times in response to incentives, based on 15 experiments carried out in the western USA. Their idea was to isolate systematic difference between two groups treated in different ways, eliminating individual and temporal differences. The authors concluded that consumers (private individuals, small- and medium-sized companies) reduced or delayed their consumption. The actual reductions in consumption varied from one experiment to another (from small to substantial). They increased with the savings made by households by reducing or delaying consumption. They were also higher in cases where users possessed the technology to control equipment remotely.

8 In economics, utility is a measurement of the well-being or satisfaction obtained through the consumption or, at the least, acquisition of goods or a service. The utility function assigns a utility level to each consumption set.

The experimental results from the western US showed between 3 and 6% reduction in consumption at peak times. This reduction was twice as large in cases where consumers had the capacity to control their usage remotely. However, above a certain threshold, the effects on demand were marginal.

Although some differences in pricing exist, French consumers pay almost the same price for electricity at all times of the day and any time of the year, despite significant hourly variations in the cost/price of electricity (meaning that price does not correspond to cost). This makes it difficult to estimate consumer sensitivity to energy price variations. Nevertheless, Faruqui *et al.* [FAR 10b] showed that dynamic pricing – a form of real-time pricing which increases consumer sensitivity to prices (price elasticity) – would result in a 6% reduction in demand at peak times.

Systems of this type are implemented through the use of smart meters. Léautier [LEA 14b] examined the relevance of the approach, which is only worthwhile if there is a sufficient reduction in consumer demand at the right time. This means that for consumers, the marginal value of real-time pricing (value of the reduction in consumption) must exceed the marginal cost (installation of the necessary technology). The author estimated the generated surplus using a theoretical model of demand, offer and the resulting prices, calibrated using French data. He concluded that real-cost pricing is only really useful for larger consumers.

5.3.4.1.2. Risk aversion

Another economic argument concerns risk aversion: consumers may wish to protect themselves from price variations. Guarantees and coverage may constitute a determinant factor. Clastres [CLA 11] included these elements in his consideration of the question of optimal regulation, designed to ensure the success of dynamic electricity pricing and to reduce peak consumption. In this article, the author showed that the choice of an appropriate form of regulation is one of the key steps in ensuring the success of smart grids, due to the uncertainty surrounding the gains to be made from the technology as well as, and especially, to the uncertainty surrounding consumer behaviors.

The reactions of other actors, not only of consumers, are also crucial. Electricity producers (both traditional producers and new producers for whom production is a secondary activity, notably the owners of energy-positive buildings) may also adapt their production as a function of the received margins.

5.3.4.1.3. Architectural and urban forms

In the context of the ENERGIEHAB[9] project, supported by the ANR[10], Jean-Pierre Levy provided an illustration of the "black box" of domestic energy behaviors. The team began by working on the ENL database[11] in order to identify types of energy consumers and the factors, including those relating to housing, affecting variations in consumption. They then carried out an additional semi-directive survey. This combination of two methodologies enabled the researchers to identify the main factors determining temporal flexibility of energy consumption along with the adaptation behaviors of households.

Based on these conclusions, spatialized data describing households and their homes may be used to improve our knowledge of behaviors.

5.3.4.2. *User behaviors*

Other studies have highlighted the existence of additional determiners. Kriström and Kiran [KIR 14], working for the OECD (Organization for Economic Co-operation and Development), showed that non-economic factors are most influential in the adoption of these behaviors, for example membership of environmental organizations. These results clearly show that communication and raising awareness is of prime importance. The ACCENTURE[12] study [ACC 10], which will be presented in greater detail below, supports these conclusions and confirms the importance of user behaviors.

If the grid develops, it will provide operators with indirect and direct information on the private life of users (timing and activities of inhabitants). However, the promises on offer in terms of improved energy yield and savings on individual bills appear to be facilitated by public acceptance. Accenture carries out regular surveys in the aim of defining new energy consumers according to a number of different profiles.

---

9 http://www.agence-nationale-recherche.fr/Projet-ANR-08-VILL-0006.

10 *Agence Nationale de la Recherche*, National Research Agency.

11 In France, the *Enquête Logement* (housing survey) forms the main statistical source for descriptions of the ordinary housing stock, occupation conditions and household expenditure for principal residences, covered in some detail. The results of this survey can be used to study residential strategies. Thanks to the level of detail provided, in-depth work on energy precarity has also been carried out. http://www.ademe.fr/sites/default/files/assets/documents/analyse-precarite-energetiqueindicateurs-enl-2013-rapport.pdf.

12 https://www.accenture.com/t20160811T002327__w__/us-en/_acnmedia/, Accenture/ next-gen/insight-unlocking-value-of-digital-consumer/PDF/Accenture-Understanding-Consumer-Preferences-Energy-Efficiency-10-0229-Mar-11.pdf.

5.3.4.2.1. Key results

Accenture studied a large number of consumers equipped with smart meters, and has carried out a number of investigations into consumer responses to these initiatives (9,108 respondents across 17 countries). These investigations were extended into a broader research program on consumer preferences regarding energy efficiency. The key findings are as follows:

– Perception of the effects of energy consumption on climate change is focused on fuel consumption. This is particularly true in France.

– The success of energy management contracts has been weak, even among those who know about them.

– Consumers prefer to contact public services or electricity suppliers with regard to questions about energy efficiency, but do not necessarily trust the response. Trust is lowest in France, Germany, Spain and the United Kingdom. Consumers trust environmental associations and consumer groups.

– Consumers do not want to modify their behaviors if this will result in an increased bill.

– Arguments which might encourage customers to sign up to a management system include a reduction in environmental impact, as well as the possibility of controlling comfort while spending less time doing so.

– The arguments limiting uptake are also important. These include the idea that the supplier might profit from savings generated by the user (see the notion of freeloaders – actors who derive profit from the actions of others without participating), the fact that suppliers have access to consumption data, reductions in comfort, increased billing complexity, etc.

– In France, the possibility of controlling equipment remotely appears to generate more interest than the amount on the bill. Inversely, consumers are more likely to sign up for a program in which suppliers do not have the option to limit the use of equipment remotely. Consumers are more likely to adopt programs if they retain a certain level of control. However, when consumers themselves are subjected to control, they may respond to an economic incentive as a form of compensation.

Looking at information about survey respondents (age, SPC[13], etc. consumption modes) in conjunction with the survey results, the Accenture study defined six different types of consumers:

– Proactives: most willing to reduce the consumption of major appliances, least interested in reducing environmental impact, prefer in-person contact to obtain information. A higher proportion use electricity to heat their home.

13 Socio-professional categories. In France, a range of SPC codes are used for census data and INSEE (national statistics office) surveys.

– Eco-rationals: highest interest in reducing their environmental impact, sensitive to social pressures, willing to accept reduced comfort in exchange for savings, interested in energy efficiency. Prepared to pay more for high-quality products.

– Cost-conscious: most sensitive to electricity bill savings, fairly sensitive to social pressure, more likely to be discouraged by complex billing, high level of trust toward suppliers. More likely to be female.

– Pragmatics: unwilling to accept appliance control. Highly sensitive to electricity bill amounts. Slower to adopt new technologies. More likely to be male.

– Skepticals: lowest acceptance of utility control, lowest trust in providers. Low sensitivity to bill savings and to social pressure. Prefer to seek advice from consumer organizations. Higher income, more likely to use natural gas to heat their homes.

– Indifferents: unwilling to reduce the use of major equipment in their homes, but accept control. Main inhibitors are bill complexity and time commitment. Often young men and early adopters of new technologies.

5.3.4.2.2. Findings to consider in relation to a smart grid project

The results of this research provide useful elements for reflection, irrespective of whether the final users are domestic consumers, professionals (e.g. store owners in a mall) or intermediary actors (social housing operators, managing trustees, etc.):

– bill reduction is an important parameter;

– however, other more subjective parameters such as trust or the quality of information on the subject should not be ignored;

– simplicity of implementation is a major argument;

– finally, the perception of the distribution of gains is crucial. The subject is a complex one. It seems that the adoption of an energy management system is sometimes dependent on subtle balancing acts, where consumers only sign up if they feel they have the most to gain.

Experimental economics [FRA 13] offers an interesting framework for reflection in areas where information is lacking, for either technical or legal reasons.

The purpose of experimental economics is to reproduce complex decision processes which are impossible to model, centering on individual judgments and decisions made in a context of uncertainty. Market situations are simulated using subjects who interact via an IT network, enabling communications monitoring and data recording.

5.3.4.3. *Governance*

The question of governance is a central one, a fact highlighted by the results of the Accenture survey, which focuses on factors other than price elasticity. The issue of trust is important.

With the increasing number of actors involved in decision processes, the notion of governance often comes up. The definition of this notion is debated[14]. Governance essentially denotes a movement of decentering reflection, decision-making and evaluation, increasing the number of locations and actors involved in decision or project construction processes.

Figure 5.4 shows the implementation of more flexible modes of control or regulation, with increased respect for the environment and for all of the actors involved. This form of regulation is based on an open and informed partnership between different actors. The system includes organizations and individual actors participating in an interdependent structure of direct and indirect relationships, and involves both stable and reconfiguration phases.

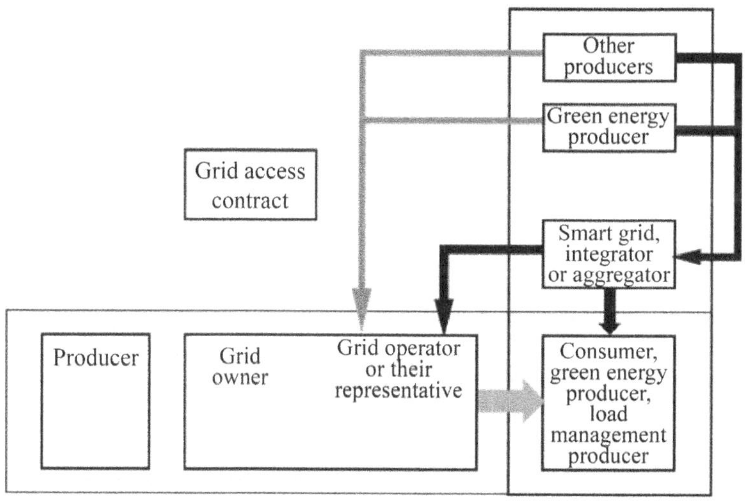

Figure 5.4. *The shift from a pyramid to a decentered approach to governance*

14 The easiest way to define governance is by opposition with the notion of government. Government implies stable political and administrative arrangements and the existence of a single power structure. Governance constitutes a more flexible, dynamic framework. When thinking about governance, we focus on the conditions involved in the coordination of different organizations, including interest groups and relationship systems (Du gouvernement des villes à la gouvernance urbaine, Patrick le Galès. *Revue française de sciences politiques*, 1995, volume 45, pp 57–95).

The "old" system, shown in gray in the diagram, was centralized and arranged in a descending structure. In a smart grid, the integrator or regulator becomes the main actor. Relationships are more complex and more egalitarian. Additional factors come into play in accelerating changes in governance, including:

– openness to competition;

– introduction of green energies;

– changes to consumer usage (electric vehicles, new household equipment, etc.);

– limitation of grid investment opportunities susceptible to reinforce the old model.

The vertical element of the diagram shows new system actors: these include other producers, including green energy producers, integrators and the options for development which are available to consumers. We thus see a conversion from a centralized macro-system to a group of localized systems.

5.3.5. *Demand management programs*

The aim of demand management programs is to reduce consumption by certain users during price peaks, representing an effective alternative to additional production at these times. The variation in demand contributes to balancing the electrical system. It can be considered to be equivalent to production and results in gains due to the reduced need for fossil fuels. This is known as load modulation and cut off.

Load modulation and cut off is thus a potential source of gains for the electrical system, notably in cases when it is less costly than a corresponding increase in production at peak times.

5.3.5.1. *Load modulation and cut off in France*

5.3.5.1.1. Operation

In France, management of permanent balance in the grid is the responsibility of the electricity transport grid operator (RTE). This balance is attained through adjustments. To minimize the total cost of adjustments, RTE selects from the offers received, activating them in order of increasing price until balance is re-established. Those actors whose offers are selected receive payment at the rate they proposed.

Historically, almost all balancing activity involved production or demand variation from major consumers (economic activities). However, since late 2007, aggregated groups of small consumers may also be mobilized as part of the load management strategy.

Load-reducing consumers are disseminated across distribution grids, leading to the expression "disseminated load reduction". To be eligible to participate in adjustment mechanisms, consumers must be coordinated by an aggregator or disseminated load-balancing operator. Aggregators submit load reduction offers to the adjustment mechanism based on the sum of the load reductions permitted by the consumers in their portfolio. To that purpose, aggregators install special units in the houses and apartments in question, allowing them to communicate and control certain electrical equipment remotely, including water heaters and electric heating systems. If a load reduction offer is accepted by the adjustment market, consumption is reduced remotely by switching off certain controllable equipment for a short period (approximately 15 min).

5.3.5.1.2. The load reduction compensation debate

The question of financial compensation for load reduction is still a subject of debate.

– Providers feel that disseminated load reduction operators should compensate them for the injected energy which is not consumed by load-reducing consumers.

– Aggregators consider that diffused load reduction should simply be paid for through the adjustment mechanism in the same way as other adjustment offers (increased electricity production, reduction in consumption by major industrial consumers, etc.). Remuneration may also be granted in compensation for reductions in comfort, in which case the question of evaluation needs to be addressed.

5.3.5.2. *Economic value of load reduction*

Many actors have concluded that the compensation-free remuneration approach leads to distortions in adjustment price and consequently to a loss of efficiency.

Let us consider the simple example put forward by Rious and Perez [PER 12, RIO 12b] (Figure 5.5). Taking a consumer and an electrical generator in a system, they compare the situations which arise when the generator group is situated on either side of the meter. Their model included:

– a consumer with expected consumption of 1 MWh;

– a regulated tariff of €90/MWh;

– a marginal cost for the electrical generator of €150/MWh;

– P, the adjustment price, which varies according to the size of the imbalance.

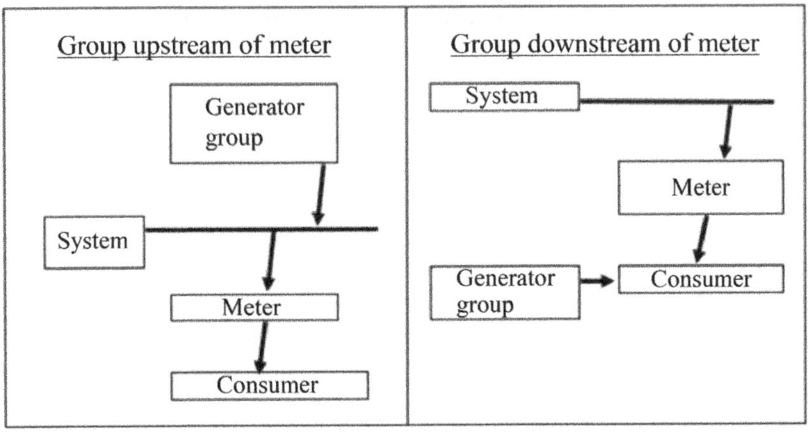

Figure 5.5. *Position of the generator group*

In cases where the generator is situated upstream of the meter, it may be considered as a means of production. In this example, the generator will only be triggered if the adjustment price exceeds €150/MWh. This ensures optimal use. If the generator is situated downstream from the meter, it may be considered as a load reduction offer. It costs €150 for the generator to produce 1 MWh, but this results in a saving of €90 which would have been paid to the provider. Generator use thus becomes worthwhile when the adjustment price exceeds €60 (150-90).

A remuneration system for disseminated load reduction in which aggregators do not pay compensation to suppliers thus results in a distortion of price signals, inefficiently favoring certain adjustment offers to the detriment of others, which would have restored balance in the system at a lower cost.

This also explains why consumers (or diffused load reduction operators) need to compensate the electricity provider (in this case, at the level of the regulated tariff, €90/MWh) each time the load reduction mechanism is activated. This compensation results in the alignment of behaviors between clients with generators situated on either side of the meter.

An offer of remuneration for diffused load reduction without compensation for providers constitutes a form of implicit subvention, encouraging the development of diffused load reduction. While this form of support is justified in societal terms, it also results in a major distortion in the energy market.

5.3.5.3. *From load reduction inhibitors to the capacity mechanism*

From a supplier perspective, there are certain factors which inhibit the development of load reduction techniques, in spite of the fact that the latter may be considered as a vector of progress:

– technical characteristics which result in imperfections in the market. This "missing money" can relate to price ceilings, the lack of consideration for cost of failure in price formulations, incomplete internalization of the social cost of CO_2 emissions, etc.;

– regulatory barriers: market rules for large market;

– technologies requiring significant investment (including investment in learning).

Several different mechanisms have been suggested to solve the "missing money" problem. Most are based on the introduction of a complementary payment in response to the availability of a production capacity (producers) or voluntary load reduction (consumers). In other words, these mechanisms ensure that the installed capacity is sufficient to cover peak consumption.

Voluntary load reduction, or the capacity mechanism, is considered to be the most effective solution for developing peak capacities. In France, article 6 of the law NOME included a new capacity mechanism designed to promote the development of sufficient capacities to respond to peak demand. The idea is to add to the energy market by creating a new good, ensuring sufficient development of the means of production and of load reduction. Following discussions between actors in the electricity system and the *Direction Générale Energie Climat* (DGEC, Energy and Climate Directorate) at the Ministry of Energy, two forms of capacity mechanism are under consideration: a decentralized capacity obligation and a loopback mechanism.

5.4. Social acceptability

5.4.1. *Introduction*

The emergence of social acceptability concerns in the context of evolutions in the energy industry came as a surprise to many. The need to justify, legitimize or defend the reasoning behind technical proposals intended to reduce energy consumption and greenhouse gases, for example, in order to gain the support of users was entirely unexpected. This real surprise is the reason consumers are often said to be conservative or even selfish, and may be perceived as hostile to the adoption of new developments which are otherwise considered virtuous.

The slow uptake of eco-actions, in spite of the likely reductions in energy bills, brought with it a number of lessons which we would do well to bear in mind. On the one hand, the experience showed that the quality of the intentions and objectives being targeted (by a policy, technical device, etc.) is not enough to stimulate public support. The level of complexity of the social substrate, due to the diversity of profiles involved, their representations, beliefs, preferences, and more generally to the socio-technical contexts of energy consumption, was largely under estimated and must be taken into account. On the other hand, it highlighted qualities which must be present in the proposals made to target publics: at the very least, they must provide a clear idea of the modalities of implementation and be free from normative expectations in terms of practice.

In the context of promoting an approach to managing and deriving maximum value from electrical energy storage systems within habitat, these lessons imply that we should pay particular attention to conditions which are likely to increase the adoption of voluntary load reduction among target publics, and to the qualities of the contract expressing the transaction between the mechanism and electrical energy consumers. First, however, we must define the sociological content of the notion of acceptability.

5.4.2. *Conceptual frameworks: points of reference*

5.4.2.1. *Definition of social acceptability*

The definition used here is that given by Véronique Yelle [YEL 13], a researcher at the Université de Laval in Québec[15]. While this definition is not universally accepted, it is the most widely cited in books and articles on the concept of social acceptability. Our subject here is the concept of social acceptability rather than the problem, which relates more closely to the situation of actors facing resistance from target publics.

> **DEFINITION.–** "Social acceptability is the aggregation of individual judgments with regard to the acceptance (or non-acceptance) of a practice or condition, whereby individuals compare it to possible alternatives in order to determine its desirability. It is transmitted by significant groups within society sharing the same judgment with regard to this practice" – Véronique Yelle.

This definition highlights several important elements:

– Acceptability is a collective process with a social character resulting from the accumulation of individual decisions, based on shared values and beliefs. Depending on the direction taken by individual judgments, and under the influence of emerging

15 Véronique Yelle is a researcher working in the Canadian Forest Studies Center.

social dynamics (e.g. potential discords with opinions which are, or are perceived to be, held by the majority), the choice process may or may not result in acceptance. For this reason, the term "adoption process" may also be used to designate the dynamics involved in social acceptance.

– The "alternatives" mentioned in the definition result from variations in strategy. Everything depends on the stability of objectives, opportunity or windfall effects, and the reasoning capacity of each actor. This also means that the intrinsic complexity of the cognitive exercise may result in one or more possible strategies, and sometimes in a poor choice.

– "Significant groups" may be informal (friends, peers, colleagues, neighbors, etc.) or intermediate groupings (syndicates, associations, etc.). They are "significant" as they form a point of reference for the actor (in terms of membership and/or influence) and as their own choice has an impact on opinion.

5.4.2.2. *The limited rationale of actors*

The choice of acceptance does not, as it were, emerge from nowhere. It is founded on reasoning carried out by an actor on the basis of four parameters: objective(s), resources, constraints and strategies. These concepts and their interactions are presented below.

– **Objective(s)**, i.e. the aims of the actor (money, pleasure, promotion, regard, distinction, information, protection, comfort, etc.). An actor may have several objectives, over the short, medium and/or long term, and these may even be contradictory.

– **Resources**, i.e. which may contribute to attaining the objective(s) (skills, information, allies, money, network, etc.). It is the objective which determines the nature, scale and value or relevance of resources.

– **Constraints** are the exact opposite of resources, corresponding to parameters which go against the attainment of the objective (rivalries, competition, lack of skills, isolation, handicaps, ignorance, etc.).

– **Strategy** corresponds to the way in which relevant resources are used to reach a given objective as a function of the constraints which must be overcome or neutralized. The cognitive exercise of reasoning on the basis of relevant objectives, resources and constraints corresponds to a rationale on the part of the actor, which is unique (due to individuality and the content of their reasoning). Hence, we speak of a "limited rationale", as each actor has their own unique perspective.

Table 5.2 is an example of a simple but very effective analytical table, used to understand the reasoning of an actor. Our example concerns the case of an integrator and producer of photovoltaic energy.

Objectives	Resources	Constraints	Strategy
– Maximum profitability of the project once operating – Identify and finalize "good" projects (sufficient critical size, easy installation of PV panels, etc.)	– Possibility of adjusting proposal for different potential sites – Specialist agencies to evaluate risks/countries	– Uncertain interest from clients who need to be persuaded – High levels of uncertainty on regulations and prices	– Integration of all aspects involved – Add energy efficiency objective to production – Identify projects with high initial profitability to limit risks – Prioritize large installations to obtain scale economies – Inject quality via energy efficiency

Table 5.2. *Example of an analysis grid for strategic reasoning*

Now that we have considered the mode of reasoning used by actors, it is important to note that the concept is not limited to individuals; it may also apply to organizational entities such as groups of individual actors with shared objectives, resources, constraints and strategies. In this case, we may speak of collective actors. Similarly, we might use the word "consumers" in a general sense to refer to a single economic agent profile, but we would then need to distinguish between private individuals and companies, because their consumption contexts (nature, scale, capacity to bear cost, etc.) are not the same.

5.4.2.3. *Consumer involvement is not a new development*

Technical advances in terms of energy mechanisms need an increasing participation of consumers (in programming, making savings, load reduction, etc.). This is reflected in a semantic change: in French, a shift has been made from "*consommateur*" to "*consomm'acteur*", while in English, the new term "prosumer" relates more specifically to consumers who also produce energy.

The description of the thought processes connecting objectives to strategy, as shown in Table 5.2, casts some doubt on the validity of the semantic shift. All agents are full actors in the system, as all agents think before acting. In addition, there is the potential for interdependency between agents via their respective objectives (which are not always compatible), resources and relevant constraints (a competing agent is a constraint).

The reasons for the semantic shift from *consommateur* to *consomm'acteur* are particularly visible in the context of the implementation of a decentralized electrical system. The diversification of sources (green energy, co-generation, etc.), the variability of production and technical constraints do not imply a change in status for consumers – who have always been actors in the system – but does offer them the possibility of extending their involvement by becoming producers themselves (e.g. by installing PV systems) and/or by participating in regulation by signing up to a voluntary load reduction contract.

5.4.2.4. *The influence of interlinked social dynamics*

The reasoning behind judgments of acceptability takes place in a context involving multiple layers of parameters. This explains the systemic nature of the social acceptability process. This is not simply a matter of personal preferences, and so the notion of social psychology is relevant, considering individuals acting and deciding within their social milieu.

<u>Contextual parameters:</u>

– individual reasoning, using the structure shown above (objectives, resources, constraints, strategy);

– the social context of production, which includes three dimensions:

- macro: regulations, market, price, incentives, social culture, national politics, etc.,

- meso: geography, territory, operators, local technical constraints, etc.,

- micro: organization, operation, needs, resources, values and representations of the actor, etc.

In this context, values, representations and attitudes carry considerable weight. Beliefs, worldviews and preconceived ideas contribute to the initial positive or negative attitude toward a proposal which may or may not be accepted. The meso- and macro-levels are equally important, but their influence is subtler, being less obvious and more debatable.

5.4.3. *Studies of social acceptability*

In the previous sections, we described conceptual and operational tools which may be used for understanding actor reasoning in one set of circumstances. Applying these notions to the case of a decentralized electrical system, and more precisely to the question of the acceptability of voluntary load reduction proposals, we need to distinguish between two different consumption profiles: household consumption and corporate consumption.

For each profile, we need to identify the components of the reasoning grid in order to understand the dynamics involved in supporting or opposing a load reduction proposal. This exercise boils down to an investigation of motivating factors and of the conditions which are necessary for a proposal to be accepted, making it easier to model the elements involved in the decision process for the two different profiles.

The different forms of electrical consumption correspond to a socio-technical context of energy consumption. This notion is intended to combine the technical context of consumption (building and equipment) with the social context (occupation, activities and social dynamics of consumption).

To make it easier to explain and understand this situation, we need to consider adoption factors for the two consumption profiles, notably:

– the social context of energy use;

– the influence of the value system, knowing that this system itself is in a state of constant transformation.

5.4.3.1. *Conditions for acceptability within companies*

5.4.3.1.1. Compatibility with the system of activity

Every organization, commercial or otherwise (industries, service companies, public services, etc.) is characterized by a particular combination of working situations, regularity of activities and public involvement with their equipment. First, these "activity systems" largely condition the ways in which energy is used (heating, electricity, water, etc.), and have a knock-on effect on the capacity to accept a technical supply system in which the energy "cards" (permanence and cost) will be redealt. Second, these systems determine the capacity for voluntary load reduction (flexibility of the consumption/supply relationship).

Components	Content	Impact/energy demand
Purpose	Open to the public? Food storage? Sample storage? Operating hours?	Heating requirements? Power requirements? Continuous supply? Supply security? Lighting requirements? Safety constraints? Regulatory demands?
Personnel	Number? Diversity? Posture? Physical activity?	Times of presence? Stability? Receptive to instructions?
Working conditions	Indoor/outdoor? Day/night? Mobile/static?	Heating/lighting requirements? Working conditions? Regulatory demands?
Regularity of activities	Operating hours? Regularity?	Time management? Programming? Anticipation of needs?

Table 5.3. *Components of a system of activity*

The voluntary load reduction capacity of an economic entity results from a combination of technical, organizational and operational factors, although it should be noted that there is no automatic connection between these factors and a decision of acceptability. The assessment made by the actor responsible for the decision is also important.

5.4.3.1.2. Environmental effects on the value system

Interest in the cultural dimension of company activities dates back to the 19th Century, but the theoretical approach is much more recent, beginning in the 1980s in tandem with the development of management theory. At this time, it focused strongly on the identification of thought processes and means of action shared by the members of an organization. The underlying aim was to identify cultural elements which are spontaneously pro-business, in order to promote them, and those which are less helpful, which it would be better to adjust.

This approach to the cultural dimension, as the product of a history, organization and operation within an entity, has resurfaced in recent years with the promotion of corporate social responsibility (CSR). The French Ministry of Ecology, Sustainable Development and Energy defines CSR (in French, RSE) as the contribution made by companies to furthering sustainable development. Broadly speaking, CSR corresponds to a visible demonstration of a desire to support individuals and the environment which goes above and beyond the basic regulatory obligations. The motivations behind this type of commitment often vary (ethics, distinction, management, etc.).

In France, a law known as Grenelle 2, adopted in 2016, obliges companies with more than 500 employees to establish a CSR policy. Below this threshold, companies are free to choose their own approach to environmental matters in general, and more specifically to energy. The regulatory obligation and the pursuit of competitive advantages (reduction in energy costs, improved image, etc.) mean that companies are often keen to increase control of their electricity usage, notably via voluntary load reduction mechanisms.

5.4.3.2. *Acceptability factors in domestic environments*

5.4.3.2.1. Lifestyle compatibility with load reduction

Household electricity consumption is distributed across a range of different applications (food, leisure, hygiene, cleaning, laundry) and, in some cases, heating. Consumption is linked to the way in which accommodation is lived in (presence, equipment, occupation), and is relatively constant due to the routines built around schedules, lifestyles and consumption practices (Figure 5.6). These electricity consumption dynamics are set to increase due to the essentially individualized and weakly controlled nature of use.

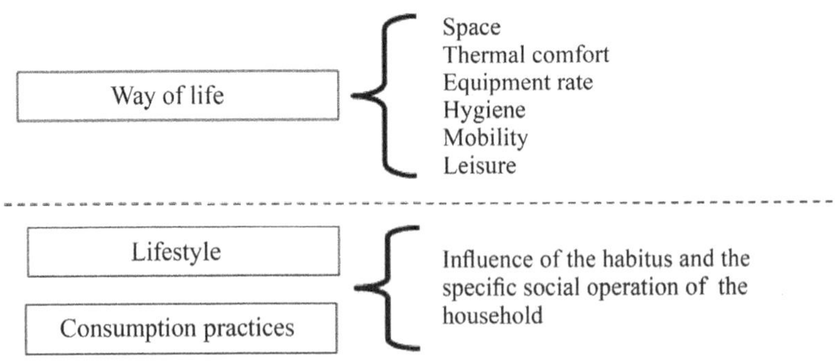

Figure 5.6. *Social dynamics and energy use*

The constant elements of use go some way to explaining why it is so difficult to change both energy consumption and bills. In the same way as activity systems impose certain constraints in the professional sphere, these constants create a social framework; the voluntary load reduction principle, which requires a certain level of self-discipline in limiting electrical consumption at times of peak need during the day or the year, must be compatible with this framework. Acceptance is not simply a question of principles, but also of feasibility, and needs to be considered on a case-by-case basis.

Nevertheless, it is possible to establish an initial list of lifestyle variables and usage modes which may play a key role for the acceptance of load reduction:

– **Electric heating**: alternative technical solutions may be used (fireplace, convectors, oil, etc.) depending on time constraints, the outside temperature and how well the home is insulated (inertia).

– **Level of electrical equipment**: this defines a degree of technical dependence which adds a further element of complexity when attempting to modify consumption practices through the introduction of load reduction techniques.

– **Nature of equipment**: some equipment requires a permanent supply and an interruption to this supply would cause significant problems (routers, aquariums, etc.).

– **Household size**: this has an impact on the level of equipment, the scale of consumption, and it may become harder to cope with the constraints imposed by load reduction as the number of people in the household increases.

– **Negotiation of modes of consumption**: the regulations for energy usage are less formal and less strict in a domestic setting. As we have seen, everything boils down to a question of lifestyle; additionally, consideration should be given to the role of negotiations in households of two or more people. Each individual has their own criteria, preferences and limits, and this may result in disagreements with regard to differing levels of vigilance in energy consumption.

5.4.3.2.2. Values supporting reasoned energy use

The French population is becoming increasingly interested in quality of life and health matters. This can be seen in the results of a study carried out by the CREDOC[16] in 1993 and again in 2013 on the definition of “happiness” (Figure 5.7). The 2013 results show an increased focus on love, social connections, peace and freedom, to the detriment of notions associated with social success (money, work, success, possessions), more clearly present in the 1993 results.

16 CREDOC: *Centre de recherche pour l’étude et l’observation des conditions de vie*, Research Center for the Study and Observation of Living Conditions.

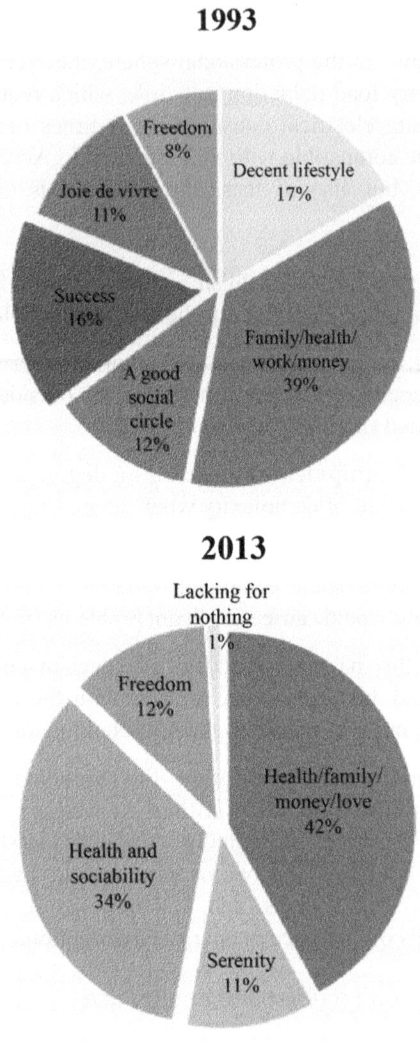

Figure 5.7. *Typology of responses about representations of happiness. Source: CREDOC, consumer surveys, 1993 and 2013. For a color version of the figures in this chapter see www.iste.co.uk/robyns/buildings.zip*

– This development has repercussions for the importance of anything associated with the environment in general, with health and well-being, and, indirectly, with reasoned energy usage. This can be clearly seen in:

- the evolution of consumption practices: households are paying increasing attention to the quality, origins and potential recyclability of products, looking for the "natural" option, promoting values of simplicity and durability[17];

- the attention devoted to developing green energy sources and to saving energy in environmental emergencies, although the motivations involved may include a mixture of ecological, economic, sanitary and/or ethical elements (a priority for over 30% of respondents[18]);

- the stable belief that energy consumption needs to be reduced: this belief is held by 74% of the French population, of whom almost two-thirds consider that energy consumption needs to be reduced "significantly" (Figure 5.8).

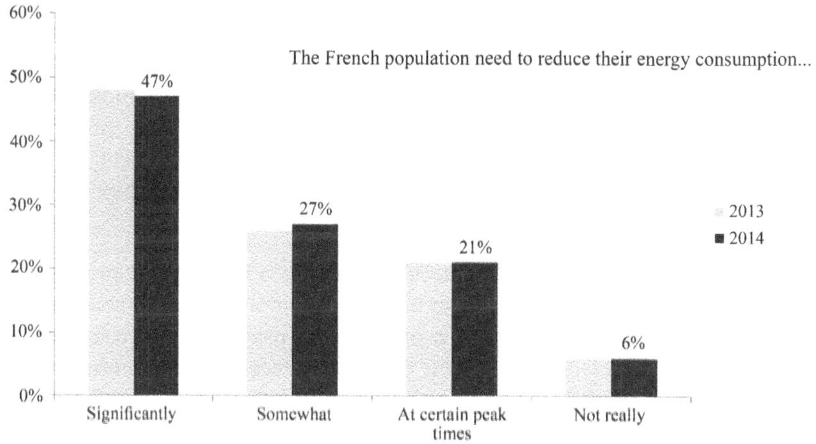

Figure 5.8. *Results of the annual Promotelec survey – CREDOC, "Habitants,habitats & modes de vie" – 2014*

The shift in values and the widespread admission of a need to reduce energy consumption should be seen in the context of the social dynamics at work on different levels (from macro down to micro). Ongoing public awareness campaigns, the increasing frequency of unusual natural events, record temperatures, car-free days intended to combat pollution in certain cities, increasing energy costs, state

17 Source: Consommation et mode de vie, CREDOC, April 2014, no. 266.

18 Source: TNS-SOFRES study on actions considered as high priority in the energy field in France in 2012.

encouragement of green transport options and eco-actions, the appearance of wind farms across the country and more, all constitute signals within the more or less immediate environment, creating an impact on the public consciousness and leading to a shift in practices.

5.4.3.2.3. The weight of the social environment

The way in which representations, attitudes and practices evolve raises questions concerning social influence. Except in cases of exceptional desocialization, actors are constantly under the influence of the social context in which they exist.

In the previous section, we listed several types of "signal" which, in combination and over time, can have an impact on practices. Here, we wish to examine the explanatory factors of this cause-and-effect relationship. In the social sciences, factors of influence are generally grouped into three categories:

– **Changes stemming from innovation**: in some contexts, technical innovation plays a dominant role in the evolution of practices. This is the case for transformations in the energy system, based essentially on modifications to the technical conditions of consumption (including the emergence of domotics) and of production. Certain actors see this transformation process as a way of satisfying some of their own motivations (curiosity, personal fulfillment, prestige, environmental concerns, financial savings, etc.), and are happy to adopt new types of equipment and new forms of energy use.

– **Concern for social norms**: this occurs if there is a risk (real or presumed) of isolation from the reference group in terms of perceptions and/or practices. Users may fear marginalization from the group or their own psychological reaction to this distance. As social norms are in constant mutation, concern for this norm acts as a hindrance to change, notably due to "conditional preferences". These preferences are dependent on the behaviors of others, or on what others consider to be socially acceptable. Many psychological and economic studies have shown that individual decisions are influenced by collective attitudes and behaviors observed within the actor's group(s) of reference[19].

– **Desire for conformity**: this is closely connected to the idea of a norm, which constitutes a reference criterion. The difference between the norm and a lack of conformity may be qualified in different ways depending on the nature of the issue: eccentricity, non-conformism, insignificance, lack of concern, deviance, etc.

---

19 CAZALS, M.P. ROSSI, P., *Eléments de psychologie sociale*, Armand Colin, Paris, 1998. LEWIN, K., *Décision de groupe et changement social*, Dunod, Paris, pp. 498–519, 1965.

5.4.4. *Theoretical application of voluntary load reduction within a reference framework*

Voluntary load reduction in energy consumption is, by its very nature, uncertain. It needs to be included in a functional framework, developed in a specific context and with reference to imperatives imposed from the outside. Evidently, adjustments may be required in the course of implementation, and these will not necessarily be problematic or traumatic. Each situation must be treated on a case-by-case basis.

To illustrate our points regarding the conditions of introduction, we will consider two different cases:

– The first corresponds to a shopping mall. This commercial setting is particularly interesting as it includes different sales units which may present a range of systems of activity. In addition, we encounter constraints specific to professional settings (working hours, personal and building security, food safety, etc.).

– The second corresponds to domestic settings, modeled on typical domestic situations. In this case, we will focus on lifestyles rather than on regulations and systems of activity, considering the rules of operation developed over time to enable harmonious cohabitation between individuals with different personalities, of different ages and different occupations.

5.4.4.1. *Case study: shopping mall*

The load reduction capacity of an entity is the result of a combination of technical, organizational and functional factors, and there is no automatic link between any of these factors and the decision whether or not to accept a system. Acceptance depends on the decision-maker, who also possesses an influencing capacity due to their ability to assess the parameters involved.

5.4.4.1.1. Modeling the degree of acceptability of load reduction systems

In this somewhat simplified view, there are four main parameters which, when taken in combination, determine whether or not a voluntary load reduction opportunity will be accepted.

Load reduction = f (Consumption E; Constraints of P; Degree of dependency; Image)

– **Consumption**: higher levels of consumption are a positive factor as the financial stakes are similarly high. However, irregular or unpredictable consumption can have a negative influence as decision-makers must operate in unstable and potentially stressful situations. It is important to determine who bears the cost of energy consumption: in terms of accepting a load reduction approach, a situation where the bill is paid in close proximity to the commercial activity is ideal (this is closely related to the degree of legal dependency).

– **Activity constraints**: our hypothesis is that these constraints always have a negative effect on acceptance; whatever form they take, their simple existence (in the form of time constraints, the use of cold storage systems, etc.) adds a level of complexity to the design, implementation and technical maintenance of the energy management mechanism.

– **The degree of dependency**: this involves several different elements that are potentially cumulative (site, brand, nature and weight of entities involved in load reduction). This factor may have a positive or negative effect, depending on the preferences of the most influential actors in a shopping mall. For example, would it be technically and financially viable to implement a load reduction mechanism if the most influential commercial entity in a mall (in this case, a supermarket) does not wish to participate? This shows the relevancy of the degree of dependency variable.

Energy consumption	Production constraints	Dependency	Image
Volume	Time schedules	Site and trading area	Communication
Cost	Heating/cooling needs	Toward the brand	Corporate social responsibility
Regularity	Lighting needs	Number of accepting entities	Brand power
Management	Machines	Weight of accepting entities	
	Security		

Table 5.4. *Acceptability factors in a company setting*

– **Image**: we assume that this factor always has a positive effect since companies, and particularly the biggest corporations, use their environmental commitments as a means of communicating with clients who are increasingly sensitive to these issues. The power of the image factor is linked to the negative impact that a brand could have on choosing not to adopt "green" technical solutions.

5.4.4.1.2. Specific considerations for commercial units

We will examine the theoretical aspects of this case using a matrix representation of factors, including positive or negative orientations depending on the nature of the economic unit. This exercise is not intended to produce general rules. We will consider the following cases:

– **A small, independent store selling non-food items**: directed by a private individual, dependent on the shopping mall in which it is located. The production constraints associated with its energy needs are simple and weak. The level of interest in voluntary load reduction (in terms of supply, usage and finance) is also limited.

– **A small branded store selling non-food items**: like the independent store, interest in change is limited. However, if the brand to which the store belongs is in favor, the balance will tip toward acceptance.

– **Laundry and restaurants**: generally independent, of medium size but with high energy consumption. Load reduction may add a further level of complexity to stressful situations, which reduces the likelihood of acceptance.

– **Large non-food stores**: higher dependence on brand than on site. Focused on commercial aims. Low inclination to acceptance.

– **Large branded store**: highly dependent on brand (legal) and site, and these dependencies are difficult to overcome. Acceptability limits are dictated by the nature of activity: high consumption in relation to the surface area, but far lower than for a food store of similar size.

– **Supermarkets**: more dependent on brand than on the site, for which the presence of the supermarket is a determining factor (size, attractiveness, energy consumption).

– **Apartment building**: located near the mall, dependent on households which require a permanent power supply. Owners may be willing to accept, but their needs have a relatively low weight in comparison with the total consumption of the commercial site.

				Electrical imperatives linked to purpose of building			
				Weak		Strong	
				Interest in voluntary load restriction		Interest in voluntary load restriction	
				Weak	Strong	Weak	Strong
Brand dependence	Weak	Dependency on shopping mall	Weak				Apartment building
			Strong	Independent non-food store		Laundry, restaurants	
	Strong	Dependency on shopping mall	Weak			Large non-food store	Supermarket
			Strong	Independent non-food store		Big brand store	

Table 5.5. *Social dynamics and energy usage. For a color version of this table see www.iste.co.uk/robyns/buildings.zip*

5.4.4.2. *Case study: socio-technical elements of the domestic context*

Energy use is difficult to model due to the very limited role of laws in the social sciences, which is far lower than in the experimental sciences. There is often a high level of uncertainty, and situations are never strictly comparable from one case to the next. However, in order to establish parameters for a supervision system, a model of electrical use before and during acceptance of voluntary load reduction is needed. In the following sections, we will describe our approach to modeling domestic electricity use. The effects of individual variations between different cases have been minimized via the neutralization of certain variables.

5.4.4.2.1. Creating domestic usage scenarios

Given the diversity of domestic situations, we have chosen to limit the modes of the selected variables. We have chosen to use two groups of variables:

– Fixed variables:

- Household composition: two parents, one of whom works outside the home, with two school-aged children.

- Description of the home:

  - o detached style;
  - o effective living space of 100 m^2;
  - o no shared walls.

- Electrical equipment:

  - o electric heating;
  - o hot water tank;
  - o refrigerator, deep freeze, washing machine, dryer, dishwasher, TV, other.

- Thermal quality of building: class D.

- Single-rate electrical subscription.

- Reference period: a working day during the week (not the weekend and not a Wednesday, when many French children do not have school).

- During the winter, using electric heating.

– Multi-mode variables:

- Main motivation to participate in load reduction:
  - o financial gain;
  - o innovation/technology/distinction;
  - o participation in reducing greenhouse gas emissions.
- Occupation of home and electrical consumption:
  - o scenario 1: morning and evening from 6 pm;
  - o scenario 2: one of the adults is home all day.
- Proposed load reduction option:
  - o permanent price reduction if drawn power remains below a certain threshold;
  - o price reduction at peak times, in the morning and at the end of the day, to encourage modification of consumption behaviors.

5.4.4.2.2. Motivation scenarios

The decision whether or not to adopt a new practice is made based on one or more objectives. In reality, a combination of aims is usually involved, but for simplicity's sake we will focus on a set of single objectives, resulting in the three modes set out in Table 5.6 (connections may be made between these motivations and some of the consumer profiles identified by Accenture, as presented in section 5.3.4.2.1: Cost-conscious, Eco-rational and Proactive).

Main motivation	Motivation scenario
Financial gain/ bill reduction	For reasons of pragmatism and/or financial preoccupations, these households are particularly receptive to proposals which would enable them to pay less, save more or maximize their buying power. They are likely to commit to voluntary load reduction on condition that the financial compensation is sufficient. Continued support will depend on their assessment of the balance between the actual compensation they receive and the level of investment required (calculation of gains + modifications to electrical consumption routines + loss of spontaneity due to need to delay consumption)
Save the environment/ social utility	First and foremost, these households wish to adopt a lifestyle coherent with saving the planet from excessive exploitation. Although their home is poorly isolated, their purchasing choices and energy consumption practices in different areas (heating, electricity, water, transportation) reflect a clear desire to limit the domestic carbon footprint. Load reduction proposals will be accepted on the basis that they reduce carbon impact on a broader scale (increased role of green energy, reduced infrastructure, etc.). The idea is all the more attractive as it conforms to the desire to live simply and frugally. The financial aspect may be taken into account but does not play a defining role. Changes to consumption habits form part of this lifestyle and the household will have few difficulties in accepting these adjustments

Appreciation of technology/ innovation	Technically oriented households are early adopters of high-tech products or devices. Combining personal enjoyment with a utilitarian element, they are happy to explore the potential of new proposals, without necessarily defining what they expect to achieve. These households play an important role in the spread of innovations. Their activities may promote new standards in equipment and usage, on condition that the new element performs well (financial savings, social distinction, thermal comfort, etc.). In the case of load reduction, which requires a greater investment in terms of behaviors (depending on the scale of the adjustments required) than in equipment, technical skills and parameter establishment, the desire to adopt a new solution is dependent on: 1) highlighting the value of green energy sources as part of the energy supply system, 2) highlighting the crucial role of households as an essential component of systems, adjusting demand to correspond to a naturally variable supply, 3) the complexity of the control system, which should not be downplayed, as this element is a source of enjoyment for technically minded households

Table 5.6. *Household motivations for engaging in load reduction approaches*

5.4.4.2.3. Occupation/electricity consumption scenarios

In a domestic context, the main variations in electrical consumption are linked to levels of occupation of the home throughout the day. We will consider two possibilities: 1) the home is occupied in the morning and the evening, with inhabitants returning at around 6 pm[20]; 2) the home is occupied all day. Usage also varies according to the times at which equipment is used, with levels of impact which differ between devices. Three types of equipment may be identified based on their impact profile:

– cyclical operation, over a duration determined by the household: washing machines, dishwashers and cooking apparatus. Impact varies depending on the selected program, temperature, etc.;

– continuous operation: refrigerator, deep freeze;

– discontinuous operation linked to occupation: lighting, media, office equipment and, potentially, heating, depending on the external temperature.

The two different home occupation scenarios result in specific electrical consumption curves (Figure 5.9). Notable features include:

– consumption continues during the night due to continuous equipment and electrical heating, although at lower intensity;

20 Traditionally, the end of the working day in France.

– a rise in power consumption for heating at around 6 am, with an acceleration in consumption due to concurrent use of multiple devices (lighting, kettle, coffee maker, hairdryer, communication devices, etc.);

– consumption falls rapidly at around from 7.30 am, to an extent which differs according to the occupation profile:

- high if the home is left empty until 6 pm,

- moderate in the case of continuous occupation, with various peaks resulting from occupant activity;

– the reduction in energy use after 7.30 am is maintained until occupants return home;

– peak power usage occurs in the evening. The difference between the two curves is because consumption is spread more evenly across the day if one person stays home.

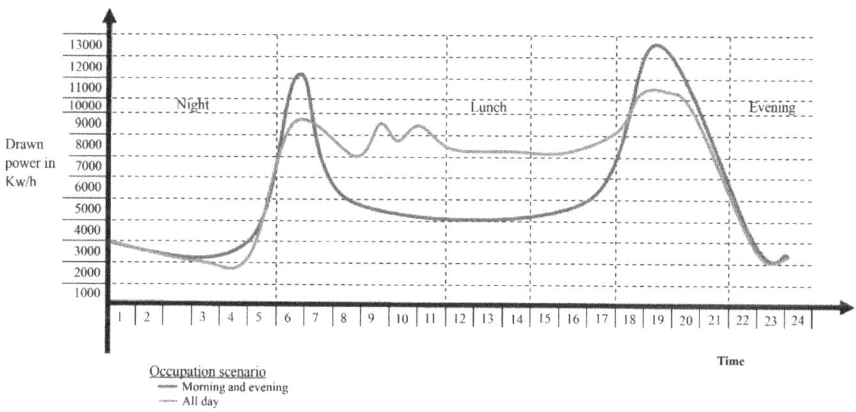

Figure 5.9. *Visualization of two consumption profiles over 24 hours on a work day*

5.4.4.2.4. Variables affecting household receptivity to load reduction proposals

Three main variables affect household receptivity to load reduction proposals:

– **The chosen pricing formula** may encourage households to reduce consumption as required. There are two possibilities:

- Permanent price reduction if drawn power remains below a certain threshold.

- Price reduction at peak times, in the morning and at the end of the day, to encourage modification of consumption behaviors.

– **Psychological and social conditions of consumption**, which includes two variables:

- The main motivation for adopting voluntary load reduction (financial gain; technical innovation and distinction; environmental preservation).

- Patterns of occupation affecting electrical consumption (morning and evening from 6 pm, or all day with one adult staying home).

– **Four variables of uncertain impact**. Current knowledge of the social dynamics of consumption shows that households' desire to moderate their consumption within a favorable context, such as that created by load reduction offers, is not enough to guarantee even the smallest reduction in electrical consumption. This uncertainty is linked to four different variables which must be included in any theoretical study of modulations in electricity usage behaviors. While the existence of these variables is undisputed, their respective scale and impact can only be verified after the effect and is practically impossible to control. For this reason, the term "uncertainties" is more appropriate for the variables below:

- *The risk of a rebound effect:* the perspective of bill reduction (through technical improvements, pricing changes, etc.) may result in the household maintaining or even increasing their level of consumption via a transfer effect between usages.

- *Household capacity to change the temporal elements of their electrical consumption:* the capacity to modify usage behaviors is essential.

- *Ability to withstand constraints generated by load reduction practices:* for example, washing laundry later may mean that clothes have to be hung at a time otherwise devoted to family activities. Changes to the timing of electrical consumption thus have an impact on any associated routines.

- *Accumulation of sensitivity factors:* nothing prevents a household from being sensitive to both bills and the environment. Our hypothesis is that this cumulative effect is most significant for technically oriented profiles.

Given the nature of these uncertainties, our hypothesis is that they are less significant and have a lower impact among environmentalists, given the level of correspondence between their own key motivations and the aims of load reduction approaches. Essentially, we consider that they are most inclined to make significant changes to their modes of electricity consumption, as their own behavior constitutes their main means of investment. Furthermore, they have a greater capacity to withstand any discomfort resulting from the re-organization of consumption schedules. The level of uncertainty is higher for the other profile types:

– Financially motivated:

- capacity to cope with practical adjustments;

- capacity to organize shifts in consumption schedules.

– Technologically motivated:

- price sensitivity;

- environmental sensitivity;

- scale of risk of rebound effect: technologically minded households are often keen to increase usage without financial constraints, notably in order to increase thermal comfort[21].

5.4.4.2.5. Modeling behavioral responses to pricing incentives for load reduction

Table 5.7 shows a summary of the behavioral modeling exercise. It was constructed using multiple parameters:

– the "main motivation" variable, with a choice of three different options (environmental concerns/financial savings/technical);

– the "pricing incentive for load reduction" variable, with a choice of two options (permanent incentive/peak period, i.e. morning and evening, incentive);

– the electrical consumption profile, based on two different occupation scenarios (morning and evening/all day);

– the four areas of uncertainty (rebound effect/capacity to organize electricity use/ability to cope with constraints imposed by scheduling changes/effects of cumulative motivations).

This exercise gives us a household sensitivity ranking, taking "main motivation" as the major input variable. Unsurprisingly, we see that the highest levels of sensitivity are found in households keen to reduce the environmental impact of their consumption. More than the other groups, households with this profile present an accumulation of sensitivity factors in favor of moderating electrical consumption, with a notable concordance between the environmental aims of load reduction proposals and the main motivation of the households themselves. Below these environmentally motivated households, we find those most interested in reducing their electricity bills (financial motivation). Members of this group are also likely to support load reduction proposals, but their sensitivity (and uptake rate) is conditional on a higher number of factors than that of the previous group, notably their capacity to organize and cope with shifts in electrical consumption schedules. Finally, technically oriented households appear to be the least sensitive, notably because of the level of uncertainty with regard to their reaction to load reduction proposals (concerning financial interest, organizational/coping capacity, the attractiveness of a

21 This transfer of costs between areas of use is the reason why technical progress alone is not sufficient to guarantee the transition toward a new energy system with reduced consumption and greenhouse gas emissions.

low-tech solution, etc.). They are most susceptible to rebound effects on their electricity consumption, as they are most likely to re-allocate the savings from load reduction in order to increase other usages, primarily for thermal comfort.

	MAIN IDENTIFIED MOTIVATIONS		
NATURE OF MOTIVATION →	**Environmental concerns/social utility**	**Financial gain/ bill reduction**	**Technological motivations/ innovation**
SENSITIVITY ASSOCIATED WITH EACH MOTIVATION →	Sensitivity: high Sensitivity factors: – reduced GG emissions – coherency with environmental commitment – reinforcement of frugal lifestyle choices	Sensitivity: variable Sensitivity factors: – bill – potential financial gain	Sensitivity: variable Sensitivity factors: – technical/"fun" element – social distinction
PERMANENT PRICING INCENTIVE	**Environmental concerns/social utility**	**Financial gain/ bill reduction**	**Technological motivations/ innovation**
Morning–Evening 	Sensitivity factors when home is occupied all day		
All day 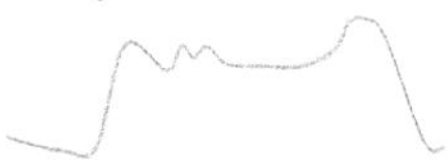	Uncertainty: – sensitivity to technology	Uncertainty: – sensitivity to environmental concerns	Uncertainty: – price sensitivity – bills – sensitivity to environmental concerns – risk of a rebound effect
PEAK PERIOD PRICING INCENTIVE	**Environmental concerns/social utility**	**Financial gain/ bill reduction**	**Technological motivations/ innovation**
Morning–Evening 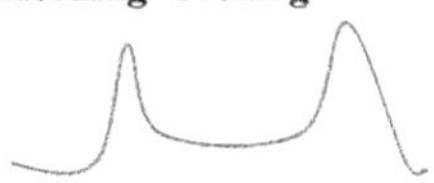	Sensitivity factors when home is occupied in the morning and the evening from 6 pm		
All day 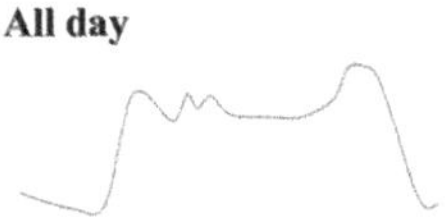	Uncertainty: – potential financial gain	Uncertainty: – capacity to adjust consumption schedule – coping capacity/ domestic constraints in evening	Uncertainty: – coping capacity/ domestic constraints in evening – risk of a rebound effect

Table 5.7. *Modeling behavioral responses to pricing incentives for load reduction. For a color version of this table see www.iste.co.uk/robyns/buildings.zip*

5.4.5. *Quality of the load reduction contract*

Up to now, our examination of the acceptability issues around load balancing has focused on the conditions in which proposals may be received, and has been limited to a comparison of expected consumption changes (delayed and/or reduced consumption), the main social parameters of consumption (major motivations and occupation rate of the home) and simple remuneration mechanisms. In addition to this, aspects relating to contract quality also need to be taken into account. These include the clauses and operating modes set out in writing in the contractual document, alongside expected or predictable effects which are not covered by specific commitments but may attract household support. By taking a broader approach to contract quality criteria, we aim to obtain a more realistic representation of consumer decision processes. Moving beyond simple financial considerations, for example, they will also consider effects on the micro level (e.g. household lifestyle), the meso level (distinguishing factors between social groups) and the macro level (e.g. GG emissions or effects on the employment market).

This presentation follows the approach set out earlier in the context of contract theory. Our conclusions result from a delicate balance of different interests, providing sufficient levels of satisfaction for each party to enable an agreement to be reached.

5.4.5.1. *Key factors in a contract*

The value of a contract is always relative in that it depends on the importance each household accords to aspects explicitly or implicitly linked to the contract in question. While it is not possible to calculate a value, it is possible to identify the most widespread factors taken into account.

Table 5.8 shows a dozen different factors, grouped into four categories:

– **Profitability**: the balance between expected reduction in bills (based on a consumption simulation) and the cost of engagement.

– **Engagement constraints**: variable depending on the contract period and the penalties for early, or temporary, breach of contract.

– **Advice and monitoring of consumption**: households may require support in adjusting their electricity usage. Profitability may be dependent on this support. This theme also covers the question of data ownership, which often features in acceptability discussions.

– **Positive externalities**: the final aim of load reduction is to reduce greenhouse gas emissions via a reduction in consumption. Effects may also include job creation, lower energy dependency, etc.

Variables determining proposal quality
Variables relating to usage costs – Expected reduction in cost of consumption – Evolution of total cost (subscription + consumption + services) – Etc.
Variables relating to the actual content of the load reduction contract – Period of engagement – Financial cost of early or temporary breach of contract (fines) – Production and/or consumption monitoring – Energy saving analysis – Consumption management support – Etc.
Variables relating to elements external to the household in question – Greenhouse gas reduction – Positive impact on employment – Contribution to energy independence – Etc.

Table 5.8. *Key factors in a load reduction proposal*

5.4.5.2. *Managing risks and uncertainties*

The modification of the technical conditions of the electricity supply brought about through load reduction practices raises questions regarding a number of unknown elements, and these can have an effect on factors which render the contract more or less attractive. This effect may take one of two forms:

– **Existence of a "risk" associated with a calculable probability**. This primarily concerns technical aspects (adjustments, reliability, hacking).

– **Existence of an "uncertainty", with a non-measurable probability of occurring**. For example, the effective reduction of greenhouse gas emissions depends on the origin of the electricity being consumed or saved just as much as on the effective management of electricity consumption.

Variables	Criteria	Nature
Environmental	Effective greenhouse gas reduction	Uncertainty
	Evolution of consumption	Uncertainty
	Actual reductions in electrical consumption	Uncertainty
Technical	Adjustment	Risk
	Reliability	Risk
	Hacking	Risk
Financial	Evolution of cost of entering into contract	Uncertainty
	Evolution of electrical consumption	Uncertainty
Regulatory	Evolution/conditions of production	Uncertainty
	Evolution of price parameters	Uncertainty
	Evolution/impacts on health	Uncertainty

Table 5.9. *Risks and uncertainties associated with load reduction proposals*

The combination of risks and uncertainties reduces the attractiveness of load reduction proposals. Adoption will depend on the desires of each potential client in terms of results (financial/environmental/technical) and on the level of uncertainty they are willing to accept in the pursuit of their objectives.

5.4.5.3. *Anticipating the concerns of potential clients*

The adoption of a load reduction proposal is based on a cost–benefit type analysis. In their reflection, target consumers will examine explicit aspects presented in the articles of the contract, identify the limits and risks involved, and will assess these elements in connection with their domestic electricity consumption practices in order to identify any adjustments which would be required. This exercise involves a certain level of complexity and is specific to each individual prospective client. However, the complexity involved can be reduced to a given number of data elements, as target clients always have a main motivation. As the aim is to generate maximum support for the project, it is important to anticipate elements of the proposal which may play a decisive role.

Table 5.10 summarizes the most crucial points affecting acceptance of the proposal. These have been expressed as questions, reflecting the questions which potential clients themselves may pose. Elements of a possible response to each question are also provided.

Decisive points	Advice/type of supporting argument
How can we be sure of the financial interest of signing up?	Between the cost of subscription, the price per kWh and the variation in consumed units, the financial aspect is not the easiest element of the proposal to understand. In order to make an informed choice, we need to analyze the costs involved, particularly those stemming from consumption, which is the main parameter targeted by load reduction. The aim is to adjust practices so that consumption takes place during the cheapest periods. Households should be supported in this using simulations, either with the assistance of an advisor over the telephone or through direct access to a simple simulation tool
How can we alter our practices?	Consumption practices are the most uncertain aspect influencing the financial interest of engagement. At least some form of support should be available, for example a guide, with the same information available on the service website. The simulation tool suggested above could also prove useful in this regard.
How can we limit the rebound effect?	The financial savings promised by load reduction mean that there is a real risk of rebound in terms of electricity consumption, without customers necessarily being aware of the effect happening. It is important to inform customers of the nature of the risk and advise them on means of avoidance
Can we take back control if we need to? **Will there be a charge for this?**	Load reduction is the main constraint. In order to fully respect load reduction periods, it is necessary to anticipate consumption. This is not necessarily easy and it is not always possible to respect these periods. Technical and business model constraints should not preclude deviations from the rule, which are to be expected given the ways energy is used (in both domestic and professional settings), which are subject to unpredictable events. Of course, limits need to be set, either in terms of the scale of the draw on the network or of frequency

Table 5.10. *Plan to address sticking points for potential clients*

5.4.5.4. *Conclusion: social acceptability and the factors involved*

The diagram in Figure 5.10 shows the most influential elements set out above. It highlights the main elements at play in actor rationales, with or without the prospective user being aware of their influence.

– ***Cost of energy consumption***: this includes unit consumption, subscription cost and remuneration for voluntary load reduction. This parameter is closely linked to the political and administrative system which defines the rules of play between actors (producers, managers, markets, etc.).

– ***Intrinsic qualities of the contract:*** these form the keys to the acceptability question, as they determine possible gains and losses (flexibility, dependence, new technical constraints, etc.). These qualities are essentially expressed through three main parameters: reversibility of the engagement (immediate? With notice?), security (due to the complexity of the technical apparatus more than to that of green energy sources) and control of supply.

– ***Social norms***: this factor is most closely linked to social dynamics which may accelerate the adoption process. Their influence is subtle as it is not always explicit or perceived. Social dynamics act on value systems and representations, eventually modifying priorities for action and the perception of what is and is not a good thing to do. Image becomes particularly important at times when conservative forces are starting to give way to new values. Big brands are particularly sensitive to prevailing social norms, important for attracting clients.

– ***Energy usage***: in terms of unit consumption, usage reflects a way of operating (lifestyles in a domestic context, activity systems for companies). Repeated usage situations lead to the development of routines which are the best means of consumption for a given purpose (promoting sales, conserving food, washing, ensuring security, etc.). Adjusting practices to correspond with the constraints of load reduction represents a significant challenge given the highly rationalized nature of many of these practices.

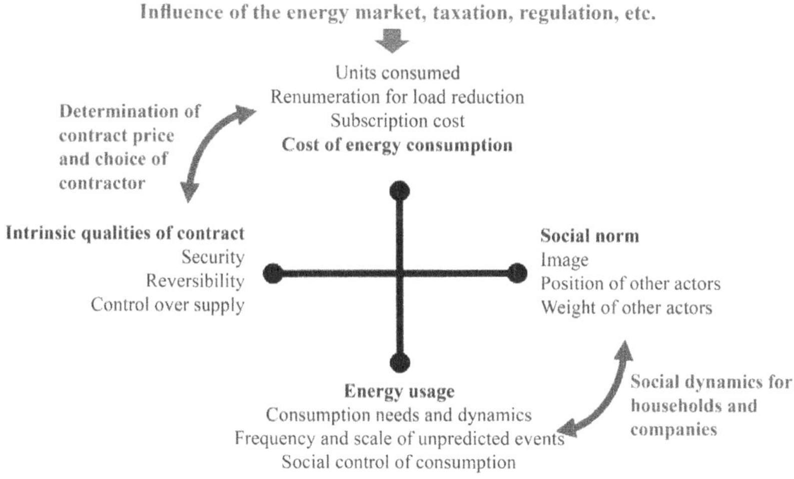

Figure 5.10. *Major elements in consumer rationales when considering a proposed load reduction contract*

5.5. Conclusion

One resource which is of critical importance for the successful implementation of smart grids, and for the reconstruction of an energy production and consumption system in the context of under-investment in grids, is information relating to consumers, who are coming to play an increasingly active role in these systems.

For economists, the question essentially relates to reactions to prices. Will customers engage in voluntary load reduction if energy prices increase? What price would have this effect on consumers? Evidently, customer reactions are dependent on household income, composition, the house itself, and the way in which households balance their own areas of expenditure.

For sociologists, the question is one of acceptability. The position of individuals in relation to their peers, support for sustainable development and levels of trust in new technologies are some of the main factors involved.

The two approaches are not contradictory, but complementary. In both cases, indicators of sensitivity to price or to institutional discourse may be partially established by carrying out qualitative customer surveys or through the use of databases.

The results tend to be fragmentary, as they relate to only a small number of participants. However, they are improving with the increasing availability of individual data, which is becoming a resource in its own right, a point of entry for energy sales and management.

In addition to energy, industrial, economic and social questions, therefore, there are also political, legal, philosophical and ethical issues to be considered, relating to the possible need to relinquish control.

5.6. Acknowledgments

The authors wish to thank the ADEME and the Nord – Pas de Calais regional authority for their financial backing for the GESEDMA project (*Gestion et échanges de services énergétiques décentralisés* – Management and Exchange of Decentralized Energy Services) through which the studies presented in this chapter were developed.

6

Energy Mutualization for Tertiary Buildings, Residential Buildings and Producers

6.1. Introduction

Within the framework of European and national policies designed to reduce greenhouse gas emissions and fossil fuel consumptions, there is an increasing push to mutualize electrical energy for groups of different actors. This notably takes the form of:

– decentralized management of interconnected electrical grids;

– mutualization of electrical energy functions and uses (consumption, storage and production) across multiple buildings or actors with different consumption and production profiles.

In this chapter, we will examine possible exchanges of electrical energy flows and services between a commercial building, such as a supermarket, and other actors such as energy producers, the public distribution grid manager, third-party consumers (e.g. residential users), a storage system using electrochemical batteries and a standby generator, operating within a distribution grid and participating in self-consumption. We will define the ways in which these exchanges may be managed, their profitability and their acceptability for all of the actors involved, including energy professionals and consumers of electricity. Our study is focused on the perspectives offered by collective self-consumption by actors within a given geographical zone, not limited by the initial distance of one kilometer defined in Chapter 5 (section 5.1).

We therefore need to develop a methodology and tools, both on a technical level to model these exchanges, define management approaches and evaluate their economic relevance, and on a socio-technical and economic level in order to evaluate motivations, hesitations to address and incentives to create in order to implement a durable mutualization and sharing project. The way in which energy is exchanged and bills are managed for third parties using public electrical networks also raises a number of legal questions.

We will begin by presenting our case study for energy mutualization between commercial, tertiary and residential buildings. The notion of an energy services aggregator will be introduced. An energy management approach based on fuzzy logic will be developed and applied to the case in question, based on the methodology presented in Chapter 1 (section 1.8). We will then establish a specification for each actor, drawing on sociology to assess the conditions of acceptability and involvement in an energy mutualization approach for each actor (see Chapter 5). An acceptability coefficient for load reduction approaches within the context of the supervision strategy will be defined. Finally, several scenarios, with and without energy supervision, will be compared on the basis of economic, environmental, self-production and self-consumption indicators.

6.2. Energy mutualization between commercial, tertiary and residential buildings, producers and grid managers

6.2.1. *Grid actors*

The case study presented in this chapter builds on that developed in Chapter 2, which concerned energy management in a supermarket. Three different actor profiles will be considered, each relating to different contexts.

Consumer/actors: this profile covers, among others, commercial premises, with energy needs which vary according to activity type, from simple needs (heating, electricity, air conditioning) in cases where there are no perishable goods to more complex situations where supply must be maintained (to maintain the cold chain, conserve fresh food, etc.). In all cases, lighting and heating conditions have an effect on purchasing behaviors and are thus subject to close scrutiny. Nowadays, for these commercial premises, energy expenditure is considered to represent an essential cost for activity to continue, but one which does not directly relate to core activity. Companies tend to focus more on management problems applied to sales activity (opening times, workplace organization, skills, purchasing, marketing, advertising, merchandising, security, etc.). The possibility that a commercial building might produce energy as a business will also be addressed in our case study.

Other consumer/actors will also be considered (homeowners or tenants of social housing). In our study, the low level of per-home consumption in comparison with that of the commercial building is offset by the weight of electricity costs in the household budgets of residential clients. Energy issues are important from a social perspective at the household level.

Energy producers: the level of uncertainty affecting the conditions of service provision has a tendency to rise, with:

– constant increases in energy production costs;

– public policies designed to encourage diversification, with an increased focus on sources which are environmentally friendly but costly to implement;

– increased consumer awareness of energy costs.

These concerns are specific to producers, but they do not affect the permanence of the social demand for a reliable energy supply. The cost of production per kWh must be optimized, and this determines the cost per kWh for final users.

The supermarket in our case study is simultaneously involved in consumption, production and storage of energy (see Chapter 2).

Grid operators: they are responsible for delivering supply to consumers; the energy mutualization question is particularly important to them for a number of reasons, including the presence of an electricity reserve in case of peaks which may then be used for demand smoothing. More generally, they must balance supply and demand in the energy system. At the same time – paradoxically – the mechanism is also a source of potential problems. At first glance, operators may appear to have a greater interest in mastering peaks in consumption than in regulating demand, which may have a reducing effect on overall demand volume and on the pricing conditions of subscriptions. For this reason, proposed services need to reconcile the different and potentially contradictory interests of grid operators.

As we saw at the end of section 5.2, there is a sociological challenge involved in understanding the specific rationale of each actor (objectives, uncertainties and strategies). In economic terms, the challenge is to assess the value of the "resources" proposed by the supermarket and by other actors. This depends on the costs of implementation and on the importance which actors accord to these resources in responding to their own needs. An energy management strategy for the storage system will be developed based on fuzzy logic, following the methodology presented in section 1.8 and implemented in Chapters 2–4.

6.2.2. *Energy service aggregator*

The mutualization of means within a micro-grid requires the involvement of a new entity, the aggregator. The first purpose of an aggregator is to emphasize a load reduction and/or energy production capacity in the best possible conditions. Currently, load reduction is the main area which aggregators can exploit in order to gain flexibility. This consists of temporarily reducing the consumption of a grid user at a specific time of day in response to an order from the aggregator, modulating the consumption of different loads or triggering a production group.

Aggregators also have a number of other aims:

– to increase the competitive power of companies through their load reduction portfolio. Companies are remunerated for services they provide to the electrical network (value is assigned to company flexibility: flexibility = $);

– to actively manage energy consumption in line with company activities, reducing the cost of consumption and participating in maintaining balance between production and consumption;

– to minimize losses within the electrical grid;

– to solve consumption peak problems by remote triggering of generator groups.

The aggregator pays clients in their portfolio to reduce their consumption and, more generally, to modulate their consumption at given times or to produce energy (for those with the capacity to do so). In order to offer load reduction services at different times of day, aggregators must attract clients from different sectors who have the capacity to reduce their load at different times.

In France, the electrical transport grid operator (RTE) is responsible for maintaining balance in the grid. This is done by means of adjustments. In order to minimize total adjustment cost, RTE selects between offers of load reduction capacity and/or electrical production in increasing price order until balance is re-established. Those actors whose offers are selected receive the payment specified in their offer.

Following a request from RTE, an aggregator will temporarily reduce the energy consumption of clients by a certain quantity in order to assist the grid operator in maintaining the balance between instantaneous consumption and production, ensuring that frequency remains stable at the national level. Each aggregator manages a portfolio of clients, and their role is to manage these clients in an efficient manner with regard to their technical constraints, increasing the profitability of their activity. The main aims of the load reduction approach are:

– to avoid power cuts or the use of expensive, pollution-producing production facilities;

– to sell non-consumed energy in the same way as produced energy;

– to control demand instead of controlling production.

In the past, almost all contributions to maintaining balance made use of production and demand reduction capacities offered by large consumers (as an economic activity). However, since the end of 2007, small consumers have also had the option to participate in load reduction methods.

Load-reducing consumers are disseminated across distribution grids, leading to the expression "disseminated load reduction". To be eligible to participate in adjustment mechanisms, consumers must be coordinated by an aggregator or disseminated load-balancing operator. Aggregators submit load-reduction offers to the adjustment mechanism based on the sum of the load reductions permitted by the consumers in their portfolio. To this end, aggregators install special units in the houses and apartments in question, allowing them to communicate and control certain electrical equipment remotely, notably water heaters and electric heating systems. If a load-reduction offer is accepted by the adjustment market, consumption is reduced remotely.

There are currently several aggregators operating in France, including but not limited to:

– Hydronext aggregates electricity from autonomous producers, mostly in the hydroelectric sphere as well as in the wind, solar and thermal spheres, which is then offered on the wholesale electricity market, exploiting the flexibility of tertiary and industrial clients;

– Actility Energy and Energy-Pool exploit the flexibility of tertiary and industrial clients;

– Voltalis offers dynamic electricity consumption management for a large group of residential and tertiary clients;

– Econometering develops and operates control and command infrastructures to aggregate flexibility capacities for the Engie group in Europe. Their clients include public or private tertiary actors, collective housing operators and real-estate managers, along with local authorities;

– Novawatt manages energy flexibility for a decentralized production park made up of gas co-generation units and small power generators.

6.2.3. *Case study: structure of the micro-grid*

Our case study relates to the management of a micro-grid made up of several actors connected to the public distribution grid, managed by the distribution grid operator (DGO). The different actors shown in Figure 6.1 mutualize their means in order to fulfill the energy demands of the micro-grid, aiming to reduce their energy costs and equivalent CO_2 emissions. Several energy sources are used to respond to instantaneous power demands, including renewable energy sources (wind farm and photovoltaic generator). A standby generator for the supermarket (an electric generator with the capacity to function using biofuel instead of diesel) is used for short periods in order to avoid exceeding imposed limits and incurring fines, and ensures that electrical systems continue to function during power cuts.

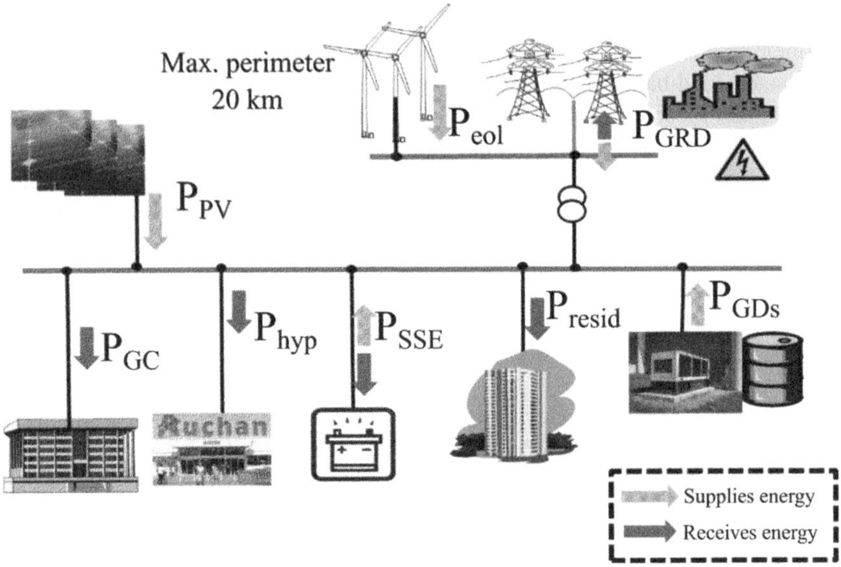

Figure 6.1. *Actors involved in the micro-grid. For a color version of the figures in this chapter see www.iste.co.uk/robyns/buildings.zip*

Three types of actors are involved in the micro-grid (Figure 6.1):

– energy producers (wind farm, PV panels and standby generator):

  - nominal power of the wind farm: 2000 kW,

  - nominal power of the PV installation: 950 kW,

  - nominal power of the standby generator: 352 kVA;

– consumers or loads (supermarket + residential building + shopping mall):

  - maximum power demand: 2400 kW;

– energy exchange actors (the DGO and an electricity storage system):

- subscribed power for energy consumption drawn from the public distribution grid: 2500 kW,

- subscribed power for energy injection into the public distribution grid: 800 kW,

- energy capacity of the storage system: 3500 kWh,

- maximum permissible power of storage system: 400 kW.

The PV installation, electrical energy storage unit and the standby generator will be integrated into the supermarket.

6.2.4. *Consumption and production profiles of actors in the micro-grid*

Figures 6.2–6.6 show typical production and consumption profiles over a week. These will be used as input data for our case study. Figures 6.2 and 6.3 represent photovoltaic and wind production respectively. Figures 6.4–6.6 are consumption models for different load types: a set of residential homes (residential development), a supermarket and a shopping mall. Note the variability of these profiles, which is predictable to a certain extent, but includes a certain level of error. This depends on weather conditions, in the case of renewable energy production, and tends to be repetitive in the case of consumption. Note also that production does not always correspond to consumption times.

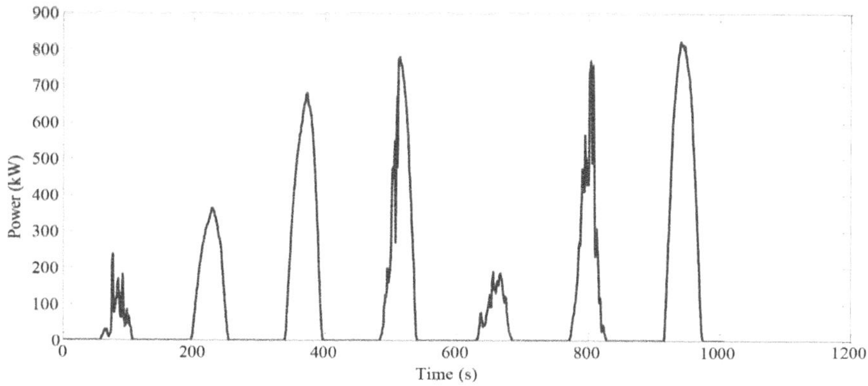

Figure 6.2. *Model of photovoltaic production over the course of a week*

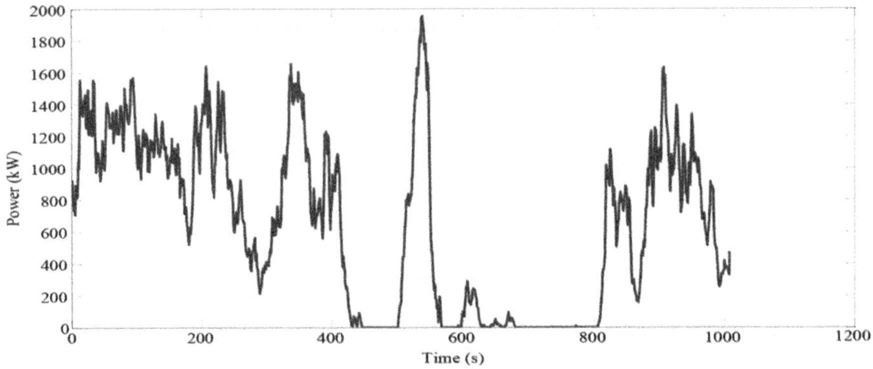

Figure 6.3. *Model of wind farm production over the course of a week*

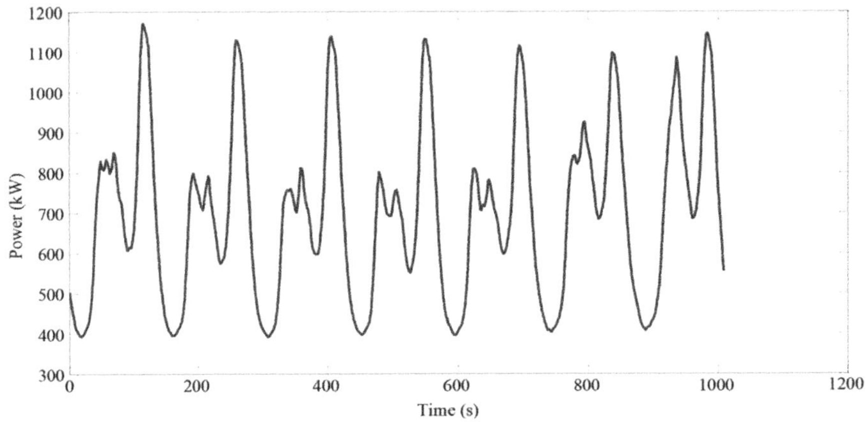

Figure 6.4. *Consumption model for a residential development*

The power profiles of the electrical energy storage system, the standby generator and the grid operator depend on the management objectives of the micro-grid. These will be discussed later in this chapter.

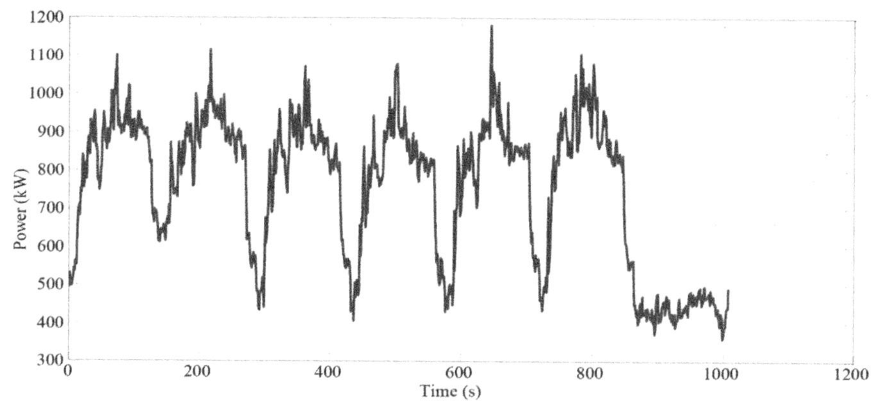

Figure 6.5. *Consumption model for a commercial site such as a supermarket*

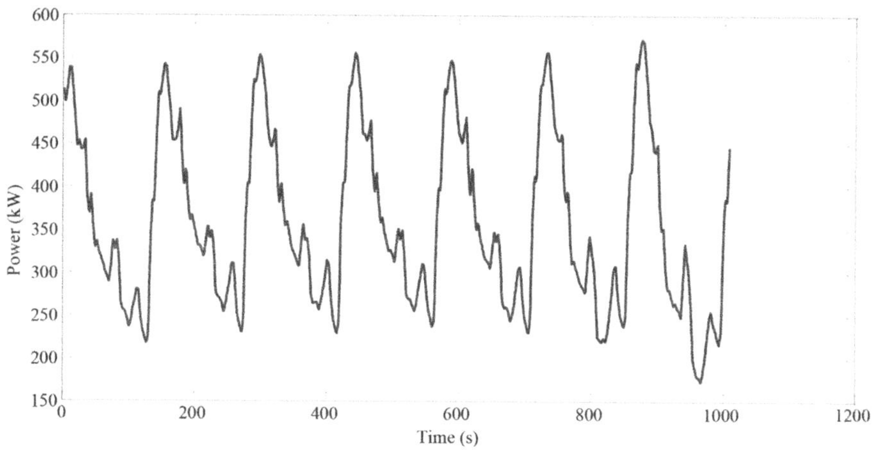

Figure 6.6. *Consumption model for a shopping mall*

6.3. Management of energy mutualization for tertiary buildings, residential buildings and energy producers

In section 6.2.3, we presented a section of a distribution grid including energy producers and consumers, an electrical energy storage system and the distribution grid operator. These actors are characterized by a high level of diversity in their consumption profiles, as well as by their economic and sociological diversity. A new desire to mutualize energy usage among groups of these actors is emerging with the promotion of local consumption of local production: this is known as collective self-consumption. The actors make up a micro-grid, or local grid, which is not intended to function autonomously; the grid operator is thus involved as an actor in its own right. Following on from the discussion of the economic and sociological implications of these changes in Chapter 5, we will propose an energy supervision strategy, demonstrating ways in which the notions of load reduction and acceptability may be integrated into this context.

The objectives, constraints and means of action involved in supervision will be described, along with principles for the energy management strategy. We will then develop a supervisor based on fuzzy logic. Simulation results will be presented for different topologies, with or without a storage system and with load reduction. These will be compared using economic and ecological indicators. The configuration of the micro-grid is shown in Figure 6.1.

The main challenge for this strategy is the association of a range of actors with different profiles and objectives. Power relationships may emerge depending on the respective priorities and weights of actors. For example, the micro-grid results in savings for the DGO in terms of future investments, but reduces revenue as less energy is transiting through the grid. A mechanism must be established in order to balance these power relationships between actors, with continued operation of the grid as a priority.

6.3.1. *Objectives and constraints of actors in the micro-grid*

Our fuzzy logic-based energy supervisor will be developed following the methodology set out in Chapter 1 (section 1.8) [ROB 15, ROB 16]. One difficulty is the association of actors with different profiles and objectives. Stage 1 of the methodology is to identify the objectives, constraints and means of action in terms of energy management for all of the actors involved. These are summarized in Tables 6.1–6.4 and are established with assistance from sociological techniques. The actors in question are:

Renewable energy sources: these include the wind farm and the photovoltaic (PV) installation. In our grid, the PV installation is located on the roof of the supermarket. The wind farm is connected to the HVB/HVA port that powers substations in the neighborhood. The objectives, constraints and means of action of renewable energy sources are shown in Table 6.1. The objectives of renewable energy producers are:

– to generate profit from production at their sites, despite the high CAPEX (CAPital EXpenditure) and OPEX (Operating Expenditure) values for their production systems:

CAPEX consists of investment spending, such as purchasing equipment for a company –for example photocopiers, printers, vehicles. It refers to immobilized assets, i.e. the working tools of the company,

OPEX is money spent in the course of operations, and includes the everyday charges involved in running a product, company or system. For photocopiers or printers, OPEX spending includes the cost of paper, ink and maintenance; for a vehicle, fuel and maintenance. In the case of a company, it may also include salaries, rent, water and energy bills (gas, electricity, etc.). For a renewable energy producer, the purchase of weather prediction data, used to estimate electrical energy production, is also a form of OPEX.

– to maintain site profitability in the face of changing legal and normative frameworks for self-consumption and production;

– to adapt to weather conditions affecting production.

The grid architecture promotes self-consumption and gives added security to renewable energy producers by adapting to natural variations in production through the use of a storage system. In our case study, local consumption is designed to prioritize use of the battery rather than the standby generator, for environmental reasons. The standby generator is used as a last resort in the grid management strategy.

Actors	Renewable energy producers	
	PV	Wind
Actor objectives	Maximize production Promote self-consumption	Maximize production Promote self-consumption (via the local grid)
Constraints	CAPEX/OPEX	CAPEX/OPEX
	Legal/normative framework for self-consumption	Legal/normative framework for self-consumption
	Weather conditions/variability of production	Weather conditions/variability of production
Means of action for the supervisor	Production clipping	Production clipping
Owners	Supermarket/independent producer	Independent producer

Table 6.1. *Renewable energy sources: objectives, constraints and means of action*

"Flexibility" actors: these include the electrical storage system (the battery) and the standby generator, usually used as an emergency power supply for the supermarket. These actors respond to requirements for additional power and/or economic imperatives. The objectives, constraints and means of action for the storage system and standby generator are shown in Table 6.2. The objectives of the standby generator are very different, as the aim is to limit usage, prioritizing self-consumption of production from renewable sources in order to reduce environmental and financial impacts. The electrical storage system is used to optimize the cost of electricity circulating within the grid, but is limited by its technical capacity and high financial impact. Our real-time fuzzy logic supervisor provides support for the grid during periods of peak consumption, prioritizing self-consumption using the services provided by these "flexible actors".

	Sources	**Storage**
	Standby generator	**Electricity storage system**
Actor objectives	Limit usage under normal conditions	Optimize energy cost for local grid
	Provide emergency power supply	Ensure storage is available
		Optimize autonomy and regulation of local grid
Constraints	Technical, economic and environmental constraints	CAPEX/OPEX
		Life expectancy
		Limited capacity (kWh) and power (kW)
Means of action for the supervisor	Set point power for production	Set point power for production or storage
Owner	Supermarket	Supermarket

Table 6.2. *Electricity storage system and standby generator: objectives, constraints and means of action*

Loads: these include the supermarket, the shopping mall and residential clients. The objectives, constraints and means of action for these actors are shown in Table 6.3. The objectives of all of these consumers are identical. They take account of the quality of supply and aim to reduce their expenditure on electricity.

Demand fulfillment may be affected by availability and changing energy costs (pricing periods). Load reduction proposals established by the energy supervisor will lead to a reduction in overall energy spending during periods when electricity is most expensive (at peak times and if the subscription threshold is exceeded).

Actors	Loads		
	Tertiary		Domestic
	Supermarket	Shopping mall	Private individuals
Actor objectives	Meet requirements (clients/equipment)	Meet requirements (clients/equipment)	Meet requirements (equipment)
	Master or reduce energy expenditure	Master or reduce energy expenditure	Master or reduce energy expenditure
Constraints	Energy cost and availability	Energy cost and availability	Price variations (local or public) applied by the distribution grid operator
	Difficulty in predicting the load curve	Difficulty in predicting the load curve	Reference periods in the price grid Energy availability
Means of action for the supervisor	Load reduction/cold storage where acceptable	Load reduction where acceptable	Load reduction where acceptable
Owner	Supermarket	Independent owner	Independent owner

Table 6.3. *Loads: objectives, constraints and means of action*

The distribution grid: the distribution grid provides supplementary power to the neighborhood surrounding the supermarket. Distribution grid operators are subject to more significant technical constraints. They must ensure that a high-quality supply is maintained, minimize transport costs and adapt to sudden changes in consumption. There are limitations resulting from the random aspect of load consumption and the unpredictability of renewable energy production. The local grid, as a client of the DGO, may help to resolve these issues. The objectives, constraints and means of action for this actor are summarized in Table 6.4.

Actors	Distribution grid
	Distribution grid operator (DGO)
Actor objectives	Maintain high-quality supply (stable frequency and voltage)
	Minimize transport costs
	Rapid response to demand
Constraints	Variable availability of sources in the grid
	Unpredictable aspect of loads
	Grid availability (installation capacity)
Means of action for the supervisor	Promote self-consumption (to reduce the load on the distribution grid)
	Load reduction in the micro-grid (selling services to the distribution grid)
Owner	Distribution grid operator (DGO)

Table 6.4. *Distribution grid: objectives, constraints and means of action*

From these tables, we see that several different objectives should be prioritized as a function of pre-existing elements and to establish a compromise between actors in the local grid:

– priority 1: maximize use of locally produced renewable energy (environmental objectives, self-consumption);

– priority 2: reduce energy bills (reduce subscription power, consumer energy from the electricity grid at the cheapest times).

These priorities will be taken into account when developing our supervisor over the following sections.

6.3.2. *Supervisor structure: input and output variables*

Initially, our supervisor will be developed without a load reduction capacity. It features three input variables and one output variable (Figure 6.7):

Input:

– Variable $P_{\_req}$ represents the power requirements of the grid and is defined as follows:

$P_{\_req}$ (kW) = consumption (loads) – production (PV, wind).

This variable concatenates the powers of the different actors in the system. The aim of maximizing renewable energy use and self-consumption translates as a minimization of $P_{\_req}$.

– Variable SOC(t) determines the state of charge of the storage system, and is used to avoid saturating or, on the contrary, completely discharging the system. The state of charge (SOC) of the battery is given as a percentage of maximum capacity.

– The pricing period variable, used to reduce energy bills.

Output:

– Variable $P_{\_sse\text{-}ref}$ gives the power to be stored or released by the electricity storage system.

An additional output variable relates to the possibility of triggering the standby generator at the supermarket. However, it will only be used *in extremis* to maintain a power supply for the micro-grid or to avoid significant excess costs, such as fines for exceeding the power subscription threshold.

The supervisor takes account of different pricing periods (peak, off-peak and shoulder), as in Chapter 2. The price bracket selected for the connection to the public distribution grid is tariff A5, a basic option, Long Use tariff, due to the presence of semi-industrial actors requiring a supply in excess of the subscribed power (see Appendix 1). Our micro-grid is considered as a single, large consumer.

As in Chapter 2, our supervisor will be developed using functional graphs, membership functions and operational graphs in order to define fuzzy rules and identify indicators.

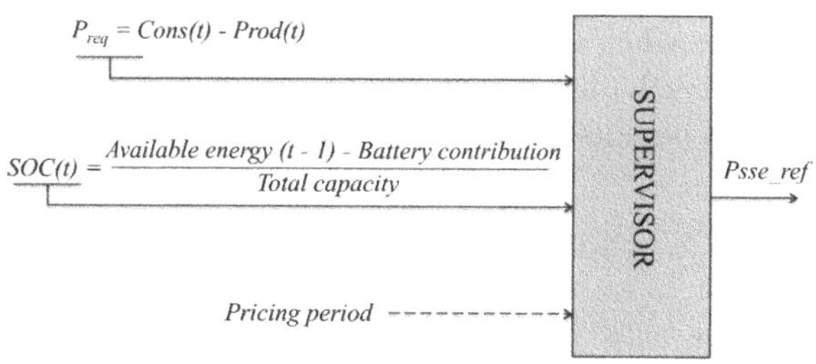

Figure 6.7. *Supervisor structure*

6.3.3. *Functional graphs*

As an example, we have chosen to present the functional graphs for the system during the shoulder period.

Figure 6.8 shows the general functional graph for the shoulder period, read as follows. The starting system N1 is the initial system, containing the management objectives to respect. When the SOC becomes high, during the shoulder period (SP), the storage system should be discharged in response to demand (N2). Inversely, if the SOC is low, then a secure capacity must be maintained by permitting the battery to store energy (N3). Once these actions have been completed, then we return to the normal state (N1).

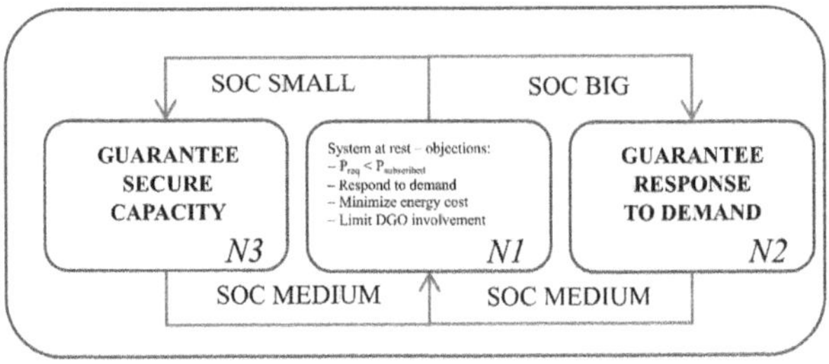

Figure 6.8. *Shoulder period – general functional graph*

The different levels of the functional graphs can be modified with the addition of a P_req coefficient (positive when additional power is being supplied by local renewable sources), as shown in Figure 6.9. We thus obtain our first more specific instruction for the electricity storage system. The graph is read as follows: levels N1, N2 and N3 correspond to the battery charge levels defined in the structure of the general graph in Figure 6.8 (Big, Medium and Small). For each battery state, if requirements are entirely covered by the renewable energy supply (N X.1), then P_req is zero and the battery does not contribute. Energy is stored in cases where P_req is negative, i.e. when there is surplus production and capacity is available (N1.3 and N3.3). Inversely, when additional energy is required (P_req) and where the SOC of the battery permits, discharge occurs (N1.2 and N2.2).

6.3.4. *Membership functions*

In fuzzy logic, the class membership of inputs is not binary, but is between 0 and 1. Taken together, the class membership degrees of an input variable should be equal to 1 (100%), as each variable is made up of a sum of classes in varying proportions. Membership functions are created based on system characteristics.

a) Fuzzification of the SOC variable

The storage system is a lithium-ion battery. This battery has the following characteristics:

– maximum charge/discharge power: 400 kW;

– constant battery yield hypothesis: 85%;

– max. SOC (State of charge): 100%;

– min. SOC: 25%;

– maximum battery capacity: 3500 kWh;

– initial energy stored in battery during simulations: 500 kWh;

– constant inverter yield hypothesis: 98%.

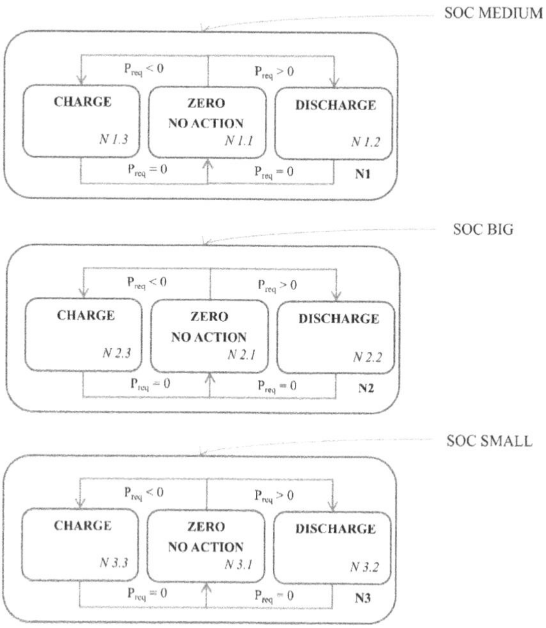

Figure 6.9. *Shoulder period – Functional graph – Presentation of levels N1, N2 and N3*

An electricity storage system such as a lithium-ion battery has a state of charge from 25% (security level to avoid reducing battery lifespan) to 100%.

This variable is fuzzified into three classes, BIG, MEDIUM and SMALL, giving us non-Boolean transition states.

The values of 25% and 100% are established in relation to the minimum and maximum SOC values for the battery. Figure 6.10 shows the fuzzification of the SOC variable.

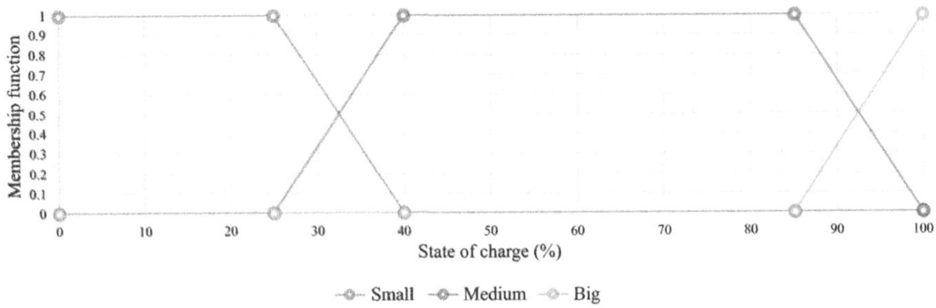

Figure 6.10. *Fuzzification of battery SOC. For a color version of the figures in this chapter see www.iste.co.uk/robyns/buildings.zip*

b) Fuzzification of the P_req variable

The P_req coefficient is the difference in power between loads and production systems. When P_req is positive, this indicates that the loads on the grid require additional energy, as the local electricity production systems are not sufficient. This energy will be obtained from the battery as a priority. Energy from the grid may then be used as a secondary solution. Figure 6.11 shows the fuzzification of the P_req variable. Several classes have been used in this case, defining all of the possible consumption states: ZERO, EXCESS, NORMAL REQUIREMENT and HIGH REQUIREMENT:

– The HIGH REQUIREMENT class applies to cases where the subscribed power is exceeded, with a membership degree of 1 from 2510 kW and up (slightly positive value of subscribed power) to a maximum consumption value (e.g. in excess of 5000 kW).

– The NORMAL REQUIREMENT class is for cases where the power draw on the grid is clearly positive while remaining below the subscribed power. The degree of membership is 1 over [+5; +2500].

– The ZERO class applies when the power of the sources and loads are balanced. This class is centered about 0 with a variation of +/– 5 kW. The name is ZERO, rather than NONE, as the terms are normalized.

– The EXCESS class covers cases where the renewable energy sources in the system produce more than is required by the loads. The degree of membership is 1 over [–5000;–5] (for 0, the ZERO class applies) from a maximum production value (e.g. approximately –5000 kW).

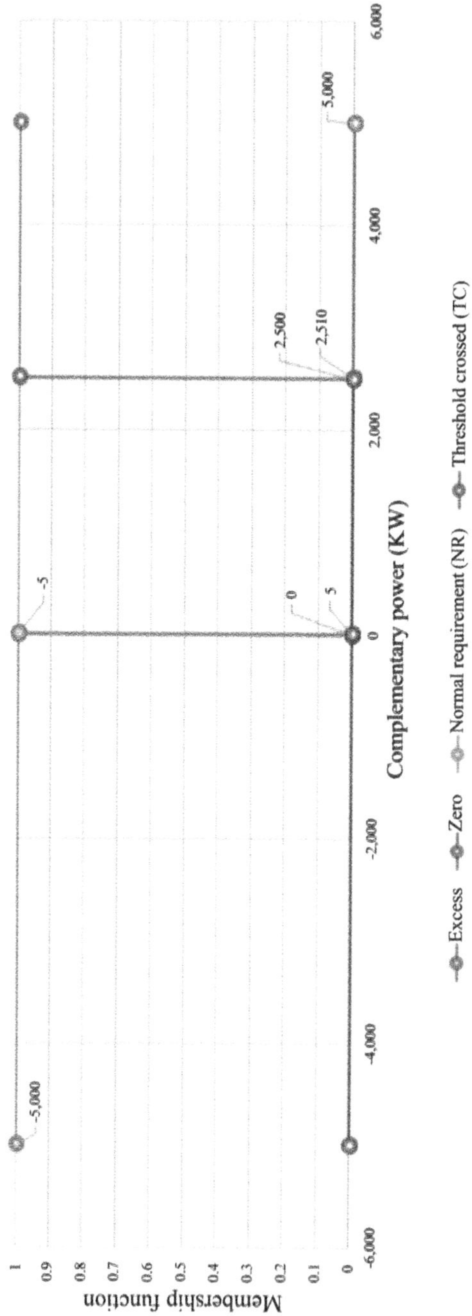

Figure 6.11. *Fuzzification of the P_{req} variable*

The decision to place class boundaries at +/–5 kW and 2510 kW was made empirically. Quasi-optimal values have been identified through different simulations (triggering battery recharging, lower price, etc.). The results for this variable are practically Boolean (like the times for pricing periods).

The pricing period is also a variable, but this takes a Boolean form. Fuzzy laws will be determined separately for each period, as in Chapter 2.

c) Fuzzification of the output variable: reference power of the battery

The reference power (Psse_ref in kW) is the set point power to exchange with the battery. In our chosen theoretical model, the lithium-ion battery gives a constant yield of 85% with a power of between –400 and +400 kW. A positive sign indicates that the battery is supplying power to a point of reference – in this case, the micro-grid.

Ongoing changes in the environment surrounding the micro-grid (technical, economic, etc.) regulates its operation. Figure 6.12 shows the fuzzification of the battery reference power variable. State ZERO, denoting the absence of energy, varies between +/– 10 kW and is centered, as far as possible, about zero. SMALL/BIG CHARGE and DISCHARGE states are also used in this mode in order to refine the reference power regulation mechanism.

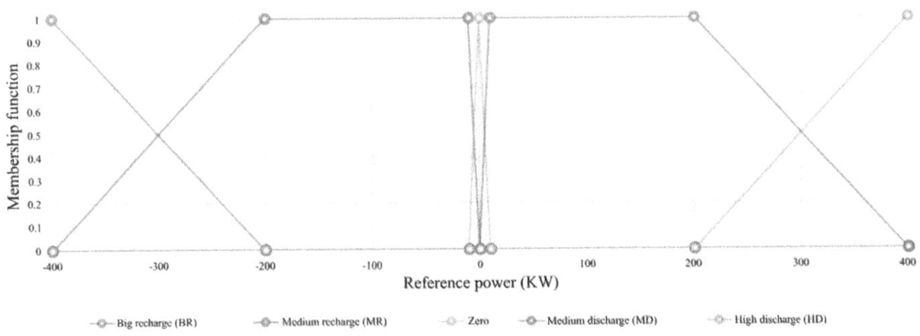

Figure 6.12. *Fuzzification of the battery reference power variable*

The degrees of membership for the BIG and SMALL recharge/discharge values are established in relation to the maximum power which may be exchanged by the battery (400 kW). The values of +/– 200 kW represent a middle state between ZERO (no instruction) and the HIGH states.

The electrical generator is also represented by an output variable, and will be used only if the subscribed power limit is exceeded, following a Boolean approach.

6.3.5. *Operating graphs*

The operating graph is obtained by combining the functional graph with membership functions. As an example, the operating graphs for shoulder periods are shown in Figure 6.13 (N1, N2 and N3).

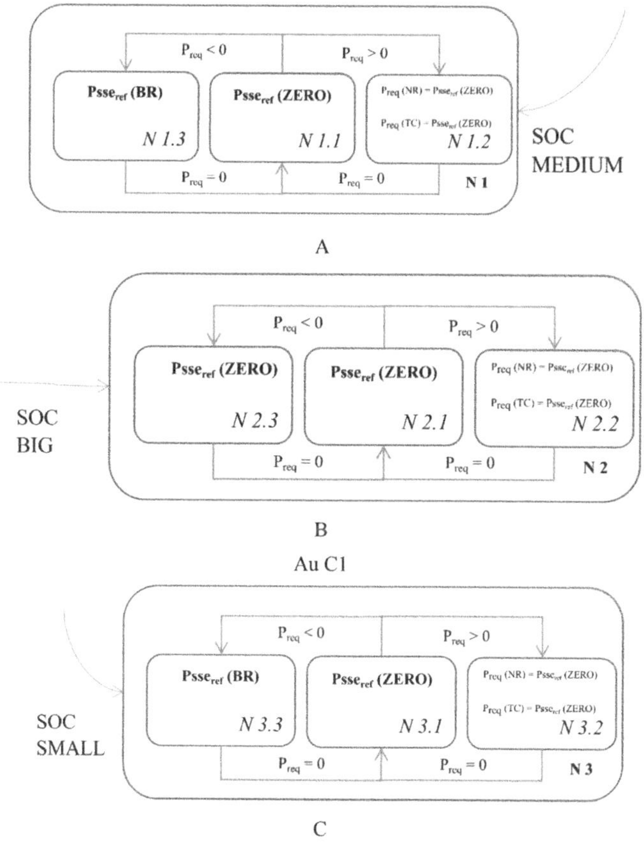

Figure 6.13. *Levels N1 (A), N2 (B) and N3 (C) of the operating mode for shoulder periods*

6.3.6. *Fuzzy rules*

Fuzzy rules are deduced from the operational modes associated with membership functions. They may therefore also be deduced directly from operational graphs. All classes are shown to avoid any ambiguity. Using four classes for P_req and three classes for the SOC, we obtain 12 possible rules per pricing period. The fuzzy rules for each pricing period are presented in the following sections.

a) Peak and shoulder pricing periods

The fuzzy rules for these periods are identical and are shown in Table 6.5. The cost per kWh is relatively close, taking the highest fixed subscription cost (TLU in Appendix 1, Table 1).

Peak period		
$\mathbf{P_{req}}$	**SOC**	$\mathbf{Psse_{ref}}$
Excess	Small	Big recharge (BR)
Excess	Medium	Big recharge (BR)
Excess	Big	Zero
Zero	Small	Zero
Zero	Medium	Zero
Zero	Big	Zero
Normal requirement (NR)	Small	Zero
Normal requirement (NR)	Medium	Zero
Normal requirement (NR)	Big	Zero
High requirement (HR)	Small	Zero
High requirement (HR)	Medium	Big discharge (BD)
High requirement (HR)	Big	Big discharge (BD)

Table 6.5. *Fuzzy rules for peak and shoulder periods*

The operating principles for these rules, from top to bottom in Table 6.5, are as follows:

Rule no. 1: in case of excess energy at peak times with a small SOC, battery capacity should be used to store a maximum quantity of renewable energy. Renewable energy sources take priority when charging storage in order to maximize self-consumption and respect environmental objectives.

Rule no. 2: in case of excess energy at peak times with a medium SOC, battery capacity should be used to store a maximum quantity of renewable energy. Renewable energy sources take priority when charging storage in order to maximize self-consumption and respect environmental objectives.

Rule no. 3: in case of excess energy at peak times with a big SOC, the battery should not be instructed to charge as its SOC is maximum. The surplus will be injected into the grid.

Rules no. 4–6: state ZERO represents a transitional phase between charging and discharging. Whatever the value of the SOC, a null instruction is sent to the electricity storage system. As correction is carried out in real-time, any small variation in consumption will result in the micro-grid exiting this transitional state.

Rules no. 7–9: if additional power is required for the grid, in normal requirement mode and whatever the SOC of the battery, a null instruction is sent. The battery will not be discharged as the grid energy level is relatively high.

Rule no. 10: if additional power is required for the grid, if requirements are higher than the subscribed power and the battery level is small, then a null instruction is sent. The battery should not be discharged when the SOC is small, nor should it be recharged, as energy drawn from the electric grid at this time is expensive and the grid itself is limited.

Rules no. 11 and 12: if additional power is required for the grid, if requirements are higher than the subscribed power and the battery level is medium or Big, then the battery is discharged as far as possible, as energy drawn from the electric grid at this time is expensive and the grid itself is limited. The economic objective is thus respected.

b) Off-peak pricing period

The fuzzy rules for this pricing period are shown in Table 6.6.

Off-peak period		
P_{req}	**SOC**	$Psse_{ref}$
Excess	Small	Big recharge (BR)
Excess	Medium	Big recharge (BR)
Excess	Big	Zero
Zero	Small	Zero
Zero	Medium	Zero
Zero	Big	Zero
Normal requirement (NR)	Small	Big recharge (BR)
Normal requirement (NR)	Medium	Big recharge (BR)
Normal requirement (NR)	Big	Zero
High requirement (HR)	Small	Zero
High requirement (HR)	Medium	Zero
High requirement (HR)	Big	Zero

Table 6.6. *Fuzzy rules for off-peak periods*

The operating principles for these rules, from top to bottom in Table 6.6, are as follows:

Rule no. 1: in a state of excess energy and with a small SOC, big recharge will be used to store cheap, low-carbon energy in large quantities in the available capacity. Renewable energy sources are prioritized in order to maximize self-consumption and respect the environmental objective.

Rule no. 2: in a state of excess energy and with a medium SOC, big recharge will be used to store cheap, low-carbon energy in large quantities in the available capacity. Renewable energy sources are prioritized in order to maximize self-consumption and respect the environmental objective.

Rule no. 3: in a state of excess energy and with a big SOC, a null instruction is sent as the storage system is already at maximum charge.

Rules no. 4–6: state ZERO represents a transitional phase between charging and discharging. Whatever the value of the SOC, a null instruction is sent to the electricity storage system. As correction is carried out in real-time, any small variation in consumption will result in the micro-grid exiting this transitional state.

Rule no. 7: during periods when complementary energy requirements do not exceed the subscribed power and when the SOC is small, big recharge will be selected, as the price of electricity from the grid is advantageous.

Rule no. 8: during periods when complementary energy requirements do not exceed the subscribed power and when the SOC is medium big recharge will be selected, as the price of electricity from the grid is advantageous and the economic objective is respected.

Rule no. 9: during periods when complementary energy requirements do not exceed the subscribed power and when the SOC is big, a null instruction will be sent as the storage system is already at maximum capacity and the economic objective is respected.

Rules no. 10–12: during off-peak periods and in cases of high requirement, as the cost of energy is at its lowest, electricity is drawn directly from the grid without recharging, as stored energy would be more expensive than the lowest price of energy following storage. The SOC has no influence on the instruction in this case, which is always null during off-peak periods and when the subscribed power is exceeded.

6.3.7. *Indicators*

A list of performance indicators for the supervision system may be used to evaluate whether economic, environmental and energy objectives have been met:

– Economic indicator: the unit cost of energy in the micro-grid, expressed in €/kWh. In our scenarios, this cost takes account of energy drawn from the grid with and without exceeding the subscription threshold; energy from renewable sources absorbed by the micro-grid, and energy from these sources which is not absorbed and is either stored (taking account of storage costs) or injected into the distribution grid; and the use of a standby generator.

– Environmental indicator: the contribution to CO_2 emissions resulting from the energy consumed by the micro-grid, expressed in kg CO_2/kWh. This takes account of the instantaneous electricity mix, and thus of time periods (http://www.rte-france.com/fr/eco2mix/eco2mix-co2), the battery (gray energy) and the standby generator where applicable.

– The summary of consumption for different actors.

– Self-consumption rate: this is defined as the part of production which is self-consumed and is equal to the relationship between the production consumed on-site and the total production of the site.

– Self-production rate: this is defined as the part of production which is self-produced and is equal to the relationship between production consumed on-site and the total consumption of the site.

6.4. Case study

6.4.1. *Characteristics of the micro-grid*

In simulating the different scenarios described below, we used the following characteristic powers for the sample grid shown in Figure 6.1:

– maximum load power (supermarket, shopping mall, residential buildings): 2400 kW;

– nominal power of wind farm: 2000 kW;

– nominal power of photovoltaic installation: 950 kW;

– nominal power of standby generator (diesel type): 352 kVA;

– maximum incoming power for the storage system: 400 kW;

– energy capacity of the storage system: 3500 kWh;

– power subscription to local grid, incoming: 2500 kW;

– power subscription to local grid for injection into the distribution grid: 800 kW.

The micro-grid is permanently connected to the public distribution grid.

The production or consumption profiles used for different actors over the course of a week are:

– the production curve of a photovoltaic installation (Figure 6.2);

– the production curve of a wind farm (Figure 6.3);

– the consumption curve for residential homes (Figure 6.4);

– the consumption curve for a supermarket (Figure 6.5);

– the consumption curve for a shopping mall (Figure 6.6).

6.4.2. *Scenarios*

We simulated three possible scenarios for a full week using the local grid shown in Figure 6.1, the local grid characteristics given in section 6.4.1 and the supervisor presented in section 6.3.2. These scenarios are:

– scenario 1: scenario without energy management;

– scenario 2: impact of renewable energy sources on our case;

– scenario 3: contribution made by the supervisor.

6.4.2.1. *Scenario 1: no supervisor*

In this scenario, there is no supervisor to carry out energy management. There is no form of electrical energy storage. Our objective is to study the grid without supervision. Renewable energy production follows the profiles defined in the previous sections. For each simulation, only the overall consumption of loads on the micro-grid (cumulated consumption profiles) will be increased, in 10% increments up to two times the initial consumption or load of the grid in question. The electrical generator is present and will produce electricity to avoid crossing the distribution grid subscription threshold at peak and shoulder times, something which happens in real-world situations on industrial sites.

Figure 6.14 shows the basic overall production and consumption profiles for the grid in question in relation to the power subscription from the distribution grid. A summary of the results of our simulation is given in Table 6.7. In this table, the environmental impact takes account of pollution from the standby generator. The self-consumption and self-production percentages only concern renewable energy.

This first scenario shows a simple example of self-consumption and self-production.

For the initial load and initial production (Coeff_LOAD set at 1), the self-consumption capacity is approximately 98%. This means that almost all of the energy produced by renewable sources in the micro-grid is consumed locally, within the grid. The environmental impact is at its lowest at this point in the scenario (39.6 kg CO_2/MWh). The self-production capacity is approximately 41%, meaning that approximately 41% of the energy consumed by the micro-grid comes from local production, and the rest is drawn from the public grid to which the micro-grid is connected.

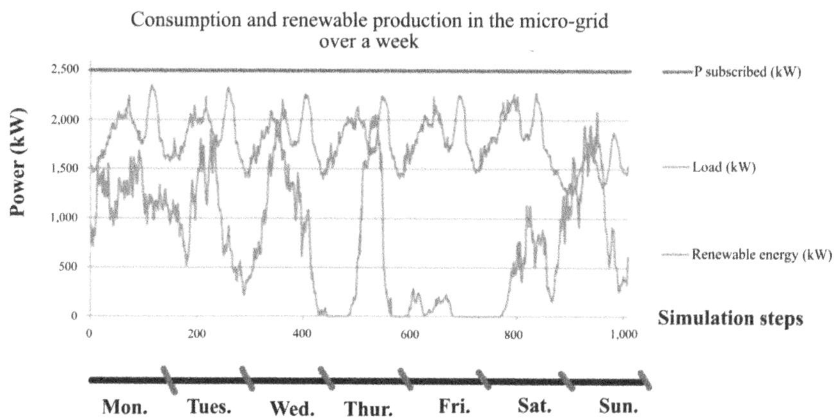

Figure 6.14. *Loads and renewable energy production in the micro-grid over the course of a week*

No storage, no load reduction, DT = 1, using diesel						
2015	$Coeff_{LOAD}$	kg CO_2/MWh	€/MWh	Self-consumption ENR (%)	Self-production ENR (%)	Diesel consumption (kWh)
ENR × 1	1.00	39.6	80.7	98.36	41.17	–
	1.10	40.6	80.1	99.3	37.79	–
	1.20	42.8	80.2	99.84	34.83	610.34
	1.30	46.6	81.2	99.98	32.19	2,053.40
	1.35	49.5	82.3	100	31.00	3,370.80
	1.40	52.6	83.7	100	29.90	4,909.30
	1.50	58.3	87.1	100	27.90	8,030.10
	1.60	62.7	90.8	100	26.16	10,794.00
	2.00	70.3	108.1	100	20.93	17,881.00

Table 6.7. *Summary of results for scenario 1*

With a load twice the size of the initial load and unvarying renewable energy production (Coeff_LOAD 2), all of the energy produced by the renewable sources in the micro-grid is consumed locally (self-consumption capacity of 100%). The self-production capacity is approximately 21%. The decline from the previous point is due to an increase in consumption while local production remains constant. Approximately 79% of the energy consumed is therefore drawn from the public distribution grid and from the standby generator, a source of pollution, which produces more and more electricity for the micro-grid during peak and shoulder periods in order to avoid exceeding the subscribed power thresholds and to limit energy costs. As a result, the environmental impact of the micro-grid increases, from 39.6 kg CO_2/MWh for the initial consumption values to 70.3 Kg CO_2/MWh when this consumption is doubled. Similarly, as more and more energy is being drawn from the generator and from the distribution grid, the price of electricity increases from €80.70/MWh to €103.10/MWh as energy produced by the standby generator is more expensive.

Note that for load coefficients 1.1 and 1.2, additional energy is drawn from the distribution grid, and this is cheaper than using the renewable sources in the grid or the standby generator. For this reason, the price of energy is lower, even with a small contribution from the standby generator.

6.4.2.2. *Scenario 2: impact of renewable energy sources on our micro-grid*

In our second scenario, we analyzed the environmental and financial impact of renewable energy sources (85% wind, 15% PV) by increasing renewable production compared to our initial values. The renewable park was either removed (x0), retained (x 1.1 times previous production potential) or doubled (x 2.2 times previous production potential). Again, the supervisor was not used in this scenario as the subscribed power limit was never exceeded. The characteristics of the electricity in the grid were actualized for each simulation step (cost and CO_2).

No load reduction was used, and the standby generator remained present. The total load was the cumulative load of all consumption profiles. The subscribed power from the grid was well above that required by the micro-grid. The renewable-source energy sold to the electrical grid was not covered by a compulsory purchase contract and was sold at market price. A summary of the results of this simulation is shown in Table 6.8.

Renewable production coefficients	kg CO_2/MWh	€/MWh	Self-consumption ENR (%)	Self-production ENR (%)	Diesel consumption (kWh)
0 (zero)	52.3	76.3	0	0	0
1 (normal)	39.6	80.7	98.4	41.19	0
2 (overproduction)	38.4	95.2	74.7	62.5	0

Table 6.8. *Summary of results for scenario 2*

Taking the zero renewable energy value as a starting point, we see a reduction in environmental impact of approximately 24% for normal production and approximately 26% with overproduction thanks to the low levels of CO_2 emissions from these sources. However, the high cost of operating these systems in 2015 led to a 5.8% increase in financial impact for normal production, and a 24.8% increase for overproduction. The overall energy cost using renewable energy was higher than that of electricity from the public distribution grid in 2015, contrasting with the change in environmental impact.

Self-production is:

– in case 1, approximately 41% with self-consumption of approximately 98% for normal renewable energy production;

– in case 2, approximately 63% with self-consumption of approximately 74% for overproduction of renewable energy.

The increase in renewable energy production in the micro-grid is not beneficial as the cost of energy increases by 17.8% between normal production and overproduction, and the environmental impact falls by only 3%. This is due to the decline in self-consumption, from approximately 98% to 75% (renewable production does not always correlate with consumption and is injected into the distribution grid). In the latter case, surplus energy is not consumed at the time of production, and 25% of this production is sold to the electricity grid. This operation is not financially viable, as the energy was sold on the spot market and the purchase price was not high enough.

This leads us to conclude that renewable energy production installations must be carefully dimensioned to reflect self-consumption capacity in order to combine a low financial impact with high environmental benefits.

6.4.2.3. *Scenario 3: contribution of the supervisor system*

This scenario features the fuzzy logic supervisor along with a storage system. For each simulation, only the overall load is increased in 10% steps up to a level of twice the initial consumption of the micro-grid. The standby generator remains present and is used to avoid crossing the subscribed power threshold during shoulder and peak periods if the storage system is unable to supply enough energy.

Energy is stored during off-peak periods, then discharged during shoulder and peak periods if the maximum subscribed power threshold is reached.

A summary of the results of our simulation is given in Table 6.9. In this table, the environmental impact takes account of pollution from the standby generator. The self-consumption and self-production percentages only concern renewable energy.

2015	$Coeff_{LOAD}$	kg CO_2/MWh	€/MWh	Self-consumption ENR (%)	Self-production ENR (%)	Diesel consumption (kWh)
ENR × 1	1.00	39.6	80.7	98.41	41.19	–
	1.10	40.6	80.1	99.30	37.79	–
	1.20	41.4	79.9	99.84	34.83	12.21
	1.30	42.4	80.1	99.98	32.19	84.21
	1.35	43.4	80.7	100	31.00	421.09
	1.40	44.6	81.7	100	29.90	914.19
	1.50	48.6	84.6	100	27.90	2,796.00
	1.60	54.9	88.8	100	26.16	6,315.10
	2.00	66.7	106.9	100	20.93	15,307.00

Table 6.9. *Summary of results for scenario 3*

The subscribed power is set at $P_s = 2500$ kW. For the initial consumption value and up to 10% additional load, the fuzzy logic supervisor is not activated as the load remains below the subscribed power. From 20% additional load, the energy management system is activated and energy is discharged from the battery. This reduces energy costs as the subscribed power is not exceeded and use of the standby generator is limited. With an additional load of 35% or more, the subscribed power may be reduced by approximately 26% to $P_S = 1850$ kW, thanks to the action of the supervisor, in order to maintain an energy price identical to that found at the start of the simulation, i.e. approximately €81/MWh (financial impact: – €3.14/MWh). At this point, consumption from the standby generator falls from 3370 kWh (Table 6.7) to 421 kWh (Table 6.9), a reduction of 87.5%.

The self-consumption indicator increases as the load increases. This shows increasing integration of renewable energy sources. However, the self-production indicator, which shows the level of dependency on the distribution grid, falls (showing increasing dependency on the grid) as consumption rises. This is because electrical energy production from renewable sources does not increase, self-consumption has reached 100% and the load continues to increase.

The evolution of the price indicator is explained by the fact that the price of renewable energy was higher than grid prices in 2015 and by the increase in the consumption of energy produced by the standby generator. However, the increase in cost is lower when a supervision system including storage is used (Table 6.9) than when these elements are omitted (Table 6.7).

With an additional load of 35%, the fact that energy is taken from the battery rather than the grid or the standby generator during certain periods reduces the environmental impact by approximately 12% (from 49.5 kg.CO_2/MWh (Table 6.7) to 43.4 kg.CO_2/MWh (Table 6.9)).

Hence, the integration of a supervisor based on an electricity storage system and fuzzy logic regulation limits the production of electricity by the standby generator in cases where the subscribed power is reached by discharging energy from the battery (Figure 6.15) and reduces the cost of electricity in the micro-grid (Figure 6.16). It enables subscription power to be reduced, even in cases of increased consumption. The blue lines in Figures 6.15 and 6.16 correspond to the case where a supervision and storage system is present, in contrast to the red lines, which are higher in all cases.

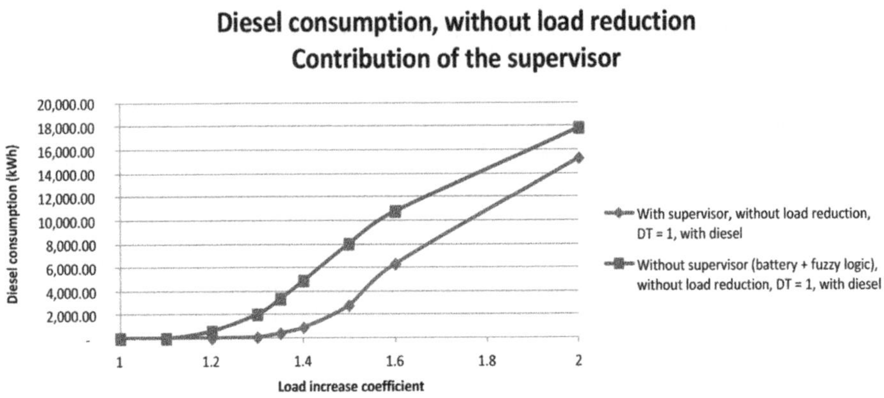

Figure 6.15. *Supervisor influence on diesel generator production*

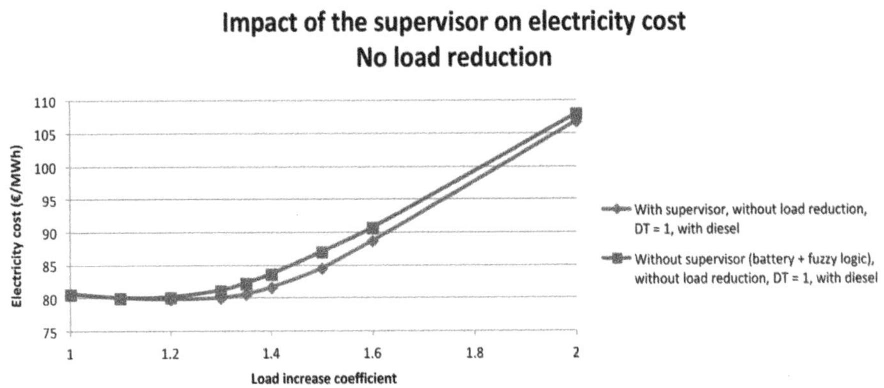

Figure 6.16. *Influence of the supervisor on electricity costs*

6.5. Load reduction

6.5.1. *Load reduction principle*

In all of the scenarios presented above, a load reduction system may be introduced in order to verify the economic and environmental advantages.

Load reduction is the ability of an entity to reduce or delay consumption of electrical energy. It is notably dependent on a societal variable. A proportionality coefficient will be included in order to calculate the load reduction power for an actor.

To encourage load reduction, the French ministry of ecology, sustainable development and energy published a decree on January 11, 2015 establishing the value of a bonus to be paid to load reduction operators. For load reduction activities on consumption sites with a subscribed power less than or equal to 36 kVA, the reference value is set at €16/MWh at peak times (7 am–11 pm) and €2/MWh during the off-peak period (11 pm–7 am) [ARR 15].

For load reduction activities on consumption sites with a subscribed power of over 36 kVA, the reference value for the bonus is null [ARR 15]. Thus, only residential consumers (*tarif bleu*, below 36 kVA) will receive the bonus. For large consumers, such as industrial and tertiary sector users, load reduction forms part of other mechanisms and is linked to reduction market with capacity requests, participating in balance management for the electric grid over different timescales [ADE 17].

6.5.2. *Introduction to load reduction and acceptability*

According to a study carried out by Accenture [ACC 10], a specialist management, technologies and outsourcing consultancy, there are six main profiles for electrical energy consumers (Figure 6.17), which we presented in Chapter 5 (section 5.3.4.2.1). These can be used to calculate acceptability coefficients.

Figure 6.18 shows the degree of support for an energy management program among different consumers: positive (green), indifferent (blue) and negative (white). These are average values for the surveyed population [ACC 10].

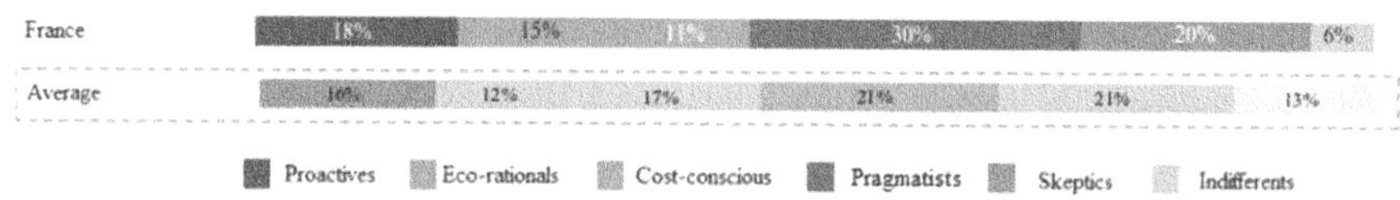

Figure 6.17. *Estimation of the distribution of consumer profiles within the French population and across the 17 developed countries around the world covered by the survey [ACC 10]*

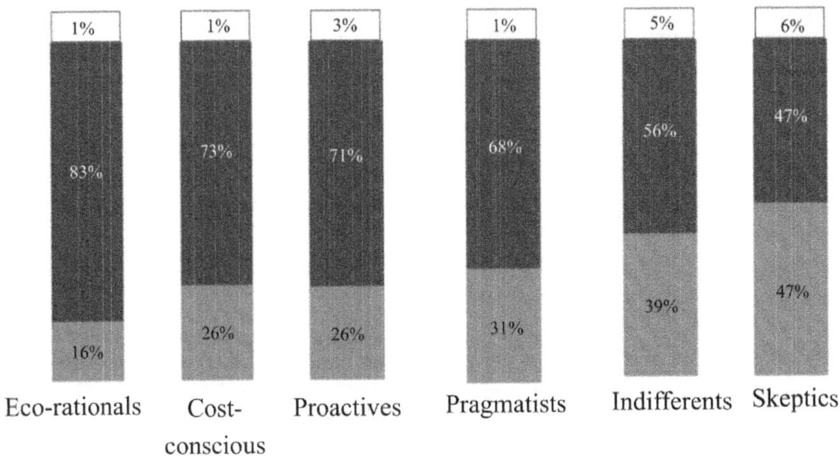

Figure 6.18. *Adoption rate of an energy management program as a function of energy consumer profiles: positive (green), indifferent (blue) and negative (white)*

The percentage of consumers in favor of an energy management program will be used as acceptability coefficients for energy supervision.

For the residential development, these percentages were averaged using the percentage distribution of profiles across the French population (Figure 6.17).

For the shopping mall and supermarket, a cost-conscious profile was selected as the advantage for these actors is of a financial nature.

This approach gives us average acceptability rates of 68.12% for the residential development and 73% for the shopping mall and supermarket.

A new variable is added to the supervision system in the form of a global acceptability coefficient for load reduction, varying from 0 (no acceptance, no load reduction bonus) and 1 (total acceptance of load reduction in return for a bonus payment). This influences the power capacity freed up by the actor (Figure 6.19).

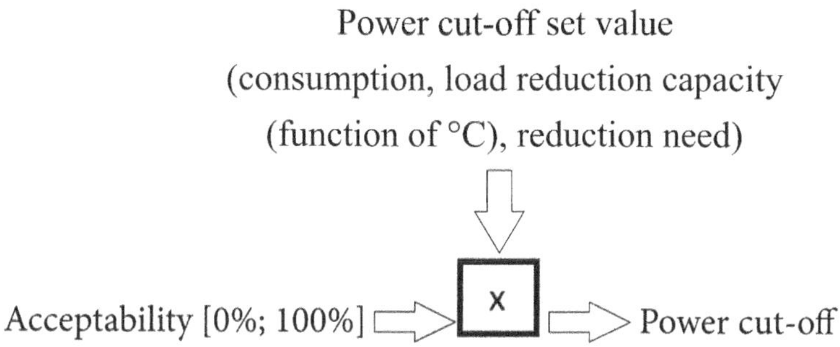

Figure 6.19. *Use of acceptability as a proportionality coefficient in calculating the power made available through load reduction*

The results of Accenture's statistical study were used to simulate the behavior of a French population in accepting load reduction. It would also be possible to install a unit for consumers in the micro-grid giving a reduction value from 0 to 1, modifiable according to consumer preferences.

The shopping mall alongside the supermarket is the first candidate for load reduction, as the supermarket owns the standby generator and the battery. The supermarket is not the first candidate as a form of payback for the services rendered to the local grid.

The residential development is the last candidate in the load reduction chain due to the lower level of acceptance.

6.5.3. *Simulation of energy management with load reduction*

Figures 6.20 and 6.21 show the impact of the potential inclusion of load reduction capacities on the use of the standby generator and in terms of electricity cost as a function of increasing charge. The green and purple curves show results without load reduction, without storage and with storage, respectively. The blue and red curves represent cases including load reduction, without storage and with storage, respectively. In terms of standby generator usage (Figure 6.20), a significant reduction is obtained by combining the contributions of storage and load reduction. A comparison of the red and purple curves, which are very similar, shows that load reduction techniques produce a similar effect to a storage system in our simulated case. Figure 6.21 shows that load reduction has a positive impact in all cases, reducing electricity cost from the point when the additional load rises above 40% of the original reference load.

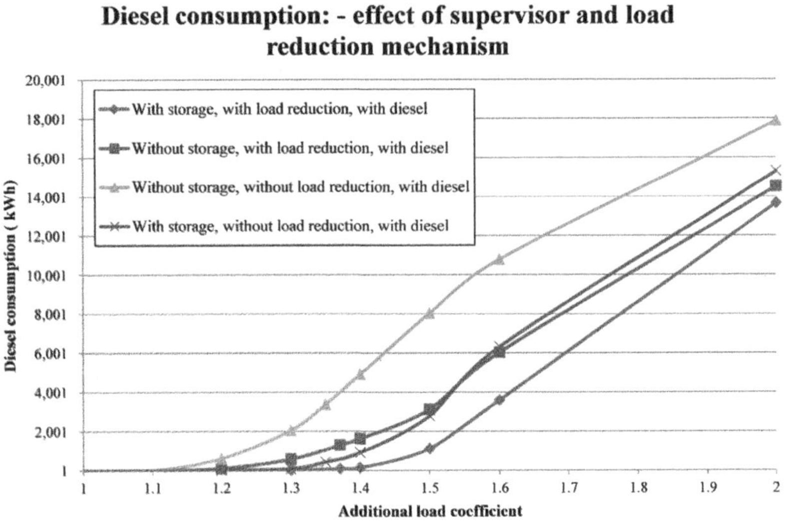

Figure 6.20. *Effects of including potential load reduction capacity on standby generator usage as a function of increasing load*

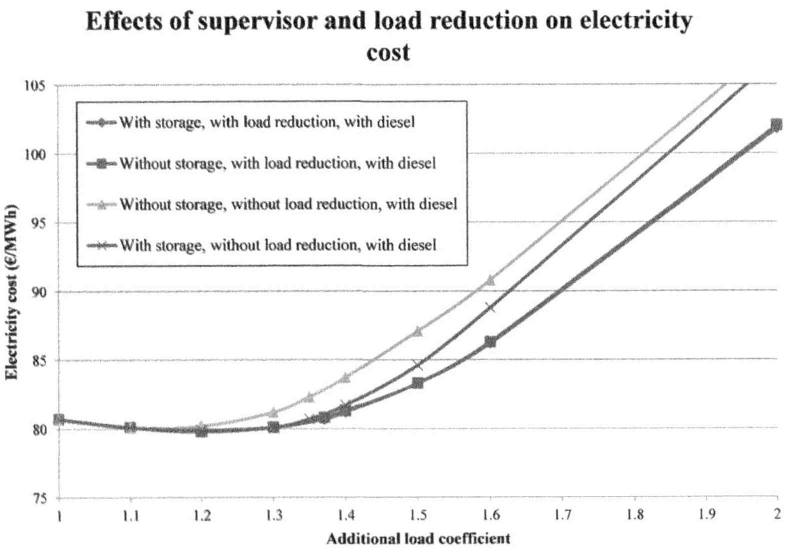

Figure 6.21. *Effects of including potential load reduction capacity in terms of electricity cost as a function of increasing load*

6.6. Conclusion

In this chapter, we have presented several simulated scenarios, based on a real-world case and on interviews with the actors involved, highlighting their constraints and requirements. These scenarios demonstrated that energy management with renewable energy production, electrical energy storage and load reduction within a micro-grid can be used to reduce energy cost by maximizing the consumption of electricity produced within the micro-grid itself, following a self-consumption approach.

Given the element of uncertainty in relation to household behaviors, we used a single acceptability criterion alongside the conclusions drawn by the Accenture survey [ACC 10]. The power relationships resulting from the differing objectives and constraints of actors within the micro-grid mean that the presence of an aggregator is crucial, establishing the priorities of the micro-grid and representing the structure in negotiations with external entities, notably concerning economic and legal matters.

The results presented in this chapter could be refined further by taking account of the depreciation of investments and of maintenance costs, particularly for the storage system and the renewable energy production systems. Moreover, the pricing principles used here are subject to change with the increasing liberalization of the energy market (choice of supplier and of regulated tariffs, or through direct contact with the energy market) and in connection with the development of renewable energy sources, particularly in the context of self-consumption. Nevertheless, the principles used to develop a supervision strategy in this chapter remain relevant, and may be applied to other energy and regulatory scenarios.

6.7. Acknowledgments

The authors wish to thank the ADEME and the Nord – Pas de Calais regional authority for their financial backing for the GESEDMA project (*Gestion et échanges de services énergétiques décentralisés* – Management and Exchange of Decentralized Energy Services) through which the studies presented in this chapter were developed.

6.8. Appendix 1

TARIF VERT A5 moins de 10 000 kW							
Versions		Prime fixe annuelle (€/kVA)	Hiver			Eté	
			Pointe	HPH	HCH	HPE	HCE
A5 OPTION BASE	TLU	76,68	7,396	6,017	4,603	4,609	2,916
	LU	56,4	10,774	6,893	4,762	4,678	2,979
	MU	44,64	14,692	8,035	5,101	4,756	2,975
	CU	31,32	22,112	10,114	5,485	4,756	2,819
Energie réactive (c€/kvarh)[(a)]							
Coefficients de puissance réduite	TLU		1,00	0,67	0,27	0,23	0,23
	LU		1,00	0,76	0,40	0,37	0,34
	MU		1,00	0,75	0,36	0,33	0,28
	CU		1,00	0,78	0,52	0,46	0,42
Calcul des dépassements	Comptage ($k_3k_2k_1$)	Electronique 4,55 €/kW		KN(PMAX-P) 1,51 €/kW		K(PMAX-P) 37,94 €/kW	
	Coefficients par poste		1,00	0,67	0,27	0,23	0,23
Hiver	: de novembre à mars inclus						
Été	: d'avril à octobre inclus						
Pointe	: 2 h le matin et 2 h le soir de décembre à février inclus						
Heures creuses	: 8 h par jour et dimanche toute la journée						

(a) L'énergie réactive est facturée selon les conditions de la formule tarifaire d'acheminement appliquée au site.

Ces prix sont à majorer de la TVA, de la contribution au service public de l'électricité (CSPE), de la contribution tarifaire acheminement (CTA), et le cas échéant en fonction de la puissance souscrite, des taxes sur la consommation finale d'électricité (TCFE) instituées par les communes (ou syndicats de communes) et départements ou bien de la taxe intérieure sur la consommation finale d'électricité (TICFE), ainsi que de tout nouvel impôt, toute nouvelle taxe ou contribution qui viendraient à être créés.

Figure 6.22. *Energy pricing. For a color version of the figures in this chapter see www.iste.co.uk/robyns/buildings.zip*

This table shows the pricing structure used by EDF for non-domestic consumers falling below the 10,000 kW threshold, known as the "green tariff" (note that the color bears no relation to environmental concerns – the company also has a "blue tariff" and a "yellow tariff" for different profiles).

A number of different versions are shown, with an annual subscription cost per kVA, peak, shoulder and off-peak tariffs for the winter season (November–March), and peak and off-peak tariffs for summer (April–October). Details of the penalties for exceeding thresholds are also shown.

7

Centralized Management of a Local Energy Community to Maximize Self-consumption of PV Production

7.1. Introduction

The needs of society are in a state of constant transformation. Over the past decade, we have seen changes in the behavior of populations who continue to aspire to increased levels of comfort, but who are also increasingly aware of the impact of their lifestyle on the environment. This has resulted in a continuous rise in electrical energy consumption worldwide (in France, this specifically takes the form of an increase in peak consumption) and in the desire to develop behaviors that are virtuous from an environmental perspective.

Case studies often focus on habitat. This context features characteristic consumption requirements, along with demands in terms of air quality, and a clear desire to reduce pollution, particularly within a wider context of neighborhoods or towns. In France, the building sector (residential and tertiary) accounted for 24% of greenhouse gas emissions in 2015, and is currently the greatest load on the electrical grid, accounting for 69% of electrical consumption in France [COM 15, MIN 16] (Figure 7.1). In 2016, electricity represented almost 25% of total energy consumption in the residential sector [MIN 16].

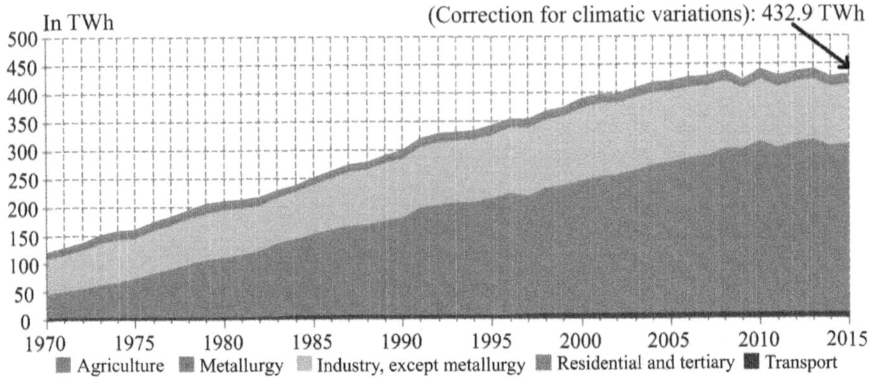

[1]Corrected for climatic variations.
***Sources:** SOeS calculations, based on results of an electricity transport and distribution survey; RTE; Enedis; Rica*

Figure 7.1. *Final electricity consumption by sector in 2015 [MIN 16]. For a color version of the figures in this chapter see www.iste.co.uk/robyns/buildings.zip*

In countries where nuclear power does not play an important role in the energy mix, in terms of reducing electricity consumption, buildings represent a crucial element in reducing dependency on petrol and the risks caused by greenhouse gas emissions in terms of climate change. In the construction industry, there is an increasing focus on constructing energy-efficient buildings of high environmental quality, with a move toward positive energy. "Positive energy" buildings (RT2020 thermal regulation for housing built from 2020 onward) are intended to produce more energy than they produce over a year, with the aim of increasing energy autonomy. In concrete terms, from January 1, 2020, all new tertiary and residential buildings in France must be energy-positive (Chapter 1, section 1.4.2). The increase of renewable sources of electrical energy in the residential sector is the main measure taken in order to reduce long-term atmospheric production and consumption. In 2015, mean annual energy consumption in the residential sector was 240 kWh_{ep} (primary energy)/year/m^2 in France. The RT 2012 thermal regulations established an objective of 50 kWh_{ep}/year/m^2 for new residential buildings. In Europe, the goal is to attain 40 kWh_{ep}/year/m^2 in 2025 and to reach energy-positive status by 2050.

At individual house level, electricity consumption is low, and the area available for the installation of electrical production units is also limited. The electrical power of installed units is therefore relatively low. Over time, the generalization of electrical production in habitat will result in an electrical system based on distributed production, i.e. multiple low-power generators, in place of a small number of large-scale production units currently used. Given that local consumption and local production from renewable and/or intermittent and/or unpredictable sources (such as PV production, domestic wind power and cogeneration) do not correlate, instantaneous electrical power production must be used in other sectors, such as transport, recharging the batteries of electric vehicles. The need to manage these new energy flows according to local need or external demand has led to the development of "smart" systems, and consequently "smart" buildings (see Chapter 1, section 1.5). Smart buildings form one of the pillars of a future smart grid with the capacity to offer production and storage services and to participate in reducing consumption (demand response) at peak times in order to support the wider electrical system.

Sources responding to the needs of the construction sector are small and modular (from a few kW to a few tens of kW) and are connected to the low-voltage (LV) grid. The majority of systems in this area are based on electrical and thermal co-generation or on residential and tertiary photovoltaic (PV) generators. Considering the possibility of a 100% renewable electricity mix, the ADEME estimates that the largest production capacity would come from PV panels installed on roofs, accounting for 34.8% of electrical production capacity (68.3 GW out of a total requirement of 196 GW) in the most socially acceptable scenario [ADE 2016]. In this situation, buildings would become some of the biggest producers of renewable energy.

PV systems have now reached a certain level of maturity, although further improvements in energy performance are still expected, and increasing use will lead to a significant reduction in cost. The connection of a large number of distributed generators and electricity producers to the grid in recent years has resulted in significant structural changes to the electrical energy supply system. The number of new PV installations connected to the grid in France considerably increased after new purchase prices for PV energy were published in July 2006; by 2016, the figures were beginning to stabilize (Figure 7.2).

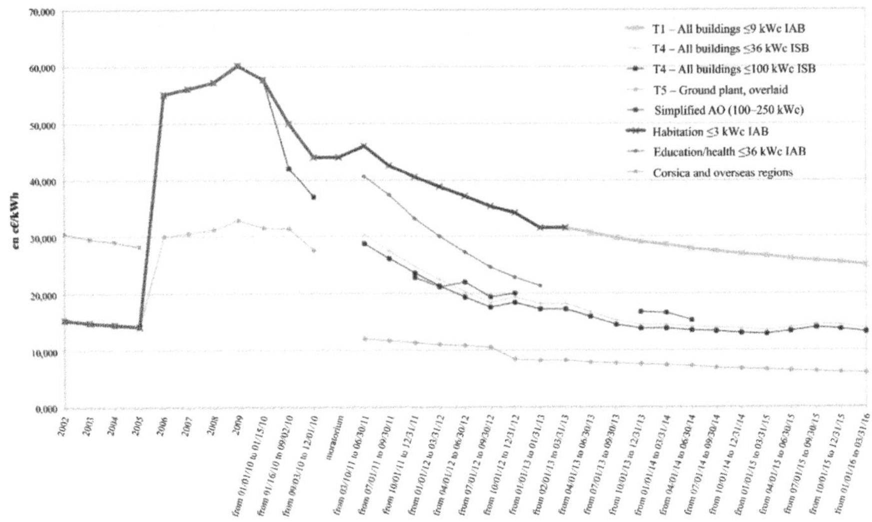

Figure 7.2. *Evolution of the main purchase prices for PV energy in France [HES 18]*

In 2014, 1,706 MW of PV production capacity was installed in the French distribution grid; this figure also appears to be declining [RTE 17a] (Figure 7.3). Furthermore, 85% of photovoltaic producers connected to the low-voltage distribution grid generally have a nominal power of less than 3 kW.

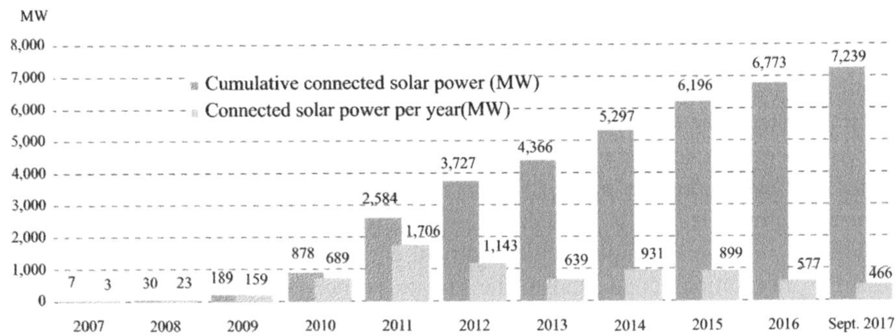

Figure 7.3. *PV facilities connected to the grid managed by ENEDIS in France [RTE 17a]*

This increase in decentralized PV production has generated numerous problems in the electricity system at local level. As [ROB 12c] explains, they must be integrated into the normal operations of the electricity system in order to expand. For example, in a residential grid where all households are equipped with PV panels, total production may exceed consumption, leading to a breach of the maximum permissible voltage. In this case, one solution is for the grid operator to offload part of the PV production: no value will be gained from this PV power. If production is based on 100% renewable sources, this creates new demands in terms of the technical and economic management of grid automation and management processes in order to maintain efficiency. Knock-on effects may be observed in voltage levels, grid frequency, fuse triggering due to modifications in power transit, grid stability and fault repair times.

The fluctuating and unpredictable nature of power generation from renewable sources means that these partially controllable or non-controllable generators are unable to provide the services needed for the grid to operate correctly. Furthermore, current grid regulation systems are not suited to the additional regulation power demands or to the new dynamics of use. Operators need to look for flexibility in new areas in order to ensure that the grid continues to operate correctly. These include:

– load reduction in cases of production shortage or low voltage (demand response);

– total or progressive production offloading in cases of over-production or high voltage;

– dynamic regulation of the reactive power of the producer;

– storage.

In a smart building, depending on the power of flexible loads and the renewable production capacity, the dynamic adaptation of flexible loads according to local energy availability constitutes a means of maximizing self-consumption. Storage is particularly useful in operational terms due to its capacity to modify the consumed or generated power flows of active or reactive power at the point of connection [ROB 15]. In the specific context of energy mutualization in a neighborhood, two forms of installation may be considered:

– centralized storage for the whole neighborhood: large size, high power, high energy;

– distributed storage with units in each building, which would be private and smaller in scale.

Given the environmental motivations, both applications should aim to pool this source of flexibility in order to maximize use and minimize investment costs.

In this chapter, we will consider the distributed storage option. This not only presents additional difficulties, such as coordination, but also enables the sharing of investment costs among producers. For example, in 2015, Tesla released an individual lithium-ion battery, the Tesla Power Wall, intended for the domestic electricity storage market (13.5 kWh, priced at 7,060 euros in France in 2018).

Globally, there is a trend toward reduction of incentivized purchase prices for renewable energy (feed-in tariff). It will therefore become less profitable to sell all PV production in coming years. The transition has already been made in Germany, where a regulatory tariff has also been introduced to subsidize self-consumption of renewable energy for nominal powers under 30 kW [RIF 09]. The abolition of fixed electricity prices has increased client interest in smart consumption control, with price changes during the day acting as an incentive to correlate domestic consumption with available domestic PV production.

The integration of storage systems into renewable energy production systems makes it possible to modulate power exchanges with the distribution grid. This means that producers can play a more active role in satisfying the needs of their own home and/or in supplying services to support the operation of the grid.

Four types of stakeholders, or actors, are involved in the new electricity systems being considered here:

– conventional and renewable energy producers directly connected to the distribution and transport grid;

– dispatchable industrial loads;

– buildings, including dispatchable loads, integrated renewable production systems, storage systems and electric vehicles;

– distribution and transport grid regulation systems, controlled by management centers (national and regional dispatching for transport, regional management agency for distribution).

Energy management in an electrical system of this type may use information flows between actors in order to master bidirectional energy flows, exploiting the flexible aspects of production and consumption in order to manage the system as a whole.

Energy-positive neighborhoods are made up of several buildings engaged in active management of their energy consumption and of energy flows between their installations and the broader energy system. In this context, the objective is to develop interactivity between PV production dispersed across buildings (which may be partially controlled through distributed storage), gas turbines (co-generation) and the micro-grid operator in order to respond to shared objectives. However, electricity generators are currently designed and controlled at a local level to satisfy their own specific optimization criteria (e.g. maximizing production potential) and are generally incompatible with the wider systemic optimization approaches required in a residential neighborhood.

On this level, objectives include increasing the energy efficiency of the neighborhood and reducing carbon dioxide emissions. This may be achieved by increasing the proportion of PV production in the local energy mix, as well as through optimized utilization of gas turbines. The operation and continued stability of the local electricity grid remain crucial, and this represents a constraint. Stability is maintained by balancing production capacity (supply) with the increasing use of electricity for new purposes (electric vehicles, heat pump systems, etc.). Our energy management approach aims to identify generator production profiles to attain overall optimization of an objective function for the urban grid, then adjusting points of operation over the course of the day in response to any differences that emerge. The purpose of optimization is to identify optimal solutions which:

– maximize the use of electrical production from renewable sources according to availability;

– minimize the cost of electricity produced in the neighborhood;

– minimize CO_2 emissions from conventional generators.

This chapter is broken down into ten main sections, concluding with a review of scientific contributions and research perspectives (sections 7.11 and 7.12).

Section 7.2 provides a rapid review of the way in which an electric grid operates, identifying the challenges to be met and the innovations to be made for applications in an urban setting, based on a number of known examples of urban micro-grids.

Section 7.3 concerns the use of storage in buildings with PV production. We will examine the local control mechanisms used to manage power exchanges between the building and the electric grid.

The role of the studied building within a neighborhood energy management system including several other buildings of the same type will be addressed in section 7.4. The energy management system is based on different control functions, on the principle of pre-optimization established one day ahead for day N using consumption and production predictions and on adjustments made as necessary over the course of a day to account for unpredictability.

Section 7.5 is devoted to a case of application, illustrating the way in which these concepts may be applied to energy management in a residential neighborhood. We will present the way in which an active generator may be integrated into a domestic dwelling. A group of producer-consumers, forming a residential electrical grid, interact with an aggregator over a communications network. We will then use micro-grid concepts to define control functions, localize them and organize energy management for the residential micro-grid. The main functions of energy management for micro-grids will be summarized and our strategy will be described.

Section 7.6 relates to the basic techniques used to predict PV production and consumption loads. We will analyze our data in order to estimate the different quantities of energy which are available and required.

In section 7.7, we will address the issue of energy planning for different generators, proposing a determinist algorithm for use in satisfying different constraints.

Prediction errors are inevitable and are taken into account by medium-term energy management every 30 min (same-day adjustment). This will be discussed in section 7.8.

Section 7.9 provides a description of a modified local control system for active generators, integrating control functions for short-term energy management in the micro-grid. The controller coordinates the energy resources of a home in order to satisfy the power references indicated by the central management system.

In section 7.10, the implementation of the residential grid in the L2EP: Electrical Power Management Lab experimental platform is described. Our objective is to validate the use of storage to maximize local consumption of renewable energy using experimental methods. We will compare different sizing strategies for identical production and consumption profiles, before analyzing economic and environmental impacts through the use of indicators.

7.2. Energy management issues in residential neighborhoods

7.2.1. *Electric grid management: basic principles*

The management of an electric grid is based around its material structure, generally involving a national dispatching center operator for managing long-distance transport and regional centers that manage local distribution.

The balance between demand for electricity and the cheapest offer is managed by the national dispatching center. The main challenge concerns the difficulty of making precise predictions concerning consumption. Figure 7.4 shows examples of the consumption curve predicted one day ahead, the curve predicted on the morning of day N, and actual consumption.

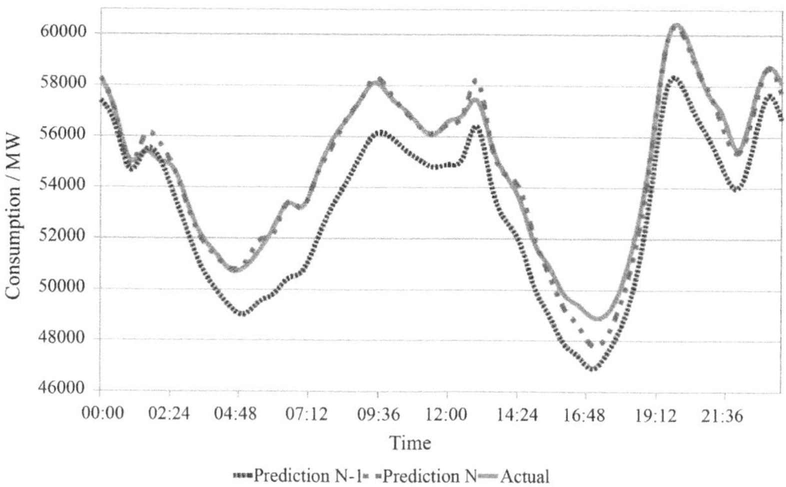

Figure 7.4. *Example of predicted and actual load curves in mainland France [RTE 17a]*

The second important function of grid management is to regulate voltage. This is normally carried out locally using adjustable transformers and through the generation or consumption of reactive power by synchronous machines or static (power electronic converter-based) compensators [BER 10]. Each electrical network has its own specific voltage levels, dependent on the power being carried, and are locally regulated by the grid operator. However, these grids currently offer little flexibility and little scope for automatic control, and regulation dynamics are either slow or non-existent.

7.2.2. *The move toward smart grids*

As the demand for electricity is variable and, except in certain specific circumstances (direct load control, load reduction, offloading, etc.), impossible to control, the electricity generation system currently needs to be adapted instantaneously to the demand in order to maintain system stability. This "instantaneous controllability" requirement for electricity production raises a number of problems linked to the technological constraints of different power plants. The output power of certain renewable energy sources (e.g. PV panels or wind turbines) is impossible to control as it is directly dependent on the availability of the primary renewable source, which is naturally volatile. Certain plants, such as conventional thermal power stations, can be started up quickly, while others require significant warm-up or cool-down periods (nuclear power). Others have a limited area of operation in which electricity yield is optimized (thermal and hydraulic). All of these specificities need to be taken into account by the energy management system in defining an operational schedule to satisfy energy demands while minimizing costs. However, the increasing presence of decentralized production (generally non-controllable) means that the way in which all grids are managed will need to evolve.

The variability and increasing unpredictability of energy transports may be accommodated by increasing the level of flexibility in energy management systems. Fast responses are needed to resolve consumption/production imbalances. There are three main ways to increase flexibility:

– acting on consumption by promoting demand management;

– acting on production by implementing controllable generators;

– acting on the grid itself through the integration of energy storage solutions.

A smart grid is an electricity distribution grid which uses new technologies to exploit these flexibilities and implement electrical energy transfers with new characteristics, such as rapid dynamics or rerouting capacities, in order to minimize distribution costs and maximize reliability. The nature of new equipment being connected to the grid, such as intermittent renewable energy production systems, and new load types, such as electric vehicles, means that these characteristics are becoming increasingly important. The "smart" aspect stems from the need to control equipment in order to optimize and coordinate the production, distribution and consumption of electricity – in short, to develop reactive energy management.

Smart meters can send and receive signals, and this data may be used in making decisions, for example, to delay certain loads. Thus, smart grids have the capacity to act on demand. The installation and use of a communications grid, including producers, distributors, consumers and local aggregators, creates an interactive aspect that can be used to improve the coordination of consumption demands and instantaneous production, with the possibility of using a decentralized storage. From this perspective, the management of electrical systems needs to evolve in order to control loads, exploit the aggregation of decentralized production and make use of energy storage capacities. This implies a decentralization of grid intelligence toward the distribution grids and more specifically toward local observation and control units, such as groups of buildings forming a local energy community and using a shared energy management system. Therefore, a dedicated infrastructure needs to be established using sensors and collection systems for production and consumption data at every level of the grid, a communications network and local analytical systems using data obtained over the short, medium and long terms. The new system will enable us to:

– use all generators at maximum yield;

– reduce losses from power lines;

– encourage the expansion of decentralized production, particularly local consumption of local production from renewable sources;

– distribute electricity at the best possible price via improved interactions between producers and consumers.

In current power system (Figure 7.5), electricity is mostly generated by large plants, supervised and controlled by the transport grid manager, who receives consumption information from transformer stations and distribution grid managers. Decentralized production cannot be controlled but must be accounted for by grid managers (and may be known as negative power). It can, however, be estimated and taken into account using weather forecasting data. If distributed production power exceeds consumption in an area of the grid, energy flows become abnormal (as the grid is not designed to cope with them), resulting in voltage regulation problems.

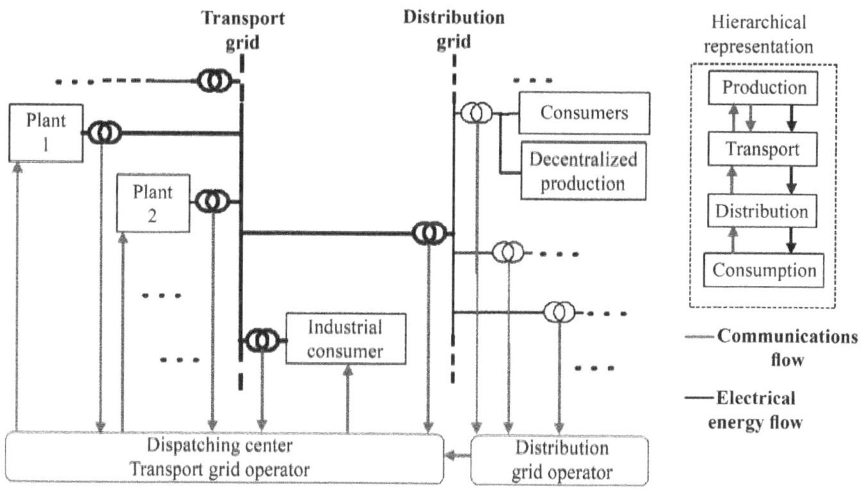

Figure 7.5. *Traditional structure of the electrical system [KAN 14b]*

In a smart grid, however (Figure 7.6), decentralized production and consumption are locally coordinated in an optimal manner by a dedicated aggregator. Each local group of decentralized generators, loads and storage devices is seen by the grid manager as a single entity, which may act as a consumer or as a producer of electrical energy [SIN 10]. Storage systems become particularly helpful in this context, particularly on the scale of a town or neighborhood.

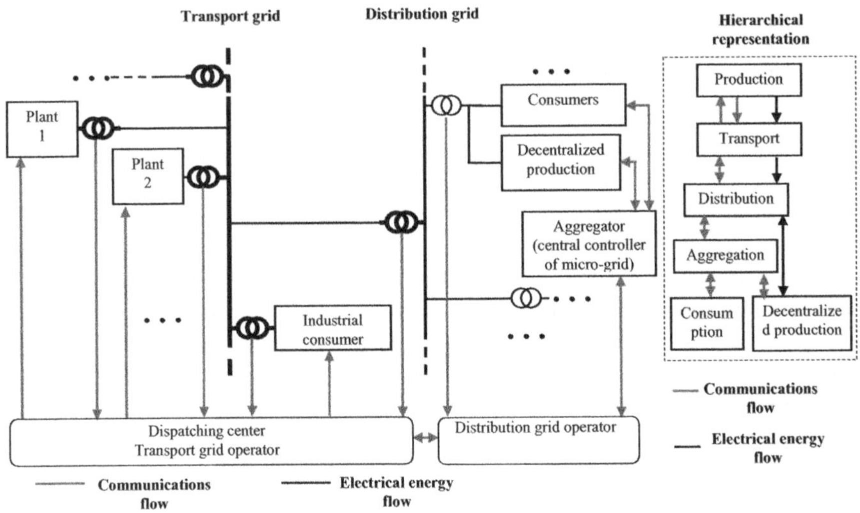

Figure 7.6. *Smart grid structure [KAN 14b]*

7.2.3. *A few applications of micro-grids for managing local energy communities*

7.2.3.1. *Local energy communities and electricity micro-grids*

According to the European Commission definition [EC 16], a local energy community (LEC) is an association, cooperative, partnership, non-profit organization or other legal entity which is:

– controlled by local stakeholders or members;

– generally, value-driven rather than profit-driven;

– involved in distributed generation and in performing activities of a distribution system operator, supplier or aggregator at local level.

There is a long tradition of LECs in Europe, notably in Germany, where there are more than 650 Stadtwerke (local public services companies providing heating and electricity), and in France, where there are more than 160 *Entreprises Locales de Distribution d'électricité* with their roots in municipal structures. The production and exchange of electricity between buildings in the same neighborhood constitutes a basic form of LEC. This form of organization may be beneficial for many stakeholders in the electricity system:

– consumers may produce and exchange some of their energy locally in order to reduce their electricity bills;

– distribution grid congestion at peak times may be reduced if decentralized producers are coordinated to comply with constraints imposed by grid operators, meaning that investment in grid reinforcement can be reduced;

– electricity transportation losses are reduced due to the short distances separating producers and consumers;

– decentralized production creates opportunities for isolated sites, where consumption does not justify the installation of high-power generators and where fuel supply is hard to come by;

– decentralized production mechanisms should increase the reliability of the power supply for critical loads.

The organization of activities linked to local electricity production and distribution following a micro-grid structure within an LEC is fascinating, due to the implementation of energy management approaches and the potential they offer for optimizing the use of local renewable energy sources. A micro-grid is a group of local consumers and small electrical energy producers connected over a low-voltage network, which may be used in a controlled and coordinated manner. From the perspective of the distribution grid organizer, the micro-grid is viewed as a single unit that may act as an aggregated producer and consumer; it may also, in exchange for payment, play the role of a small energy producer or provide system services to support the distribution grid.

A set of co-existing and interacting micro-grids within an urban location is one important element of a smart city. In general, the environmental impact of these infrastructures and of their operation is taken into account in long-term urban development projects and from a global perspective. However, many other elements are needed for a city to be truly sustainable (optimal water management, urban waste, transportation, reasonable urban density, etc.). In energy terms, smart cities aim to adopt more efficient and responsible approaches to energy consumption by the town and its inhabitants. “Smart” grid management enables simultaneous local optimization of:

– the use of different sources of supply and of energy consumption on different timescales [SMA 11];

– the integration of renewable energy sources and of new usages (electric vehicles, etc.).

7.2.3.2. *Context of the case study*

The current electricity system follows a hierarchical, centralized structure. By increasing interconnections in terms of information flows and by decentralizing management of electricity production, the operational efficiency of the entire system is expected to be improved. With this in mind, we will consider local groups of "energy buildings", i.e. generators and consumers, aggregated into micro-grids, which may be seen as either consumers or producers of electrical energy from the perspective of the broader grid.

The first experimental micro-grids, operating autonomously or connected to the grid, demonstrated the value of these systems for sustainable residential neighborhoods and sustainable cities. This application will be considered here.

Each micro-grid is supervised by a central controller which:

– monitors energy demand, source availability and price;

– optimizes operational planning for local sources and consumers;

– communicates demand or offers of excess energy to the distribution grid operator.

Strategies and algorithms to optimize the operation of the central controller still need to be developed, especially in relation to the integration of new hardware for grid regulation and to new approaches for an accurate prediction of both consumer demands and excess energy production from intermittent (passive) generators. In response to these needs, prediction-based optimization algorithms will be developed in section 7.7 using predictions to integrate active PV generators and gas turbines. Furthermore, an adjustment technique using a communication network between generators and the grid operator will be developed in order to improve precision.

To implement demand-based management of electricity transit in a grid, generators with controllable power are required. This is an issue in systems which only possess conventional wind or PV generators, where electricity production is directly dependent on the availability of intermittent energy sources. The following section presents PV generators with integrated storage systems, which can be controlled by a grid manager.

7.3. The active PV generator

7.3.1. *Current PV production*

In general, all of the energy produced by a grid-connected PV system is sold to the main grid (Figure 7.7). Photovoltaic energy is not yet able to compete with existing electricity prices [HES 18]. For this reason, in response to the need to reduce production and dependence on non-renewable energy sources, most countries have implemented economic measures to promote renewable energy sources. In the case of PV power, this usually takes the form of interesting electricity purchase prices imposed for PV production (feed-in tariff). Another, less habitual measure is to offer direct subsidies to private individuals, reducing their investment outlay.

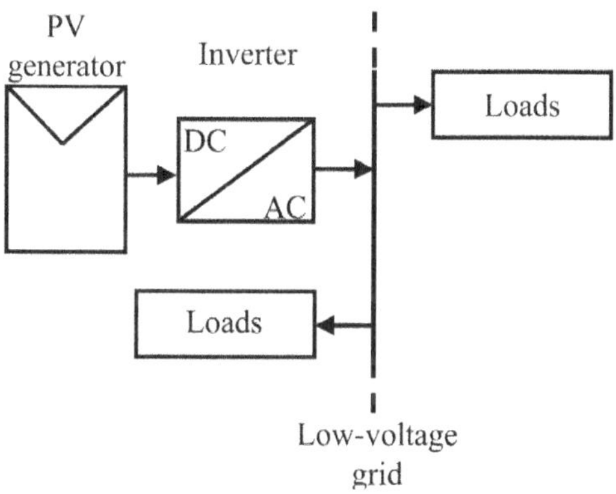

Figure 7.7. *Example of a PV system connected to the distribution grid*

7.3.2. *Limits and necessary developments*

One of the main problems of current PV generators is that their power output fluctuates and is dependent on weather conditions. Moreover, the control mechanism offers only one option, that of maximum power. This means that more power may be generated than is needed by the grid, causing electrical instability if the excess is not consumed. The maximum level of integration of passive PV generators into the grid is therefore limited, and the electricity produced by these generators is not always used.

A temporary solution is to disconnect the PV system when production cannot be locally consumed to avoid aggravating problems in the grid. This solution is effective but not satisfactory, in terms of production requirements during the night.

One way to increase the use of PV generators is to alter non-controllable (passive) systems so that they can be managed in the same way as conventional generators. These new "active" generators would be available for use in grid regulation activities (e.g. primary voltage and frequency adjustment), providing additional flexibility for grid operators.

The active generators considered here are made up of ultracapacitors, for instantaneous internal power regulation, and batteries, to guarantee long-term energy availability. As long as the state of charge of the batteries is sufficient, the local control system is able to manage internal energy transits between storage technologies in order to deliver a specified power, while providing services for the micro-grid [FAK 07, LU 10a].

7.3.3. *Cascade structure*

The cascade structure was developed from non-uninterruptible power supplies (Chapter 4) and has been used for the development of self-consumption in certain countries. A cascade conversion structure has also been used for storage systems. In this structure (Figure 7.8), batteries are connected in parallel to the inverter DC bus. The PV installation is connected to the battery bank via an electronic converter (the PV controller), which enables maximum power to be extracted irrespective of conditions (lighting, temperature, etc.).

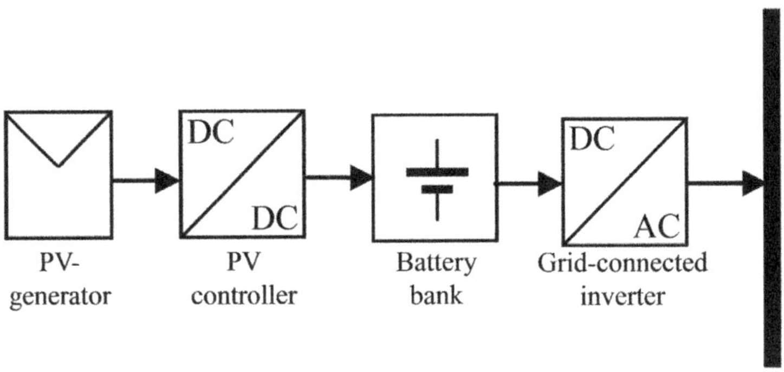

Figure 7.8. *Cascade structure of a hybrid generator [LU 10b]*

In the case of a domestic system, the electricity produced by the PV generator may be self-consumed, with any surplus sold to the grid operator. The household uses the electric grid for their domestic supply if insufficient PV power is available. Charging storage elements from the grid is not currently permitted by most legal systems so as to guarantee the "renewable" origin of energy sold to the grid.

However, in this structure, the stochastic nature of PV energy can lead to premature battery aging due to fast charging and discharging with high currents. Furthermore, the structure was designed to minimize injection of electrical energy into the grid, rather than for sharing energy within a local community. It enables individual energy optimization instead of localized optimization across a whole neighborhood; in the latter case, energy information needs to be exchanged in order for interactions to develop. In order to respond to this objective and enable the use of cheaper batteries (e.g. lead-acid) in a more efficient manner, we have chosen to use the shared continuous bus structure presented in Chapter 4 (section 4.3.4).

7.3.4. *Domestic application*

Our study concerns a residential neighborhood of homes, typically equipped with 3 kW of PV panels. Ultracapacitors are used to compensate fast variations of the power generated by the panels. One criterion used in power sizing is that it must deliver the peak PV power (3 kW) during 12 min. Similarly, one of the energy sizing criteria is that the system should be able to operate for an hour without solar energy. This local energy reserve guarantees a certain volume of energy production for at least an hour and enables the system to participate in the energy market. For the purposes of our study, we selected a 106 Ah battery bank (under 48 V, i.e. 5 kWh). The batteries and ultracapacitors are connected to a shared DC bus with a voltage of 48 by three DC/DC electronic power converters (using the architecture presented in Figure 4.10). These converters are used to control the power-exchanged with each source. The grid connection operates through a three-phase inverter. The control system sends control signals to each power converter in order to satisfy power demands from the grid owner and to provide ancillary network services.

The particularity of active generators lies in the reception of active and reactive power references (constant for each time step; P_{ag_ref}, Q_{ag_ref}) from the micro-grid's central controller. These references are then generated by using electronic power converters (Figure 7.9) to precisely monitor energy flows between different sources (storage units, PV, grid, etc.). Additionally, charge/discharge control is required for the two storage devices in order to respect state-of-charge (SOC) limits and maximum charge/discharge currents. Various techniques may be used in designing a management approach for these multi-source hybrid systems, such as fuzzy logic or neural networks [HAJ 07, MOR 06]; regulation techniques using the DC bus have also been proposed [AYA 07, MAR 05, THO 09]. A hierarchical control structure is an attractive option due to the clarity of design and organization, and this approach is still used [ZHO 15].

In our case, the control system must generate inner control instructions for the power electronic converters (choppers) of all three sources and for the grid connection converter (three-phase inverter), according to measured values and power references [FRA 96]. The control system is organized and ordered using four different control levels depending on the required response time (Figure 7.9):

– mode control;

– power control;

– automatic control for each converter (AC);

– switching control (SC) blocks.

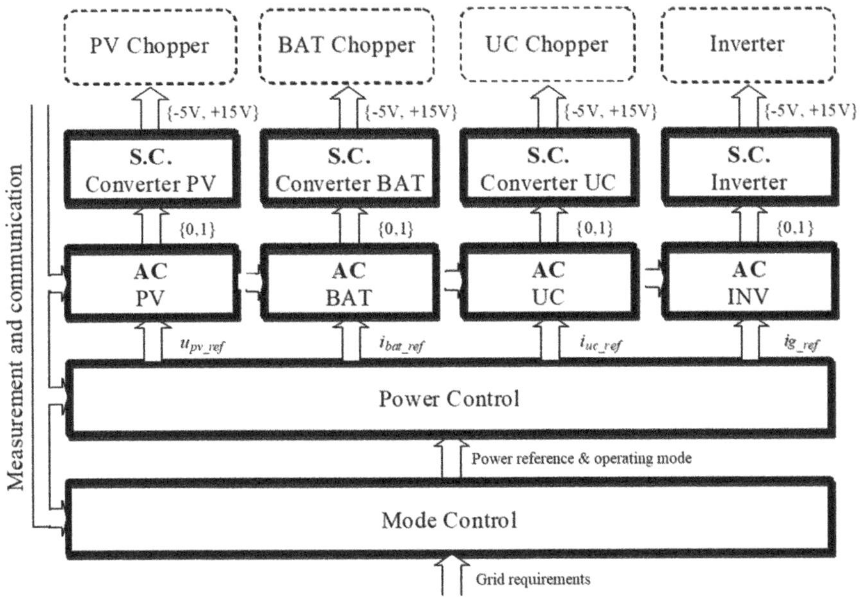

Figure 7.9. *Hierarchical control structure for the active generator [LU 10b]*

Using power references received from the grid operator (or the Microgrid Central Energy Management System (MCEMS) in the case of a micro-grid with hierarchical control) and information on the SOC of the ultracapacitors and batteries, the mode control unit calculates the power reference to exchange with the ultracapacitors and batteries, transmitting this information to the power control unit. The power control unit transforms the information into current or voltage references and transmits them to the automatic control level (Figure 7.10).

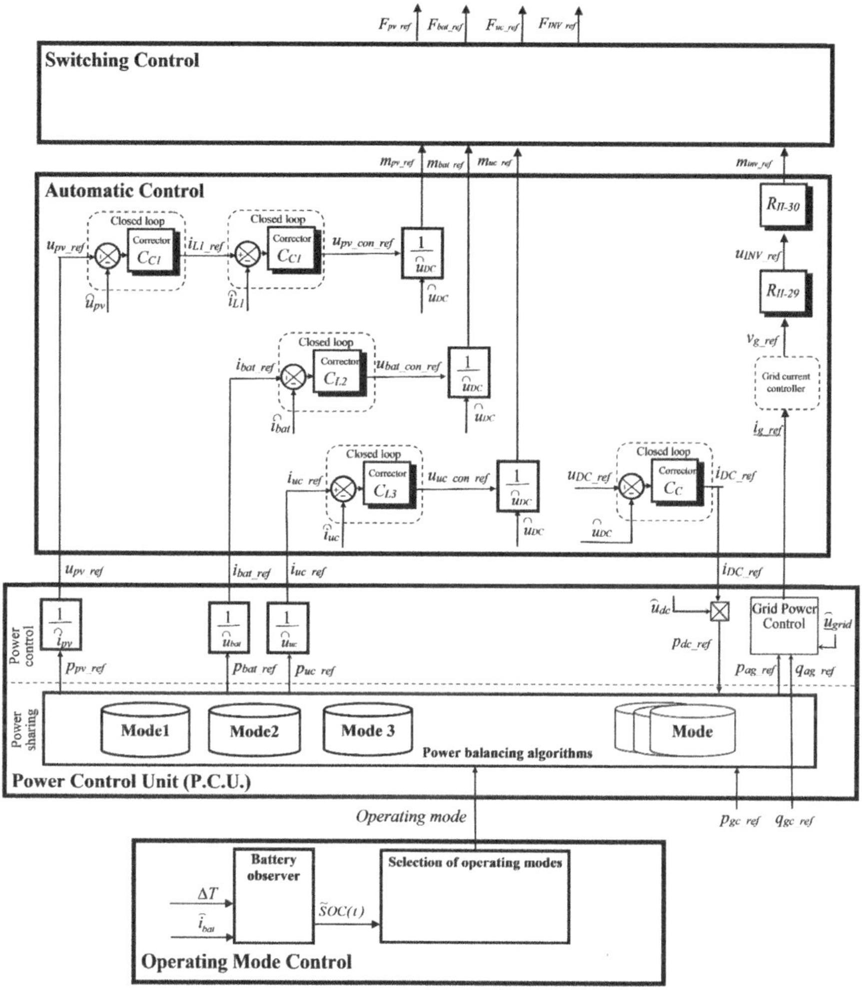

Figure 7.10. *Block diagram representation of hierarchical control structure [LU 10b]*

Each converter receives instructions from the automatic control unit regarding the current for each storage element, the voltage across the terminals of the PV panel (tracking the maximum or limited power point, depending on the selected mode), the voltage across the DC bus and the currents injected into the network. Finally, switching signals for the semiconductors are generated by the switching control block.

7.3.5. *Energy management of the DC bus*

7.3.5.1. *Energy management of the battery using mode control*

There are three possible operating modes for the active generator, selected according to the availability of each source and the SOC of the batteries: normal mode, limitation mode or disconnected mode. The mode controller makes the selection following a series of tests [LU 10a] (Figure 7.11).

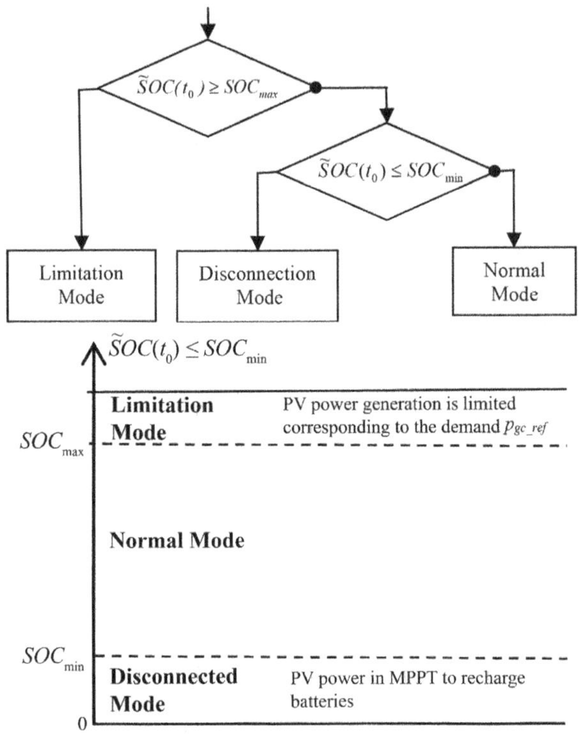

Figure 7.11. *Operating mode selection algorithm for the active generator*

If the state of charge of the batteries is acceptable (between the minimum value SOC_{min} and the maximum SOC_{max}), the PV generator can operate at full power, and the automatic control unit will monitor the maximum power point (maximum power point tracking – MPPT). This is the normal mode.

If the batteries are discharged, they can no longer act as a buffer in smoothing PV energy. The active generator moves into grid-disconnected mode, and PV energy is internally used to recharge the batteries.

If, on the other hand, the batteries are fully charged, then the active generator needs to move into limitation mode. In this case, the power generated by the PV panels must be limited (typically to the value requested for injection into the network). The operating mode selection process is presented in an algorithm form in Figure 7.11.

7.3.5.2. *Internal DC bus control and instantaneous power balancing*

Tracking the voltage of the DC bus is essential in order to guarantee the operation and stability of the hybrid generator. This voltage is expressed as a function of powers and capacity C_{DC}:

$$C_{DC}u_{DC}(t)\frac{du_{DC}}{dt}=\frac{dE_{DC}}{dt}=p_{DC}(t)=p_{UC}(t)+p_{BAT}(t)+p_{pv}(t)-p_{ag}(t) \qquad [7.1]$$

with:

E_{DC}, the energy stored in the DC bus;

p_{DC}, the power exchanged with the DC bus capacitor;

p_{pv}, the power injected into the DC bus by the PV generator;

p_{BAT}, the power exchanged between the batteries and the DC bus;

p_{UC}, the power exchanged between the ultracapacitors and the DC bus;

p_{ag}, the power extracted from the DC bus and injected into the electricity grid;

p_{sour}, the total power output from sources.

To maintain a constant internal DC voltage, the instantaneous power balance must be zero (or "balanced"). This balance must be modeled for each operating mode so that it may then be controlled. By convention, assuming that the storage elements are discharging, i.e. operating as a generator, and that power is being injected into the grid, the power balance is shown in Figure 7.12.

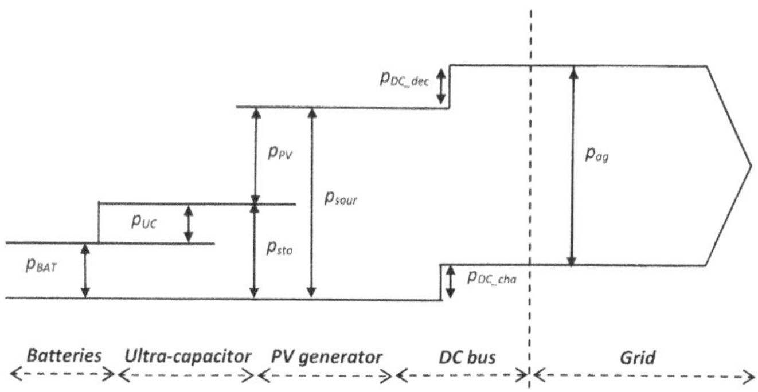

Figure 7.12. *Power balance in disconnected mode [LU 10b]*

The power exchanged with the batteries (p_{BAT}) and with the ultracapacitors (p_{UC}) makes up the total exchanged storage power (p_{sto}):

$$p_{sto}(t) = p_{BAT}(t) + p_{UC}(t) \qquad [7.2]$$

The total exchanged storage power may be used to compensate for or absorb produced photovoltaic power (p_{pv}). The resulting power is labeled (p_{sour}):

$$p_{sour}(t) = p_{sto}(t) + p_{PV}(t) \qquad [7.3]$$

Taking account of the power required to regulate the voltage of the DC bus (p_{DC}), we obtain the output power toward the electricity grid (p_{ag}):

$$p_{ag}(t) = p_{sour}(t) - p_{DC}(t) \qquad [7.4]$$

In Figure 7.7, the power of the DC bus is broken down into two terms: p_{DC_cha} if the DC capacitor is charged ($p_{DC} = p_{DC\_cha}$) and p_{DC_dec} if the capacitor is discharged ($p_{DC} = p_{DC\_dec}$).

7.3.5.3. *Power control in disconnected mode*

In certain conditions, the active PV generator must be disconnected from the electric grid and must operate in isolated mode. In this mode (not covered in Chapter 4), the active generator does not supply power to the electric grid and needs to control its own internal DC bus:

$$p_{ag}(t) = 0 \qquad [7.5]$$

In these specific cases, the power balance gives:

$$p_{DC}(t) = p_{sour}(t) = p_{sto}(t) + p_{PV}(t) \qquad [7.6]$$

The PV panels continue to operate in MPPT mode to obtain maximum electric power from solar energy.

The DC power (p_{DC}) must be equal to the required value (p_{DC_ref}) to maintain a constant voltage at the DC bus (u_{DC}). Thus, the total power from the sources (p_{sour}) may be adjusted by regulating the total power exchanged with the storage units

(p_{sto}). Estimating (or measuring) the PV power output from the panels $\tilde{p}_{PV}$, the power reference to exchange with the storage units is:

$$p_{sto\_ref}(t) = p_{DC\_ref}(t) - \tilde{p}_{PV}(t) \quad [7.7]$$

However, the reference power (p_{sto_ref}) rapidly varies as a result of PV power (p_{pv}) fluctuations. Furthermore, batteries have relatively slow power dynamics, and fast variations in power references have a negative effect on life expectancy. One approach to this problem involves using a simple method to distribute power between batteries and ultracapacitors via a low-pass filter [LU 08]:

$$p_{BAT\_ref} = \frac{1}{1+\tau s} p_{sto\_ref} \quad [7.8]$$

with τ being the time constant. This value must be set sufficiently high to take account of the slow power dynamics of the selected battery technology and of the sizing of the ultracapacitors. Using this power dispatching, the whole of the low-frequency spectrum for regulating power is provided by the batteries, with the ultracapacitors providing the remainder.

Ultracapacitors have a limited energy storage capacity due to their low energy density. However, they have extremely fast power dynamics and can supply rapidly varying powers. In this case, they are used as an auxiliary power system alongside the batteries in order to compensate for power shortages during transitions:

$$p_{UC\_ref}(t) = p_{sto\_ref}(t) - \hat{p}_{BAT}(t) \quad [7.9]$$

A synoptic diagram of this strategy is shown in Figure 7.13.

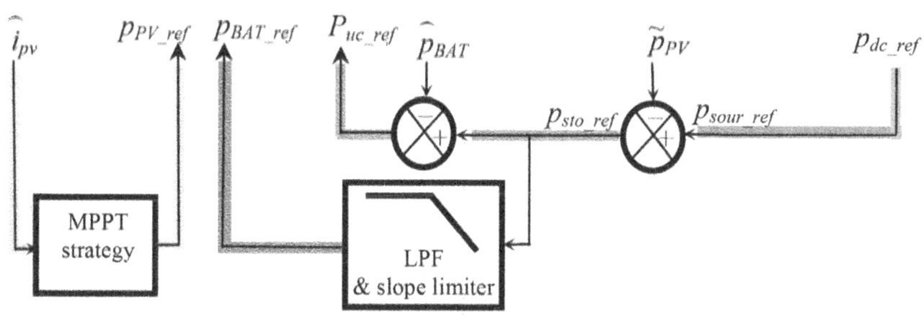

Figure 7.13. *Outline of the storage management strategy in disconnected (autonomous) mode [LU 10b]*

7.3.5.4. *Power control in normal mode*

In this particular case, the power balance gives us:

$$p_{DC}(t) = p_{sour}(t) - p_{ag}(t) = p_{sto}(t) + p_{PV}(t) - p_{ag}(t) \qquad [7.10]$$

The grid power (p_{ag}) must be equal to the power reference (p_{gc_ref}) sent by the micro-grid operator. The power reference for the storage system is thus:

$$p_{sto\_ref}(t) = p_{DC\_ref}(t) - \tilde{p}_{PV}(t) + p_{gc\_ref}(t) \qquad [7.11]$$

This power distribution strategy means that the DC bus can be regulated using power produced by the PV generator and exchanged with the storage units. A synoptic diagram of this strategy is shown in Figure 7.14.

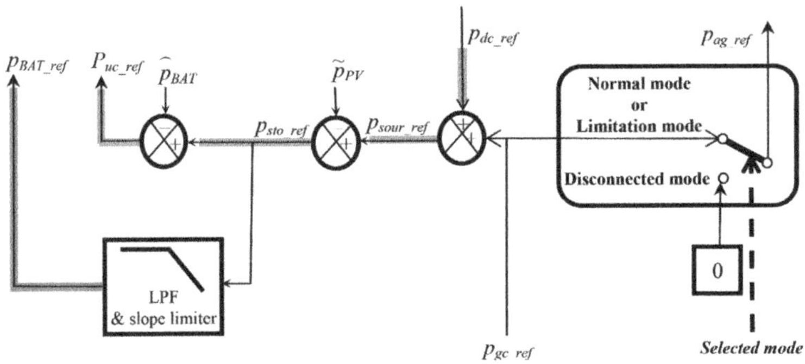

Figure 7.14. *Storage management strategy in connected mode*

Note that this approach is also used in applications where generators need to create a grid (grid-forming strategy). It is completely different from the grid-following strategy or grid-feeding strategy used in passive PV generators.

7.3.5.5. *Power control in limitation mode*

PV power may be sent to storage systems and/or the grid. If neither of these options is available, then production must be limited. This occurs when the storage units are already fully charged and when the PV power is higher than the reference power for the grid.

The power references of storage units should be calculated in relation to their characteristics. There are two restrictions.

The first is the nominal power reference, which must be taken into account for all types of energy storage system. The power exchanged with the storage unit must fall within a certain range, determined on the basis of its characteristics. To determine the need for limitation mode, we first need to consider the power exchanged with the storage unit (ultracapacitors):

$$p_{sto\_required}(t) = p_{ag\_ref}(t) - \tilde{p}_{DC}(t) - \tilde{p}_{PV}(t) \quad [7.12]$$

If this power is greater than the nominal value (p_{sto_rated}), the PV power must be limited to the following value:

$$p_{PV\_ref}(t) = p_{ag\_ref}(t) + p_{DC\_ref}(t) - p_{sto\_rated}(t) < \tilde{p}_{PV}(t) \quad [7.13]$$

The second limitation is the energy capacity limit. If the energy storage units (batteries) are fully charged, they will be unable to absorb any further PV production. If they are empty, they are unable to compensate for any power shortage occurring should the PV system fail. A synoptic diagram of this strategy is shown in Figure 7.15. The management of stored energy must take account of lost PV energy and be implemented within the central management approach for the micro-grid (described in the next section).

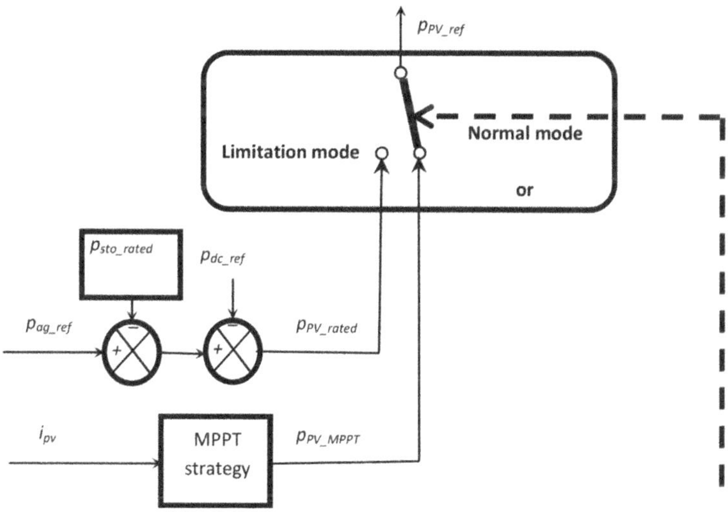

Figure 7.15. *Principle of the PV limitation mode [LU 10b]*

A summary of the implantation of these modes is shown in Figure 7.16.

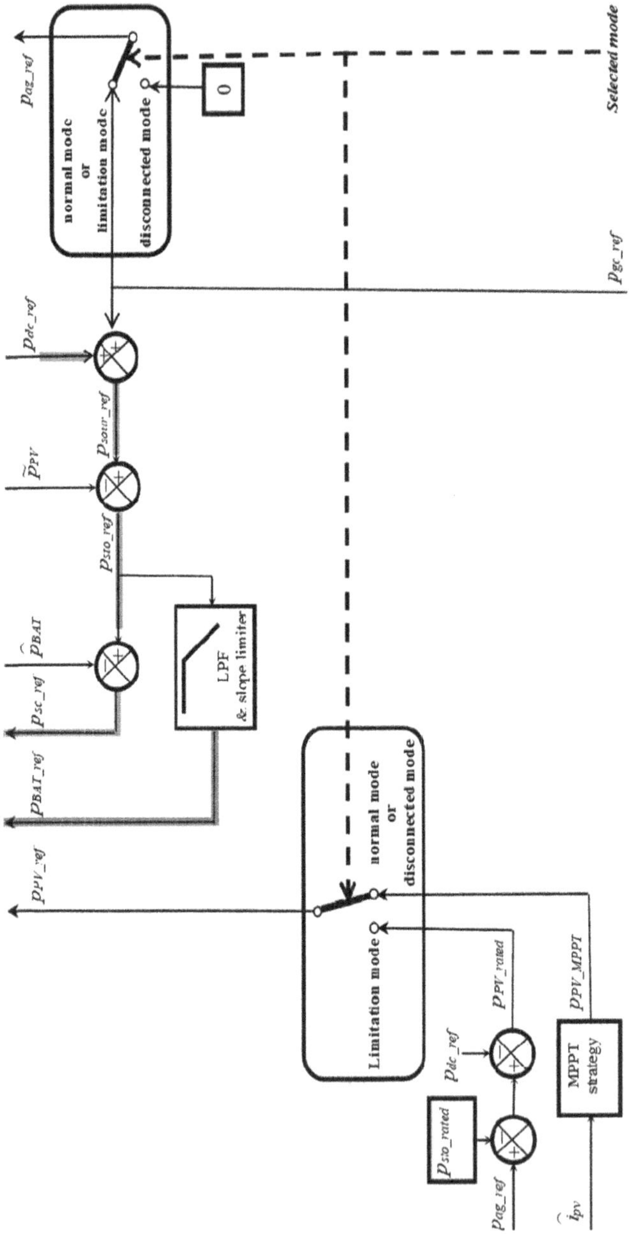

Figure 7.16. *Block diagram of distribution control [LU 10b]*

7.3.6. *Energy management of ultracapacitors*

7.3.6.1. *Principle*

Ultracapacitors are used to balance internal power in active PV generators. When both energy storage systems are available (and their SOC levels are within the correct range), the active PV generator operates in "normal" mode. In this context, power balancing algorithms can be directly used without modification.

Ultracapacitors are fast power units, and their SOC can be directly managed by balancing algorithms. The energy level of ultracapacitors can be expressed using the voltage of the units and is given as a percentage:

$$Level_{UC}(t) = \frac{E_{UC}(t)}{E_{UC\_\max}} = \frac{C_{UC} u_{UC}^2(t)/2}{C_{UC} u_{UC\_\max}^2/2} = \frac{u_{UC}^2(t)}{u_{UC\_\max}^2} \quad [7.14]$$

with:

E_{UC}, the energy stored in the ultracapacitors;

E_{UC_max}, the maximum energy which may be stored in the ultracapacitors;

u_s, the voltage at the terminals of the ultracapacitors;

u_{UC_max}, the maximum voltage of the ultracapacitors;

C_{UC}, the equivalent capacity of the ultracapacitors.

For efficiency and safety reasons, the energy level should be kept within a range between 25% and 95%, corresponding to an operating voltage between 50% and 97.5% of maximum voltage. In our study, we used BOOSTCAP 48V ultracapacitor modules in a parallel assembly with a nominal voltage of 48 V. This value gives us an operating voltage range of between 24 V and 46.8 V.

In order to prevent instability phenomena from occurring, two hysteresis cycles were used to define the storage state of the ultracapacitors ($State_{uc}$): $Empty_{uc}$, $Normal_{sc}$ or $Full_{uc}$ as a function of the stored energy ($Level_{UC}$) (Figure 7.17).

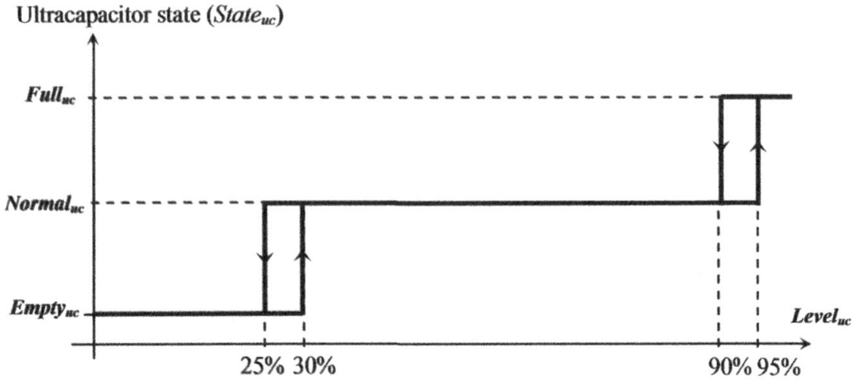

Figure 7.17. *Detection of ultracapacitor state by hysteresis [LU 10b]*

7.3.6.2. *Ultracapacitors in charged mode*

An ultracapacitor voltage in excess of 46.8 V indicates that the energy storage level in these units is too high ($Level_{uc}$ > 95%). If the battery SOC is within an acceptable range, the active PV generator will pass into "full ultracapacitor" mode. A special energy management strategy is required to bring the ultracapacitor storage level ($Level_{uc}$) back within acceptable limits, while PV power and the real power in the grid must still be able to respond to power demands. The strategy adopted in this case is to cease using the batteries in generator mode. This results in discharging of the ultracapacitors and a return to the normal state. Using the same power control algorithms for other sources, the ultracapacitors will be discharged automatically.

7.3.6.3. *Ultracapacitors in discharged mode*

An ultracapacitor voltage of less than 24 V indicates that the energy storage level in these units is too low ($Level_{uc}$ <25%). If the battery SOC is within an acceptable range, the active PV generator will pass into "empty ultracapacitors" mode. A special energy management strategy is required to bring the ultracapacitor storage level back within acceptable limits, while PV power and the active power in the grid must still be able to respond to power demands. In this case, the batteries in the system cease to be used to store excess power, and this power is then used to recharge the ultracapacitors until a normal state is re-established (Figure 7.18).

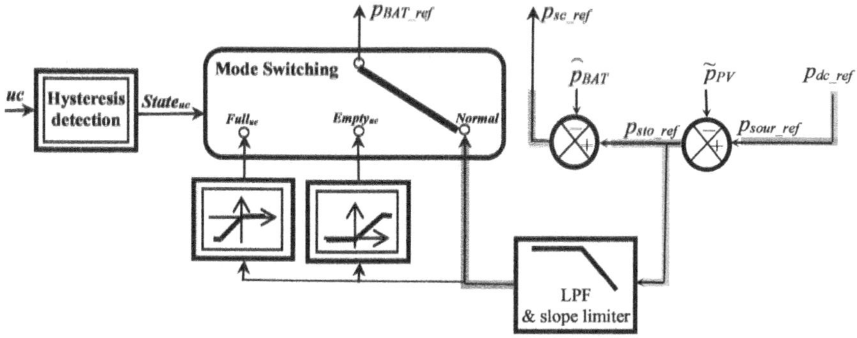

Figure 7.18. *Energy management of ultracapacitors [LU 10b]*

7.4. Micro-grid management

7.4.1. *Organization of electrical grid management*

Technical and economic optimization of an electrical system consists of minimizing the price of the energy produced and distributed to consumers and/or minimizing the environmental impact, taking account of:

– the need to balance energy production and consumption within a micro-grid over the course of a day, broken down into one-hour steps;

– constraints relating to voltage and frequency limits and to their regulation;

– constraints associated with material installations (power plants, etc.).

In order to implement the results of this optimization, management functions are generally organized according to the desired execution dynamics across two different timescales (Figure 7.19). Short-term control functions are carried out by local controllers inside the generators. A central controller (located in a dispatching center in the case of larger grids) is responsible for long-term management. In micro-grids, this entity is known as a Microgrid Central Energy Management System (MCEMS). It must have the capacity to maintain operations in isolation and to manage disconnections from and connections to the main grid.

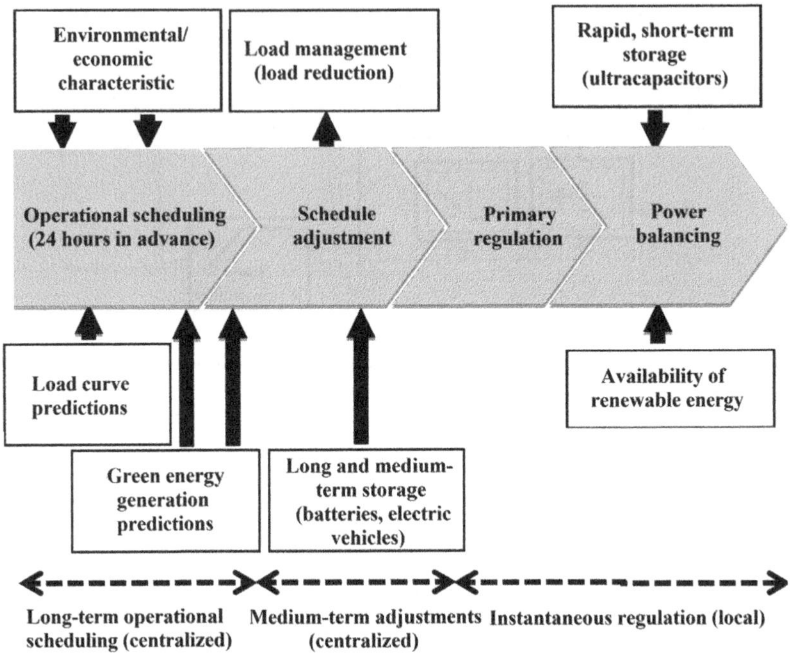

Figure 7.19. *Time classification of micro-grid management functions [KAN 14a]*

7.4.2. *Key functions*

7.4.2.1. *Long-term management or unit commitment*

One essential function of long-term management is to establish generator schedules for the next day, minimizing a given criterion (economic and/or environmental). This is known as the unit commitment problem (UCP). For controllable generators, the operator must decide which units to trigger/use and the operating point to impose, based on technical constraints and effective costs. The objective of this management activity is to determine instructions for all generators in order to comply with the physical constraints of the electrical grid, while minimizing the economic cost of electricity in the system in question. This constitutes an economic dispatching problem. The main constraints are:

– the long start-up time for certain generators, although these same generators have lower production costs;

– fossil/nuclear fuel supply;

– availability issues due to industrial activities (maintenance, etc.).

To account for slow variations over the course of a day (loads, renewable energy production, etc.), the global optima used to establish operating points for controllable generators are determined across a fixed number of time intervals, within which operation is assumed to be stationary. Separate demand and renewable energy generation predictions are used for each hourly step. To account for the uncertain element of these predictions, a power reserve may also be used. An hourly program is established for each generator, made up of real and reactive power references for each step, and sent out one day in advance.

This long-term management becomes more complex if we wish to make use of the flexible element of the smart grids, including storage, load management (e.g. via electric vehicles) and electronic transformers. We will consider the integration of the storage capacity of active generators into long-term management approaches here, designing an optimization algorithm within the context of a somewhat simplified system: an urban micro-grid.

7.4.2.2. *Medium-term management and adjustment*

Over the course of the day, as the planned program unfolds (Figure 7.19), power references may be adjusted by using a communications network to contact the generators, taking account of updated predictions. There are two key reasons for adjusting instructions. First, given that predictions become more precise closer to the time in question, adaptive optimization offers a means of taking account of a changing environment. Second, long-term energy availability, for example, the provision of reserve power or maintaining constant power within active generators, is less critical. This means that the batteries used in our case study can be smaller. In the case study in section 7.7, readjustment will take place every 30 min. "Slow dynamic" power plants (nuclear, coal, etc.) are not subject to adjustments in this way.

If a communications network is used, a centralized or decentralized management approach may be selected. The architectures in question are presented below.

7.4.2.3. *Centralized micro-grid control*

Each MCEMS can transmit demands or excess power notifications to the distribution grid operator. From the grid operator's perspective, each micro-grid thus appears as either a consumer or a producer (Figure 7.20).

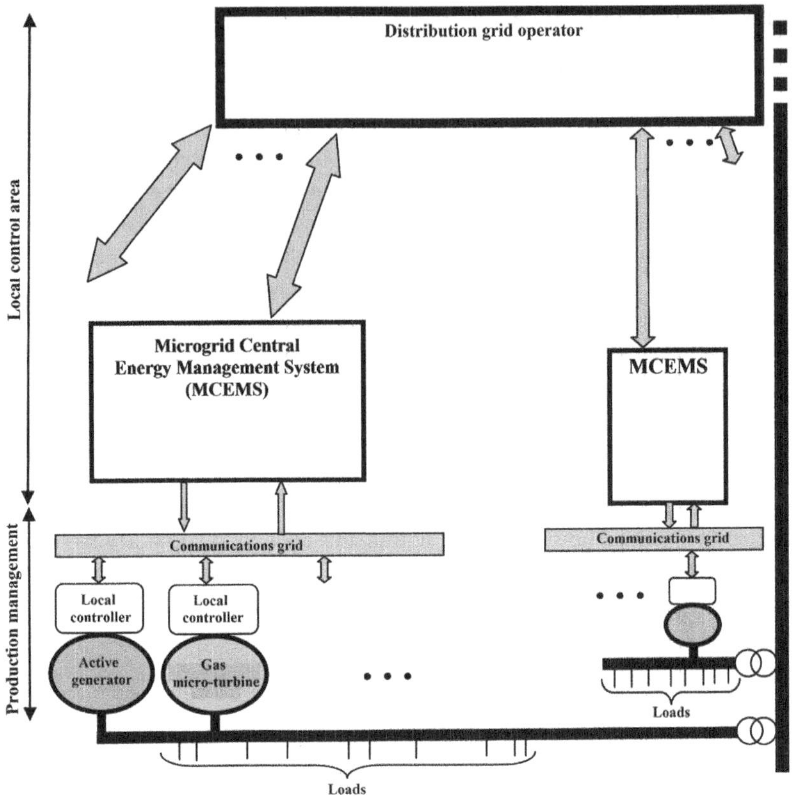

Figure 7.20. *Centralized management of a micro-grid [KAN 14b]*

For the purposes of long-term management of a local control area, a distribution grid operator may assign regulation actions to all of the MCEMS present.

Centralized organization reduces the complexity of energy management. The operation and management of the whole electrical grid is broken down into a set of less complex sub-grids with simplified management (integrated into the MCEMS). The complexity of each management algorithm means that new functions may also be provided, for example, minimization of the greenhouse gas emissions for the area in question. By observing the ways in which actual activity differed from predictions for a given day, each MCEMS may modify its own production schedule over the course of the day (medium-term optimization) in order to refine the technical and economic optima for the area or to take account of unexpected events. As corrections are made as soon as possible after the deviation, MCEMS gain the capacity to make more frequent adjustments, resulting in an improved quality of management over the medium term for the area in question.

7.4.2.4. *Decentralized management of a micro-grid by a multi-agent system*

In decentralized management structures, MCEMS are also in communication with each other [DIM 05] (Figure 7.21). The interest of exchanging information (production costs, demands, excess energy, etc.) lies in the ability to cooperate in order to satisfy grid operator demands. Using this approach, centralized decisions can be relaxed in order to further the specific interests of each micro-grid. Systems have been developed using multi-agent systems and dynamic electricity pricing based on the aggregated consumption of the LEC, creating an incentive to share the residual production of each building [CEL 17]. Individual MCEMS are often referred to as agents, and the group of these units as a multi-agent system. The need for a specific real-time communications infrastructure connecting all agents means that additional investment is required [HAT 07].

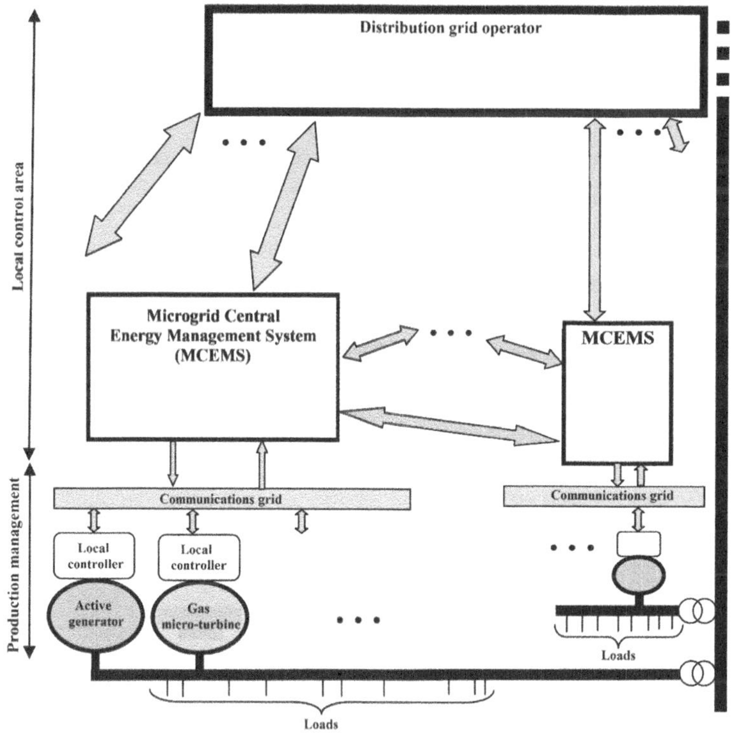

Figure 7.21. *Decentralized management of multiple micro-grids [KAN 14b]*

7.4.3. *Characteristics of local controllers for distributed production*

Electronic converters are used to connect most distributed generators to the grid. Their control systems enable modification of the powers transmitted over the grid and protect the generator from excessive currents and voltages. Passive distributed production generates electrical power according to weather conditions (as in the case of run-of-river hydroelectricity and current PV and wind systems). Active distributed production involves additional management functions (as in the case of hydraulic power, gas turbines and the active PV generators presented in section 7.3) using real and reactive power references sent by a centralized management system to the MCEMS, or as a result of negotiations with decentralized management agents.

The distribution grid operator may decide to connect a generator via a *PQ* or *Vf* node. These reflect different modes of local control.

In the case of a *PQ* node connection, sine-wave voltages are imposed on the electronic power converter, which operates as a source of current in order to generate the active and reactive power references (P_{ref} and Q_{ref}) received from the central controller (MCEMS). This converter control mode is sometimes known as grid following [SEC 13]. If the power converter includes an internal DC bus, then the voltage of this bus may be controlled by regulating the power output from either the primary sources to which it is connected (power dispatching: [LU 10b], Chapter 3) or the grid (grid following: [KAT 08]).

In the case of a *Vf* node connection, the electronic converter acts as a voltage source with an RMS and frequency within a specific range (e.g. 230 V +/–10% and 50 Hz +/– 0.1 Hz). This mode of local control is known as grid forming, as the generator must produce the electric voltage [KAT 08]. In micro-grids which are at risk of isolation, this function is essential. If the topology of the power converter includes an internal DC bus, then this element can only be controlled by transfers of power from the primary source using the power dispatching strategy.

7.4.4. *Fundamentals of power balancing*

7.4.4.1. *An essential function in grid management*

The power generated by renewable energy sources may be estimated in advance, and a generator commitment schedule is created 24 hours ahead of time with the aim of balancing demand. Imbalances between actual power consumption and generation are eliminated via regulatory actions, split between instantaneous primary regulation of frequency and a secondary regulation designed to distribute the remaining operating points of the generators.

7.4.4.2. *Primary frequency regulation with local controllers*

Frequency is the image of the imbalance between production and consumption, and thus fluctuates. Maintaining a constant frequency value via closed-loop control (with a response time varying from a few seconds to 1 min) makes it possible to adjust the power production, meaning that the level of production is constantly being adapted to the level of demand (Chapter 4, section 4.4.1.2) (Figure 7.22).

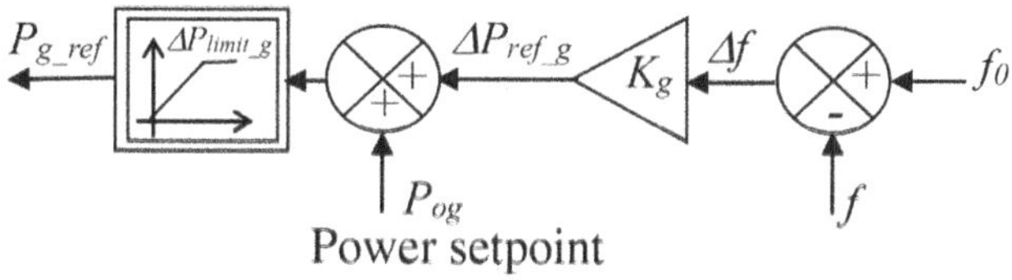

Figure 7.22. *Principle of primary frequency regulation for a generator*

The variables used are:

– P_{0g} [kW], the power reference of generator g for normal frequency;

– f [Hz], the measured frequency in the grid;

– f_0 [Hz], the normal (reference) frequency in the grid (50 Hz);

– K_g [kW/Hz], the primary regulation energy of generator g.

The maximum power reserve supplied for frequency regulation is established by limiting it to the value declared to the grid operator (P_{limit_g}). The droop value is obtained from this value using the minimum frequency value (f_{min}).

7.4.4.3. *Secondary and tertiary regulations*

Secondary regulation (Chapter 4, section 4.4.1.2) consists of calculating new power references for generators as a function of the available secondary power reserve and their operating cost. If the secondary power reserves are exhausted, then they are reconstituted by any generators which are not running at maximum power, or by new generators with rapid startup times (micro gas turbines: 30 s).

7.4.5. *Load management*

The purpose of load management is to reduce unnecessary consumption in buildings and to optimize consumption profiles, making it easier to maintain balance in the grid. Peaks in consumption may be reduced through actions designed to:

– reduce the overall demand for electricity;

– delay certain loads until a later time;

– develop electricity production capacity during times of peak consumption.

Different approaches may be used to accomplish these aims. Increasing load efficiency (e.g. by installing LED lighting, which consumes less than conventional systems) and developing distributed production result in a reduction in overall consumption. Dynamic pricing may also be used, encouraging voluntary and/or contracted reduction of certain loads at peak times in exchange for advantageous tariffs. However, load reduction often results in a loss of comfort for the users of the building in question (e.g. if electric heaters are switched off) and simply results in a shift in consumption, meaning that no energy is actually saved. The use of storage systems at peak times provides the means of satisfying local demand while delaying the load on the main grid. The effects of this approach on self-consumption will be examined later in this chapter.

7.5. Application to the context of a residential electrical network

7.5.1. *From managing domestic demand to managing domestic production*

For several decades now, operators and suppliers are engaged in partial modulation of the consumption of certain industrial and/or domestic electrical charges to respond to market constraints (economic objectives) and technical constraints (technical objectives) via demand management. This approach is based on a number of scenarios in which consumers may reduce their energy use at peak times or other critical consumption periods. It makes use of variable pricing strategies, which may be grouped into two families. The first is based on price signals, which are used to incentivize customers to change their load profiles. In this case, customers voluntarily manage their loads, disconnecting certain equipment at peak times. Higher prices are applied at peak times, making users more likely to use equipment at other times. Customers pay for the electricity they use following a predetermined specific price structure (hourly price, time of use, critical period pricing, etc.). In the second case, customers are notified of reduction requirements one day ahead of time, then remunerated or penalized according to the measured load reduction.

A third option is also available in the form of direct control of certain loads. In this case, customers enable the grid operator to remotely connect or disconnect their equipment in certain circumstances. Electric vehicles and their batteries play a particularly important role in this regard, offering flexibility and the potential to store local PV production.

Figure 7.23 shows a domestic application involving PV panels and storage units, along with the capacity to manage demand via certain controllable loads. During the daytime, this domestic application may act as an energy producer or consumer, hence the term "prosumer", which is becoming increasingly widespread.

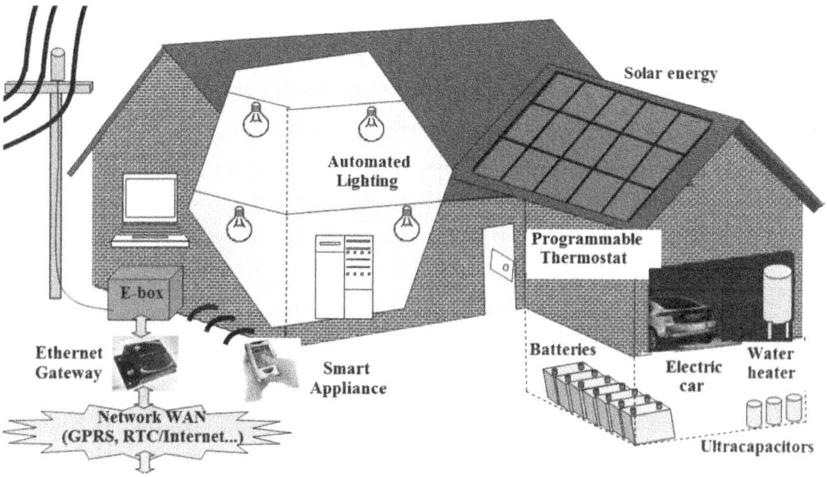

Figure 7.23. *Producer/consumer (prosumer) with load management, electricity production and storage capacity [LU 10b]*

Smart meters (such as the Linky in France), which can track energy consumption, have become increasingly widespread over the past few years. Various improvements have been made to increase consumer satisfaction, offering various means of automatically controlling certain loads. Customers now have the opportunity to adapt their consumption according to pricing periods, while grid operators can improve the quality of the power supply. The integration of smart controls at the point of connection to the grid means that all of the equipment required for energy management strategies (processor chips, networks, communications devices, sensors, etc.) is available on-site. These systems are not yet able to manage storage and PV systems, and further work is needed in this regard.

One improved form of the smart meter is the energy box (EB). These boxes receive signals from the grid operator and can reduce the energy demands of a house or increase production accordingly, as in the case of the Ohta City NEDO project [MAR 07]. Homes possess a dedicated connection for injecting energy into the distribution grid (Figure 7.24). The EB fulfills at least three functions: load control, smart metering and local energy management. Using the load controller, equipment can be programmed to automatically start when prices are low. The unit also offers other consumption facilities such as uninterrupted supply for critical loads and use of equipment for a predetermined duration. Furthermore, it has the capacity to reduce the power demanded from the grid at certain critical times by disconnecting controllable loads. The meter device communicates with the local energy management system and the load manager. Users or grid operators may also consult it remotely.

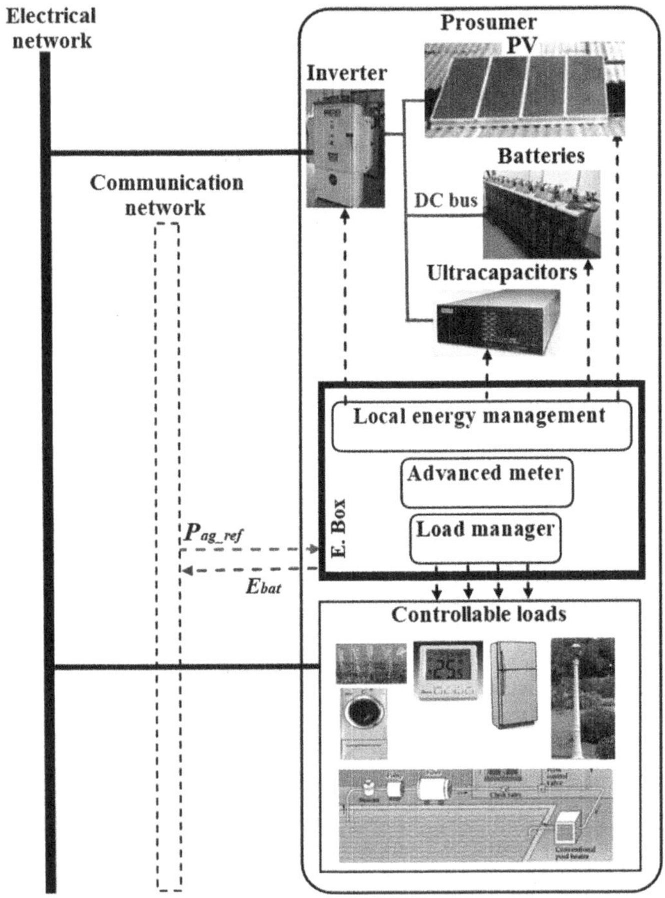

Figure 7.24. *The EB in context: controllable loads, production and storage [LU 10b]*

In this chapter, we will examine a means of connecting PV and storage systems to a shared DC bus to create an active PV domestic generator (as presented in section 7.3.4). The interest of this hybrid generator lies in the fact that it can deliver a prescribed power level ($p_{AG\_ref}(t)$) in the same way as a conventional generator (e.g. a micro gas turbine). Local energy management thus makes it possible to use PV energy in response to the needs of the grid operator even when the sun is not shining. In this case, the batteries are sized to supply the required power. This concept is referred to as an "active generator" to differentiate it from classic PV panels. Excess PV production is stored in batteries to be used as needed, and instantaneous power control is carried out using ultracapacitors. The local energy management system fulfills the same functions as a local controller in a centralized generator.

The energy box enables partial remote control by grid operators. Fast adjustments can be made in response to the state of the grid, and operators gain the flexibility to adjust power flows within a predetermined range.

7.5.2. *Residential grids and application of micro-grid concepts*

The purpose of a micro-grid in a residential context is to supply power to a group of domestic consumers. Small PV panels are installed on the roof of every house in order to:

– optimize local energy production and consumption;

– provide economic benefits via the sale of locally produced electricity and the provision of services within the electrical system;

– increase the reliability of the system, with the possibility of operating in isolated mode.

We will consider a residential grid including several prosumers using active PV generators, several houses with controllable loads and a micro gas turbine, aiming to optimize the electricity use and environmental impact of the LEC (Figure 7.25). We will aim to maximize local PV consumption within the neighborhood by exchanging energy between neighboring buildings, as well as storing surplus PV production in the battery system within each building. Particular attention will be paid to the benefits of aggregating domestic units into a shared storage system in terms of increasing the proportion of PV energy consumed in the local energy community.

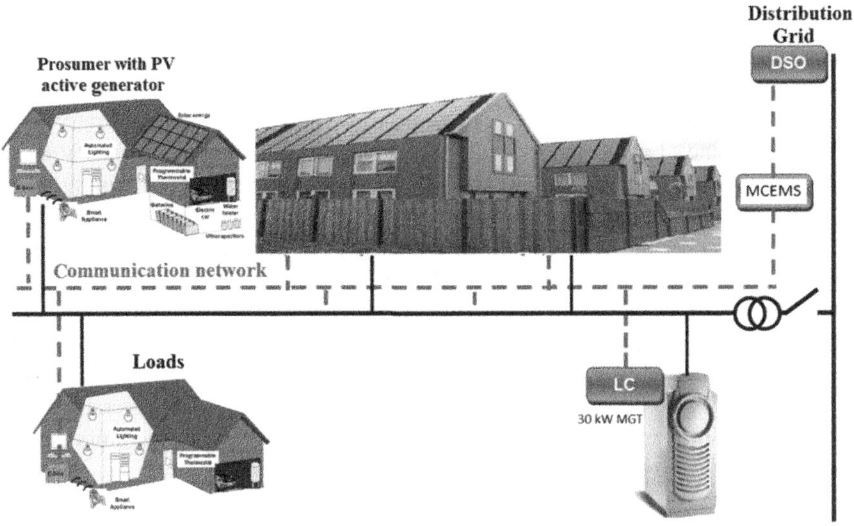

Figure 7.25. *Residential eco-neighborhood*

Given the short distances involved in residential grids, all power generators and electrical loads are connected locally, so line losses and voltage drops can be ignored.

Consumption within the grid is made up of domestic loads from several homes. These loads are modeled as constant impedances and are grouped into critical and controllable loads for the purposes of islanding in cases of autonomous operation.

The micro-grid concept may be applied to the implementation of an aggregator by coordinating domestic applications with classic production units via a centralized energy management system (EMS) (Figure 7.9). Subscribers to this micro-grid may produce or consume power following a setpoint transmitted over a communications network. Local control (as described in [LU 10b]), implemented via a domestic energy box, enables active generators to reproduce the prescribed power setpoint, taking account of the state of charge of the batteries and ultracapacitors in the system.

Centralized micro-grid management in the form of interactive organization provides a suitable means of organizing an energy management system for an eco-neighborhood. The global objective is to balance total electricity production with demand in an optimal manner [KRO 08]. This concept is particularly relevant in the context of smart grids and can be applied via a communications network including the centralized energy management system and local controllers [LU 08, ZHO 11]. The organization represents a step between the requirements of the current electricity distribution grid and the requirements that will emerge over the coming years.

The MCEMS (section 7.4.2.3) coordinates all production, acting as an aggregator between this "link" in the chain and the distribution grid operator (Figure 7.25). As ever, the global objective is to balance offer and demand in the regulated area. The centralized supervision sends out the next day's production schedule, broken down into 30 min steps, to each generator in the local park. The limited size of the area and the communications capacity available mean that the system is interactive, with an adjustment algorithm applied every half-hour. In cases where the load curve and PV production predictions for the coming hour differ from the predictions made 24 hours previously, the medium-term adjustment algorithm modifies the power setpoints. Communications between the MCEMS and the generators take place over a network using the Modbus communications protocol.

Our electrical system features two types of generator: micro gas turbine and active PV generators (which include storage and generate received power setpoints). All of the generators are connected to the grid via power electronic converters. Control of the grid connection inverter is key to managing operations of the micro-grid. The residential micro-grid has two operating modes: isolated/autonomous mode and grid-connected mode. The control algorithms used in these two cases are different [LI 09].

In isolated mode, all energy demands must be met by local electricity producers (PV generators and micro gas turbine). The micro gas turbine is used as the single master generator in this case as it is the main energy source. It is therefore implemented in Vf mode ([LU 10b] Chapter 3), i.e. this generator controls and imposes voltage and frequency (grid forming mode). The active PV generators operate in PQ mode using static control laws ([LU 10b] Chapter 3), meaning that the active and reactive powers are controlled and imposed (grid following mode) (Figure 7.26).

In grid-connected mode, the residential grid may supply or consume electrical power from the distribution grid, and voltage is imposed by the latter. In this case, both the active PV generators and the micro-turbine operate in PQ mode. A communications bus is used to distribute micro-grid management functions across all producers in a form of interactive organization (Figure 7.26). For the purposes of our case study, we will consider grid-connected mode.

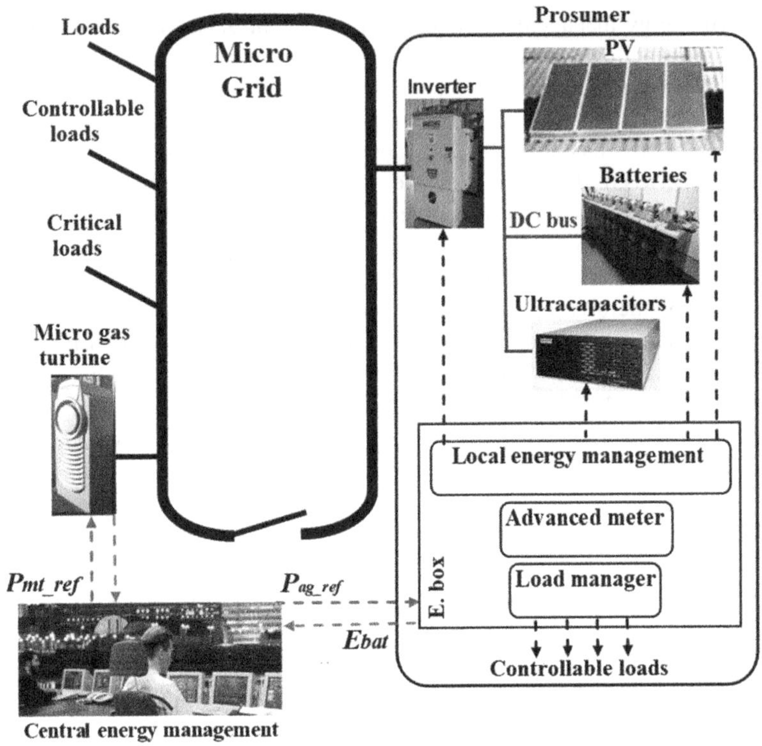

Figure 7.26. *Integration of a prosumer and micro gas turbine in a micro-grid [LU 10b]*

The central controller measures the state variables of the micro-grid and sends orders to the production sources via the communications bus. The local controllers (of the micro-turbine and active PV generators) receive power setpoints from the micro-grid's central control system. They also send out various data, such as energy production measured at the connection point. The central controller manages the micro-turbine and active PV generators in order to control the micro-grid.

The design of a centralized control strategy for this micro-grid is presented in the following section. The overall objective is to prioritize the active PV generators when supplying the residential grid, with the micro gas turbine used as an emergency backup generator. In order to simplify the presentation of theoretical developments, our micro-grid includes only one prosumer and only one micro gas turbine (Figure 7.26).

7.5.3. *Energy management of a micro-grid*

7.5.3.1. *Tasks and organization*

In general terms, the purpose of centralized management is to manage power and energy exchanges between sources and loads in the micro-grid. Real and reactive power production must be shared between sources (in this case, each prosumer) and the micro gas turbine. The central energy management system must assign real and reactive power setpoints, along with other control signals, to dispersed units, classic production units and controllable loads.

Our micro-grid management strategy operates on three essential levels according to timescale: long-term, medium-term and short-term energy management (Chapter 1, Figure 1.13). Each level is characterized by objectives, degrees of freedom, response time and temporal horizon. Depending on the objectives and performance required from the system, energy management may take place in one of two locations. Long-term and medium-term energy management is centralized for the whole micro-grid, whereas short-term management via instantaneous power balancing takes place in the EB.

The two management units must communicate, as data and information on the state of each resource (SoC, real-time power production, etc.) are essential to centralized micro-grid management. Control orders are also sent out from the central energy management system to local controllers (Figure 7.20).

7.5.3.2. *Strategy*

The centralized energy management approach must be adapted in order to accommodate active generators. The control systems for these generators are characterized by adjustment criteria for electrical energy production in response to demand, as a function of economic value and different uncertainty constraints. In designing an energy schedule, we must take account of a variety of criteria necessary for the correct operation of the grid, such as:

– generator use: the power references sent to the generators should not result in reduced performances or modify characteristics (e.g. by reducing life expectancy);

– efficiency: references should make the best possible use of generators, remaining within the correct power range for optimal operation.

Many different criteria need to be taken into account, and some of these are contradictory. The obtained plans generally reflect a compromise between performance, robustness and cost.

The strategy used here is based on load and PV production predictions. The load curve is not predetermined, but is predicted one day in advance based on data from the previous days, measurements and weather forecasts. Energy planning should aim to optimize the operation of a micro-grid in technical, economic and/or environmental terms based on the available information (grid state, generator availability, predicted consumption, etc.), and a production plan is established in response to this.

Medium-term management operates over the course of the day, correcting planned references every half-hour based on observed deviations between predicted and real data.

A number of different centralized management functions must be modified or developed. These include predictions for renewable energy production, load predictions, energy storage reserves, load smoothing, maximization of renewable energy use, CO_2 emission reduction and new energy schedules. A determinist approach will be used in designing an energy management strategy for our micro-grid, building on PV production and load demand information. In the following section, we will present prediction techniques and the way in which data is processed in designing a long-term energy management approach.

7.6. Prediction techniques and data processing

7.6.1. *Predicting PV production*

Solar energy production is hard to predict, and this is problematic in the context of an electricity system. Photovoltaic panels only supply power during the daytime, with peak power production around noon. On cloudy days, production may vary significantly. The 18 kW PV installation at the L2EP ENSAM site in Lille, for example, frequently exhibits large fluctuations in production between 4 and 6 pm (Figure 7.27).

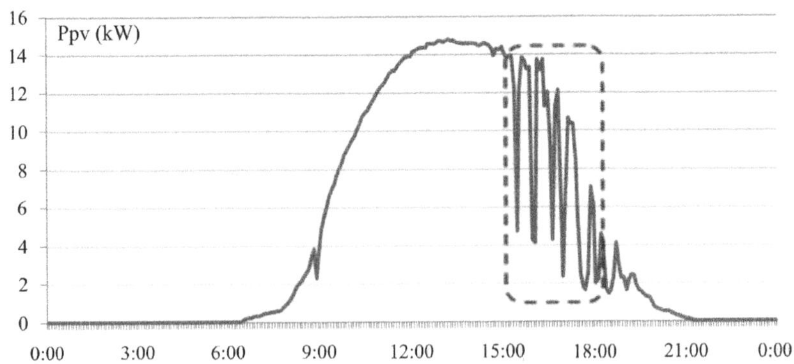

Figure 7.27. *Recorded PV energy production (Lille, France: 05/11/2008) [LU 10b]*

A further issue arises in winter, when peak energy consumption occurs at sundown. Nevertheless, using historic data, increasingly precise power predictions can now be made thanks to adaptive models which refresh predictions during the day. Certain applications use photographs of the sky to detect clouds [KLE 08], adapting predictions made the day before to reflect actual weather conditions.

In our case study, we will make use of predictions made one day ahead, along with real-time predictions of PV power production.

PV predictions for the following day are used in long-term energy management to calculate a daily operating plan for generators in the neighborhood, based on available weather forecasts and historic data. PV power production for the next day is predicted in half-hour steps (Figure 7.28).

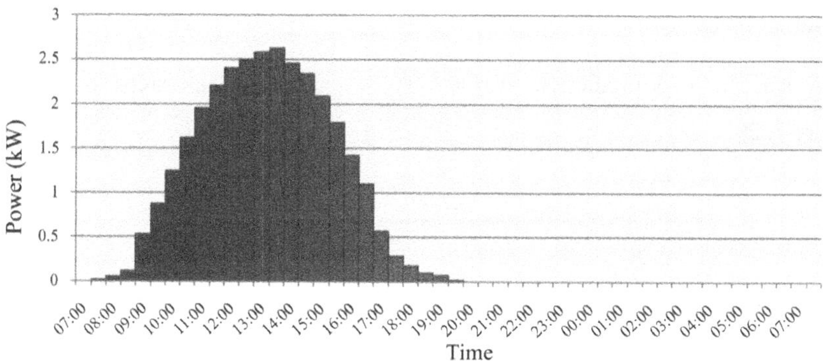

Figure 7.28. *PV power predictions made one day ahead [LU 10b]*

An approximate estimation of available instantaneous PV power is made by all PV generators, and is used by the MPPT algorithm to respond to local objectives.

7.6.2. *Load prediction*

Load predictions are used for a number of applications, including energy purchasing and development of the electricity infrastructure. In the context of our case study, load predictions are used in establishing an electricity production plan and in managing energy within the micro-grid.

Load predictions can be made over different timescales depending on requirements [FEI 08]. The nature of these predictions differs. For example, for a given region, the load for the following day may be predicted to within 1–3% accuracy. However, it is impossible to predict peak load to a similar level of precision, given the uncertainty of long-term weather forecasting. A variety of methods have been developed for short-term predictions, such as day-to-day similarity, regressions, temporals, neural networks, statistical learning algorithms, fuzzy logic and expert systems [MCM 98, MIS 08].

The residential loads involved in our case study differ from the national load curve published by RTE due to the low industrial presence. It would be unrealistic to expect predictions of residential consumption from one day to the next to be perfect. However, certain information can be used to estimate load profiles. Based on the consumption history of the homes in question, it is possible to estimate and predict load behavior within the micro-grid. There are several important factors that influence loads in an electric network: the weather (temperature, cloud cover, etc.), economic activity (loads are different during vacation periods), length of work days, etc. Our load prediction method is based on weather data and historic consumption data.

A sample load prediction profile ($\tilde{P}_{Load\_24h}$) in 30 min steps is shown in Figure 7. 29.

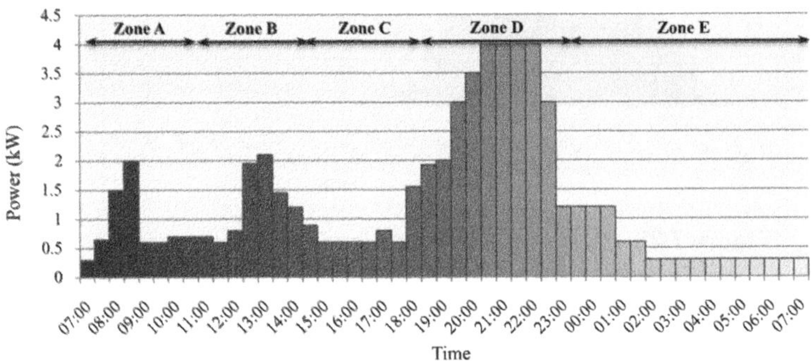

Figure 7.29. *Load predictions ($\tilde{P}_{load\_24h}$) [LU 10b]*

This type of energy consumption is typically characterized by five main periods:

– the morning peak (Figure 7.29, Zone A) linked to households waking up, rail transport, industry and the tertiary sector;

– the noon peak (Figure 7.29, Zone B);

– a drop-off in the afternoon (Figure 7.29, Zone C);

– the evening peak (Figure 7.29, Zone D) due to the end of the working day in the tertiary sector, a new increase in residential consumption and the use of transport;

– night-time, when consumption is lower (Figure 7.29, Zone E).

During this final period, consumption is limited to continuous industrial processes, public lighting, electronic devices left on standby and electrical heating during the winter.

7.6.3. *Energy estimation*

7.6.3.1. *Context*

Energy analysis is of key importance in planning energy production and in implementing energy management algorithms. The maximum energy production value for available PV panels, electrical energy consumption during the day and the quantity of energy stored in batteries are all estimated and used in calculating the amount of additional energy required from non-renewable generators (in our case, a micro gas turbine). Energy estimation based on prediction data (PV energy production and loads) is therefore essential.

For the purposes of our energy estimation, the starting point is the beginning of the day (t_0) and the duration of the day is noted (Δt) (Figure 7.30).

Figure 7.30. *Breakdown of the day used in our study*

The initial time (t_0) is generally set at sunrise. However, for cloudy or rainy days, the starting time may be adjusted because solar radiation and temperature have an effect on PV energy production. The hours of sunshine also change from day to day. These two parameters are dependent on the season and on weather conditions. Time (t_0) is thus defined as the moment when the PV panels begin producing electricity. Time $t_0 + \Delta t$ represents sundown, when production ceases to be significant.

7.6.3.2. *Estimation of available PV power [LU 10b]*

Long-term photovoltaic energy predictions are made in half-hour steps (Figure 7.32). The energy expected from photovoltaic production for each half-hour interval may be calculated using PV power prediction data:

$$\widetilde{E}_{PV\_1/2h} = \int_{t_0+nTe}^{t_0+(n+1)Te} \widetilde{P}_{PV\_24h}(t)dt = Te.\widetilde{P}_{PV\_24h}(t_0 + n.Te) \quad [7.15]$$

for $t_0 + nTe \leq t \leq t_0 + (n+1)Te$ with $Te = 30$ min and $n \in \{0,1,...,47\}$.

Total PV energy for a day is the sum of all the half-hourly calculations:

$$\widetilde{E}_{PV} = \sum_{n=0}^{n=23} \widetilde{E}_{PV\_1/2h}(t_0 + nT_e) \quad [7.16]$$

7.6.3.3. *Estimation of required load energy*

In order to establish a production plan, the energy required by the load over the same time interval also needs to be estimated:

$$\widetilde{E}_{Load\_1/2h} = \int_{t_0+nTe}^{t_0+(n+1)Te} \widetilde{P}_{Load\_24h}(t)dt = Te.\widetilde{P}_{Load\_24h}(t_0 + n.Te) \quad [7.17]$$

7.6.3.4. *Estimation of energy required for storage*

If PV power production is greater than the load requirement over the course of the day, surplus power may be sent to storage units (batteries). The energy difference for each half-hour may be estimated using the PV energy and load predictions:

$$\widetilde{E}_{diff\_1/2h}(t) = \widetilde{E}_{PV\_1/2h}(t) - \widetilde{E}_{Load\_1/2h}(t) \quad [7.18]$$

The presence of energy storage devices means that this energy difference may be absorbed or offset. The size of the storage unit determines the capacity to absorb power or compensate for a power shortfall. The nominal power of the storage system means that the energy exchanged with batteries during each half-hour step is limited. Taking a constant maximum value ($P_{bat\_\max}$), we obtain:

$$\widetilde{E}_{bat\_1/2h\_\max} = \int_0^{Te} P_{bat\_\max} dt = P_{bat\_\max} \cdot T_e \quad [7.19]$$

7.7. Day ahead operational planning and half-hourly power reference calculations

7.7.1. *Objectives*

Our case study features two energy sources: an active PV generator and a micro gas turbine. Due to the advantages of renewable energy sources (reduced greenhouse gas emissions and low usage costs), the PV generator will be used as a priority, with the gas turbine serving to compensate for any shortfall. The objective of the daily operational planning is to calculate the power reference for each generator and the necessary power reduction for each controllable load in half-hour steps. Our study does not take account of controllable loads, but further details can be found in [FOG 09]. In this section, we will focus exclusively on those management functions that directly relate to technical information. Other functions, such as prediction and economic information, will be considered to be ideal.

7.7.2. *Constraints*

7.7.2.1. *Constraints relating to the use of lead batteries*

The life cycle of a battery is generally defined by the number of complete charge and discharge cycles which it can tolerate until capacity drops below 80% of the initial nominal capacity. Figures from 500 to 1,200 cycles are typical. The aging process results in a progressive reduction in capacity over time. The life cycle of the lead-acid battery used in our study is limited; charge and discharge algorithms should therefore be used in order to extend battery life. The simplest and most obvious way of maximizing life expectancy is to ensure that the battery is always used in the conditions for which it was designed. However, other actions can also be taken to increase life expectancy. Flexible charging algorithms and a protection system to prevent overcharging help to prevent unnecessary damage.

The battery in our system is charged using PV panels, an intermittent power supply. The strategy used to guarantee the lifespan of lead batteries involves establishing a charge cycle (during the day) and a discharge cycle (during the night) with a constant charge current for each half-hour step [LU 08]. Given the intermittent nature of both PV energy and demand, ultracapacitors are used in the active generator (see section 7.3.5) and controlled in such a way that they handle all rapid power transitions. If PV production is not sufficient, the micro gas turbine is used to make up the shortfall; batteries are only discharged at night. The depth of discharge can be specified as a function of battery characteristics in order to maximize life expectancy. Stress tests are carried out with an SoC of between 10% and 70% in normal operating conditions.

7.7.2.2. *Constraints relating to the micro gas turbine*

The nominal power of the micro gas turbine (P_{MGT_max}) is 33 kW. A turbine needs considerably more fuel to start up than when operating continuously and generating constant power. It is also a significant source of pollution. Furthermore, turbines take several minutes to start up when they are cold [PIL 02].

To minimize energy losses and gas emissions each time the turbine starts, the system should operate continuously. In the case of a low power demand, the turbine is maintained at minimum power (P_{MGT_min}), producing minimum energy:

$$\tilde{E}_{MGT\_1/2h\_\min} = \int_0^{Te} P_{MGT\_\min} dt = P_{MGT\_\min}.Te \qquad [7.20]$$

7.7.3. Determinist algorithm for generator use

Based on daily predictions of power availability and PV energy ($\tilde{P}_{PV\_24h}, \tilde{E}_{PV\_1/2h}$), alongside power demand predictions and load energy ($\tilde{P}_{Load\_24h}, \tilde{E}_{Load\_1/2h}$), energy production plans are drawn up for the prosumer (P_{AG_ref0}) and the micro-turbine (P_{MGT_ref0}). The central energy management system updates power references every 30 min. Given the absence of PV production at night-time, power references are separately calculated for night and day.

7.7.3.1. *Daytime*

During the day ($t_0 < t < t_0 + \Delta t$), and for each half-hour, two different cases are considered (Figure 7.31).

Case 1: if the available PV energy plus the minimum energy production from the gas turbine is less than the energy required by the loads ($\tilde{E}_{PV\_1/2h} + \tilde{E}_{MGT\_1/2h\_\min} < \tilde{E}_{load\_1/2h}$), then maximum power will be drawn from the PV panels (operating in MPPT) and all PV power is injected into the grid. The micro gas turbine must generate any additional power that is required:

$$P_{AG\_ref0} = \tilde{P}_{PV\_24h} \qquad [7.21]$$

$$P_{MGT\_ref0} = \tilde{P}_{Load\_24h} - P_{AG\_ref0} \qquad [7.22]$$

where $P_{AG\_ref0}, P_{MGT\_ref0}$ are the initial planned references for the active PV generator and the micro gas turbine, respectively.

Case 2: otherwise, when available PV energy plus the minimum energy production from the gas turbine is more than sufficient to meet requirements, renewable energy is prioritized, and the gas turbine is kept at minimum power (to avoid a switch off and hence an overcost). The active PV generator is limited to residual power:

$$P_{AG\_ref0} = \widetilde{P}_{Load\_24h} - P_{MGT\_min} \quad [7.23]$$

$$P_{MGT\_ref0} = P_{MGT\_min} \quad [7.24]$$

where P_{MGT_min} is the minimum energy production from the micro gas turbine.

Excess photovoltaic production will be managed by the local control system in the active PV generator and stored in the batteries.

7.7.3.2. *Night-time*

Energy management during the night ($t_0 + \Delta t < t < t_0 + 24h$) depends on the availability of the stored energy in the battery system. This energy ($\widetilde{E}_{bat}$) may be estimated, or communicated to the central energy management system by the E-box. Two separate cases are considered. In both cases, the batteries must be discharged in order to be ready to charge the next day at t_0. Depending on the stored energy ($\widetilde{E}_{bat}(t_0 + nTe)$) and the nominal energy ($\widetilde{E}_{bat\_1/2h\_max}$), we obtain the energy available from the battery for the following half-hour:

$$\widetilde{E}_{bat\_1/2h\_rest}(t) = \min\left[\widetilde{E}_{bat\_1/2h\_max}, \widetilde{E}_{bat}(t_0 + nTe)\right] \quad [7.25]$$

with:

– $\widetilde{E}_{bat\_1/2h\_rest}$, the battery energy available for the next half-hour;

– $\widetilde{E}_{bat\_1/2h\_max}$, the maximum energy that may be exchanged with the battery during a half-hour period;

– $\widetilde{E}_{bat}(t_0 + nTe)$, the energy stored in the batteries.

Case 3: if the energy available from the battery added to the minimum production from the turbine is greater than the demand from the loads ($\widetilde{E}_{bat\_1/2h\_rest} + \widetilde{E}_{MGT\_1/2\_\min} > \widetilde{E}_{load\_1/2h}$), then electricity from the active generator is used as a priority, as sufficient energy was stored from earlier PV production. The gas turbine continues to operate at minimum power:

$$P_{AG\_ref0} = \widetilde{P}_{Load\_24h} - P_{MGT\_\min} \tag{7.26}$$

$$P_{MGT\_ref0} = P_{MGT\_\min} \tag{7.27}$$

Case 4: otherwise, when the energy stored in the battery plus the minimum production from the gas turbine is not sufficient to meet demand, the power reference for the active generator is calculated so that the batteries are discharged, and the gas turbine supplies the additional power required:

$$P_{AG\_ref0} = \frac{\widetilde{E}_{bat\_1/2h\_rest}}{Te} \tag{7.28}$$

$$P_{MGT\_ref0} = \widetilde{P}_{Load\_24h} - P_{AG\_ref0} \tag{7.29}$$

This covers the case where insufficient energy is supplied by the battery (Figure 7.31).

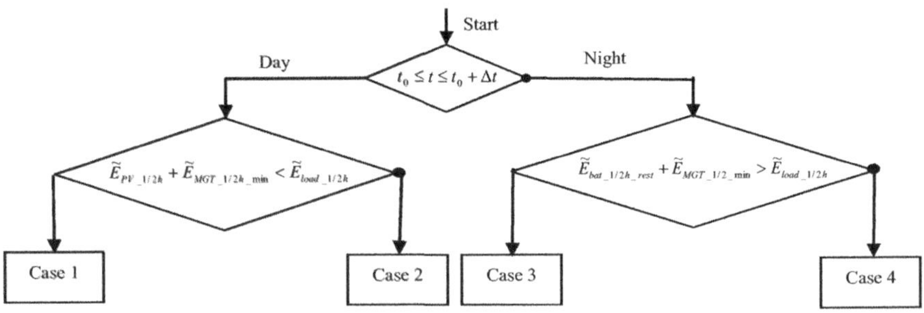

Figure 7.31. *Determination of operating modes [LU 10b]*

7.7.4. *Practical application*

To illustrate these theoretical results, we calculated a next-day energy plan based on load predictions (Figure 7.33) and estimated PV power (Figure 7.32), taking $t_0 = 07{:}00$ hours and $\Delta t = 12{:}30$ hours.

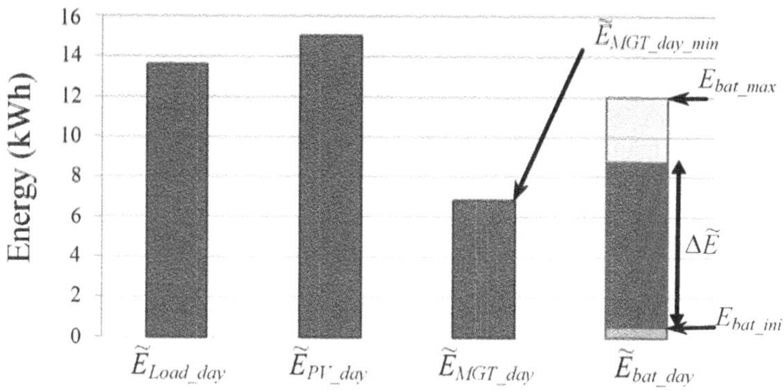

Figure 7.32. *Daytime energy analysis [LU 10b]*

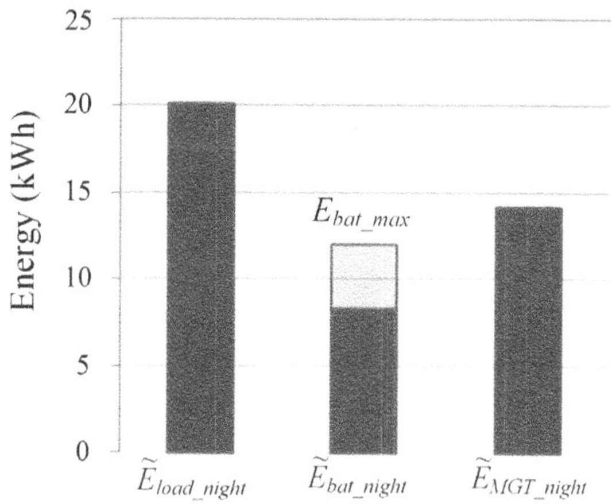

Figure 7.33. *Night-time energy analysis [LU 10b]*

Comparing the estimated energy required by the load over the course of the day in question ($\widetilde{E}_{Load\_day}$) and the estimated energy provided by the PV panels ($\widetilde{E}_{PV\_day}$), we see that excess renewable energy is available (Figure 7.33).

As the micro gas turbine needs to produce a minimum amount of energy during the whole day, the energy surplus is as follows:

$$\Delta\widetilde{E} = \widetilde{E}_{PV\_day} - \widetilde{E}_{Load\_day} + \widetilde{E}_{MGT\_day\_\min} < E_{bat\_\max} \qquad [7.30]$$

$$\widetilde{E}_{MGT\_day\_\min} = \int_{t_0}^{t_0+24h} P_{MGT\_\min} dt \qquad [7.31]$$

This energy may be stored in batteries. The energy which the micro gas turbine needs to supply to maintain operations during the night is thus:

$$\widetilde{E}_{MGT\_night} = \widetilde{E}_{Load\_night} - \widetilde{E}_{bat\_night} \qquad [7.32]$$

where $\widetilde{E}_{bat\_night} = \Delta\widetilde{E}$

As a communications network is available, the estimated value of energy stored in the battery may be replaced with a measured value, sent by the E-box at the beginning of the night ($t_0 + \Delta t$). The power references calculated using the determinist algorithm are shown in Figure 7.34.

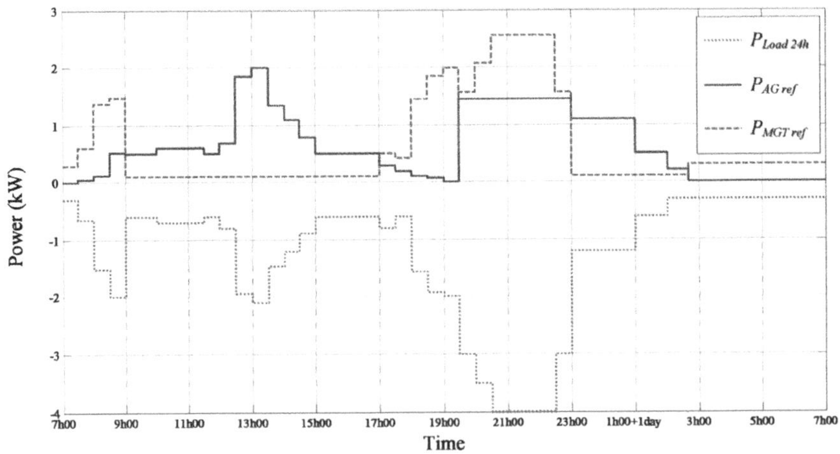

Figure 7.34. *Power references calculated for the centralized management plan [LU 10b]*

From 7:00 to 7:30 am, PV power is not sufficient, so the power reference for the gas turbine ($p_{MGT\ ref}$) is equal to the demand (p_{Load_24h}) (Figure 7.34). Between 9 am and 5 pm, the power reference of the gas turbine is set at a minimum, and the active generator provides the remaining power required by the load (Figure 7.34).

7.8. Medium-term energy management

7.8.1. *Reducing observed deviations*

The power references used in long-term energy management are calculated using load and PV production predictions. These predictions, produced one day in advance, are obviously subject to error or uncertainty, and may be corrected using a power reserve [YAN 17]. Real situations (weather conditions, demand from loads, etc.) very often differ from predictions. In practice, considering actual and predicted load ($\widetilde{P}_{Load\_1/2h}$) and predicted PV production ($\widetilde{P}_{PV\_1/2h}$) for a given half-hour time step, we obtain the following errors:

$$\Delta P_{PV\_1/2h} = \widetilde{P}_{PV\_1/2h} - \widetilde{P}_{PV\_24h} \quad [7.33]$$

$$\Delta P_{Load\_1/2h} = \widetilde{P}_{Load\_1/2h} - \widetilde{P}_{Load\_24h} \quad [7.34]$$

Medium-term energy management takes account of these differences in conditions by modifying power references at local level as a function of real situations. This approach is similar to the secondary regulation process in electrical grids. In this case, updates are applied every half-hour by the central controller of the residential micro-grid:

$$P_{AG\_ref1} = P_{AG\_ref0} + \Delta P_{AG1\_1/2h} \quad [7.35]$$

with:

- P_{AG_ref1}, the power reference for medium-term energy management;

- P_{AG_ref0}, the power reference from the daily plan;

- $\Delta P_{AG1\_1/2h}$, the power modification.

The modified reference used in medium-term energy management depends on the correction applied to the PV prediction and on the predicted load:

$$\Delta P_{AG1\_1/2h} = \Delta P_{PV\_1/2h} + \Delta P_{Load\_1/2h} \quad [7.36]$$

The power reference for the micro-turbine is also corrected:

$$P_{MGT\_ref1} = P_{MGT\_ref0} + \Delta P_{MGT\_1/2h} \qquad [7.37]$$

with:

- P_{MGT_ref1}, the power reference for medium-term energy management;
- P_{MGT_ref0}, the power reference from the daily plan;
- $\Delta P_{MGT\_1/2h}$, the power modification for medium-term energy management.

Once these power references have been corrected, the micro-grid's central controller sends the corrected information to the local controller in each generator (active PV generator and micro gas turbine) through the communications bus.

7.8.2. *Energy management to minimize the aging of batteries*

Medium-term energy management within an active generator has a direct impact on batteries. Deep discharges, under-charging and over-charging can damage batteries and reduce their lifespan. To minimize battery aging, we established that only one charge and discharge cycle should be allowed each day.

During the daytime, the power reference for charging the battery (P_{bat_ref0}) is set at a constant level over each half-hour step as a function of the power reference of the active generator (P_{AG_ref0}), local short-term PV power predictions ($\widetilde{P}_{PV\_1/2h}$) and the estimated state of charge of the batteries ($\widetilde{SOC}(t)$) during the half-hour in question. Using our proposed strategy, batteries should not be discharged during the day. The daytime charging algorithm is triggered as soon as measured PV power exceeds the reference received by the active generator:

$$\widetilde{P}_{PV\_1/2h} - P_{AG\_ref1} > 0 \qquad [7.38]$$

If the predicted PV power ($\widetilde{P}_{PV\_1/2h}$) is less than the reference received by the active PV generator (P_{AG_ref1}), then the batteries will not be charged and remain in "waiting" mode (Figure 7.35). Otherwise, they will be charged during this half-hour period, although the SoC must be verified first.

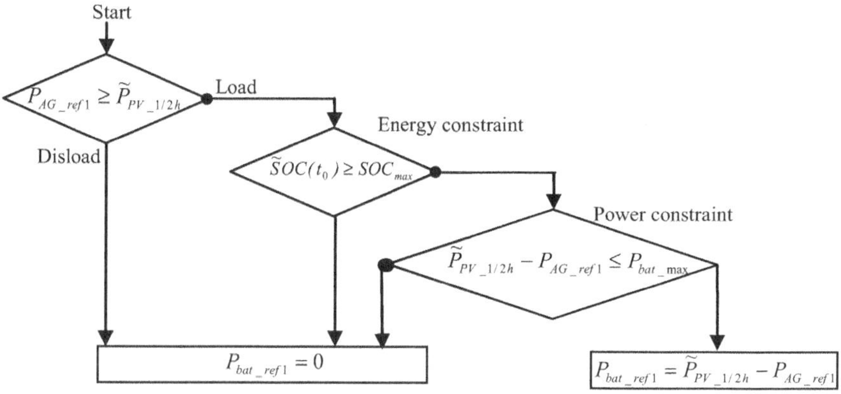

Figure 7.35. *Daytime battery charging algorithm [LU 10b]*

If the batteries are already fully charged, then they must remain in "waiting" mode.

If the surplus photovoltaic electricity (PV power remaining after the load requirement is met) is lower than the nominal power of the battery, then the battery may be charged using this maximum power. Otherwise, part of this potential PV production must be offloaded, i.e. production must be limited to the nominal power.

At night, the batteries are discharged and the discharge power reference is calculated every half-hour based on the SoC. Once again, the energy capacity limit and the discharge power limit must be respected (Figure 7.36).

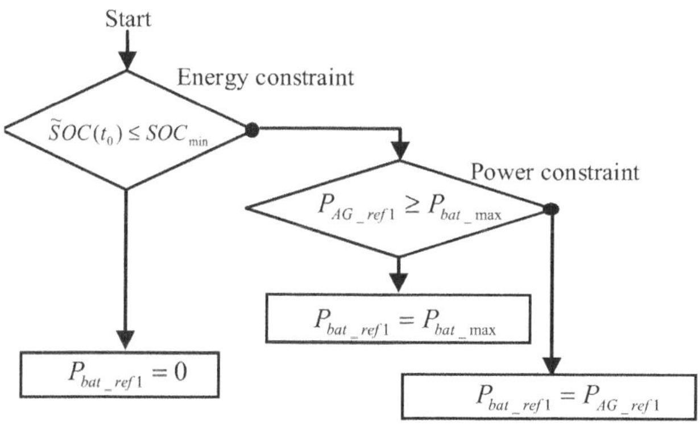

Figure 7.36. *Night-time battery discharge algorithm [LU 10b]*

7.9. Short-term energy management

7.9.1. *Primary frequency regulation*

The central energy management system sends power references (P_{AG_ref1} , P_{g_ref0} and P_{MGT_ref1}) every half-hour. These values are adjusted and planned for this duration. As unexpected variations may occur in both consumption and PV power production, primary frequency regulation is required in order to modify generator production instantaneously, maintaining the balance between production and consumption. The generators in the micro-grid share responsibility for this control function.

Primary frequency regulation and local control for a micro gas turbine are described in [LI 09]. Following a similar approach, active generators must possess primary frequency regulation capacity in order to participate in balancing [COU 08].

When the frequency deviation exceeds a pre-defined threshold, a controller is activated in order to increase or reduce power and contribute to re-establishing balance [COU 08]. A droop parameter (k_f) establishes the additional power to supply proportional to the frequency deviation (Figure 7.37).

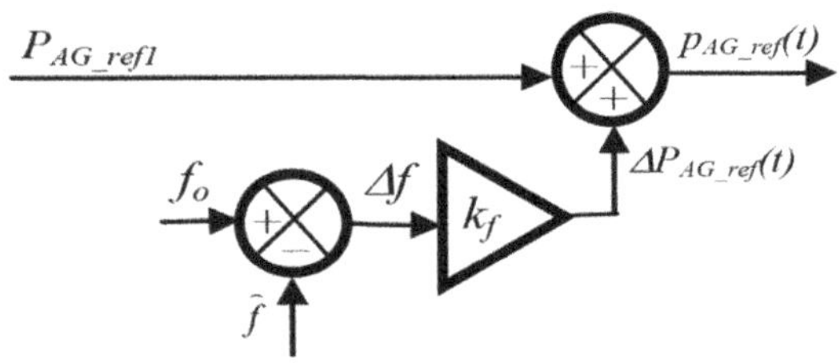

Figure 7.37. *Primary frequency regulation [LU 10b]*

7.9.2. *Power balancing strategies in the active generator*

The active generator must provide an instantaneous power reference ($p_{AG\_ref}(t)$), which is the sum of the secondary power regulation (P_{AG_ref1}) and the instantaneous power reference for primary power regulation ($\Delta P_{AG\_ref}(t)$).

For the purposes of our application, battery power is constant over each half-hour step. Instantaneous balancing power must be managed by equipment offering a high-dynamic performance. As [ROB 15a] indicates, ultracapacitors are fast dynamic storage systems with high power exchange capacities. They can therefore be used to ensure optimal battery charging and to provide power to the grid at peak times, but only for short periods (due to their low energy density).

As the power reference used in charging or discharging the batteries is constant, the power distribution carried out by the control system needs to be adjusted. A local algorithm is used to calculate power references for the ultracapacitors, PV panels and the grid connection inverter, taking account of the selected operating mode and measured values (Figure 7.38).

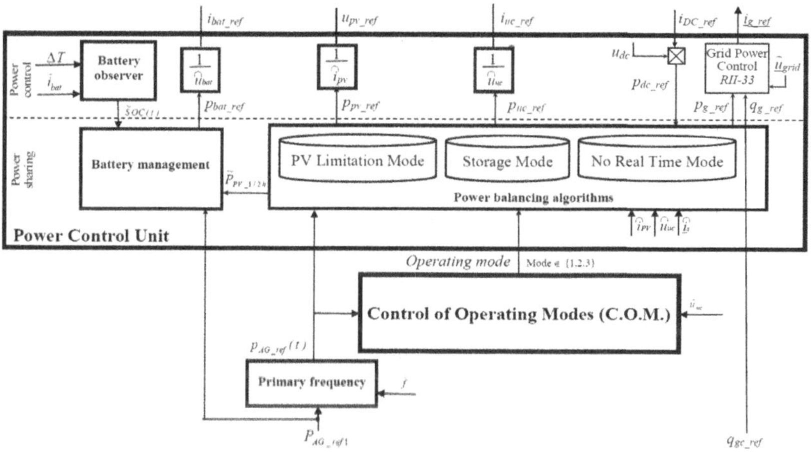

Figure 7.38. *Modified local control system [LU 10b]*

These power references are then transformed into current or voltage references and are controlled in a closed loop. The SoC is also calculated in real-time, and charging may be stopped for the last half-hour where necessary.

7.9.2.1. *Control of operating modes*

In our application, the SoC of the batteries present in all of the active generators is controlled by the central micro-grid management system in order to maximize the value of PV power. The aim is to avoid limiting PV power generation and to charge the batteries using energy that is not injected into the grid. To perform this, the batteries must be correctly sized. The other operating modes, along with the way in which modes are selected, remain unchanged.

7.9.2.2. *Energy management of ultracapacitors*

As the batteries are managed by the micro-grid's central control system, control of the active PV generator must operate via the state of charge of the ultracapacitors.

In our system, batteries are not used to charge or discharge the ultracapacitors; charging is carried out using PV production, and discharging is carried out by injecting power into the grid. Our strategy is thus modified in order to calculate the power setpoint exchanged with the grid in emergency mode (Figure 7.39).

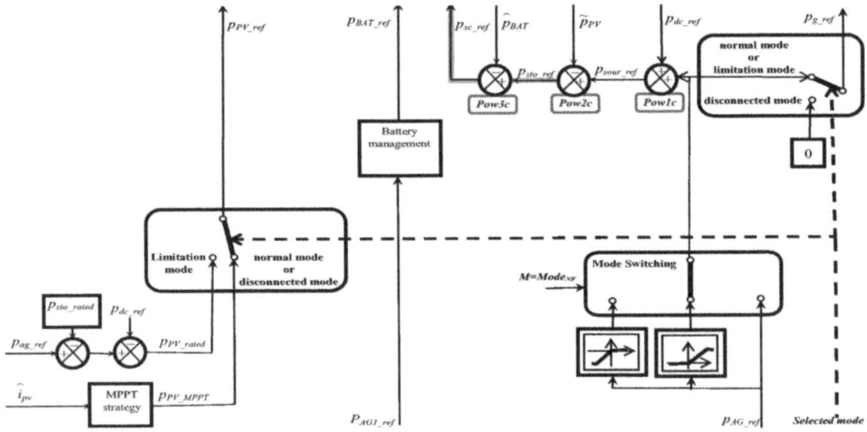

Figure 7.39. *Internal power distribution [LU 10b]*

7.10. Experimental testing using real-time simulation

7.10.1. *Benefits of real-time simulation*

The development of new equipment, such as storage systems for managing electrical grids, involves a validation phase where the hardware and immaterial control algorithms present in the system are tested. Experimental validation, using simulated models in non-real time and then in real-time, is important in order to validate the sizing, the control algorithm and achieved performances before creating an industrial prototype (Figure 7.40).

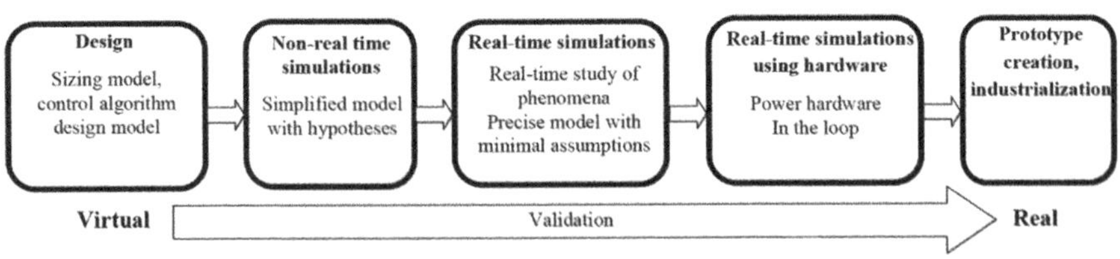

Figure 7.40. *Stages in experimental validation [BAC 08]*

Using real-time simulation, a usage environment is simulated or emulated in real-time in order to test experimental hardware on a 1-to-1 scale. This is known as Hardware In the Loop (HIL) validation. A variety of situations may be recreated in order to safely evaluate the behavior of the prototype. Dedicated processing units, generally operating in parallel and with simulation times shorter than the lowest time constant in the simulated system, are required.

Control or supervision algorithms are tested by implementation on their own processor or controller (DSP, FPGA, micro-controller, PC, etc.), interfaced with the real-time simulator through analog/digital converters. This process is known as an HIL controller (Figure 7.41(a)).

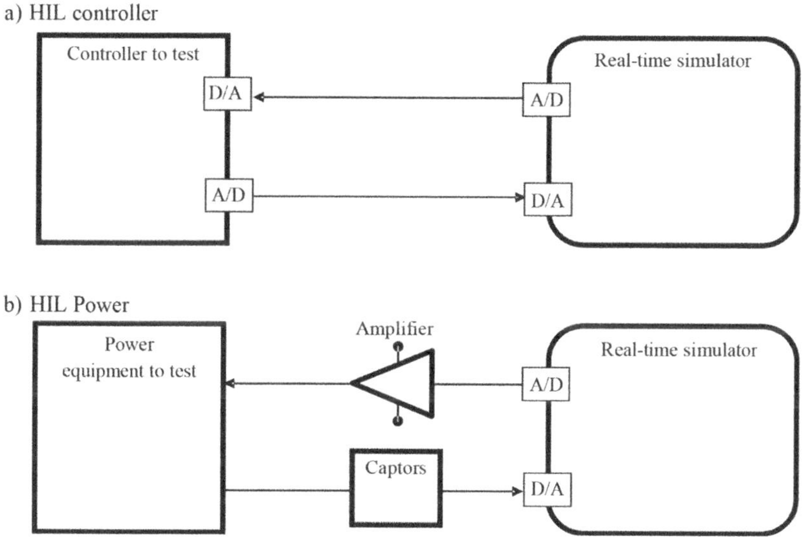

Figure 7.41. *Comparison of an HIL controller and HIL power [KAN 14b]*

In the case of hybrid or Power Hardware In the Loop (PHIL) simulators, the environment is simulated digitally, and some of the calculated electrical values are converted into analog signals (generally +/–10 V, <50 mA), amplified and re-created to enable measurement or application by the prototype (Figure 7.41(b)).

7.10.2. *The Electrical Power Management Lab*

The "Electrical Network" team at L2EP created a specific platform (Electrical Power Management Lab) in order to test innovative electrical grid management algorithms in a real situation. A real-time hybrid RT-LAB® simulator, developed by

Opal-RT Technologies (Montreal, Canada), is used to simulate different energy sources and different network architectures in real-time (Figure 7.42).

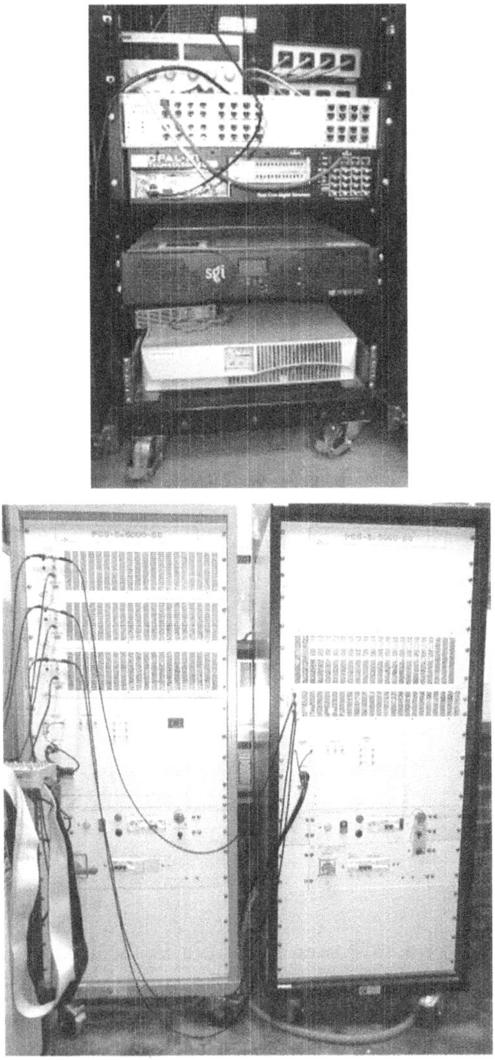

a) Parallel core RT-LAB® processor *b) Power amplifier (2 × 15 kVA)*

Figure 7.42. *Hardware for real-time PHIL simulation*

The simulator includes a processor with seven parallel cores and may be interfaced with real equipment through three power amplifiers for HIL and PHIL applications.

The platform includes a supervision system (PcView), which enables:

– visualization of data and the state of the electrical grid;

– centralized management (MCEMS), remote control and regulation of energy production sources and controllable loads (Figure 7.43).

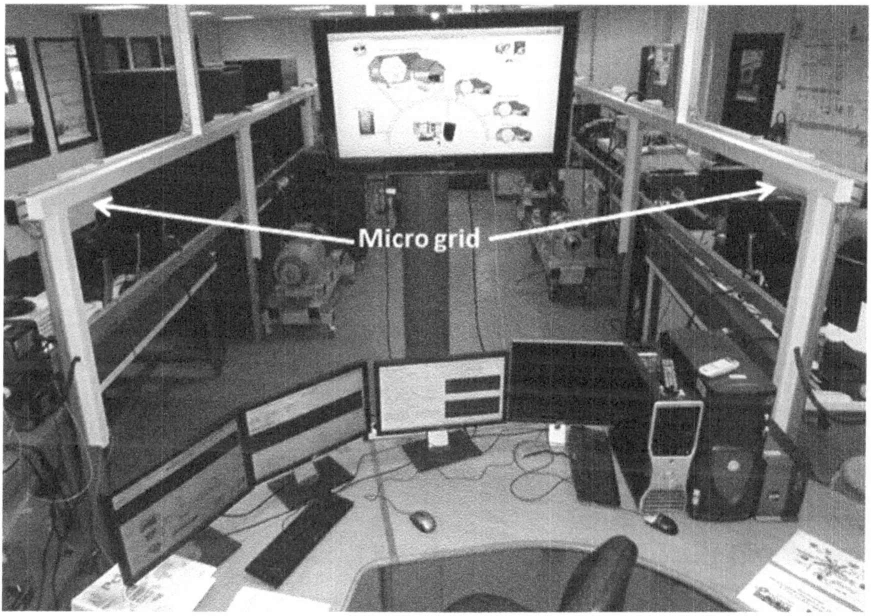

Figure 7.43. *MCEM SCADA (MGCC)*

Communications between different elements of the platform take place using an OPC-type protocol.

7.10.3. *Experimental implementation*

A micro-grid was set up around a central supervisor (Figure 7.43). Our objective was to validate centralized management using real-time simulation in order to verify the coordinated operation of different sources in the micro-grid (active PV generator and micro gas turbine). The choice of experimental verification mode was problematic, as identical conditions (load profile and PV production) needed to be reproduced for various storage sizings in order to identify their impact on the rate of PV usage in the grid. This issue was resolved by using real-time simulation to validate our algorithms.

The chosen hardware layout is shown in Figure 7.44. Test equipment was split into two groups: power elements (generators, loads, etc.) and control units (local controllers, central management, etc.).

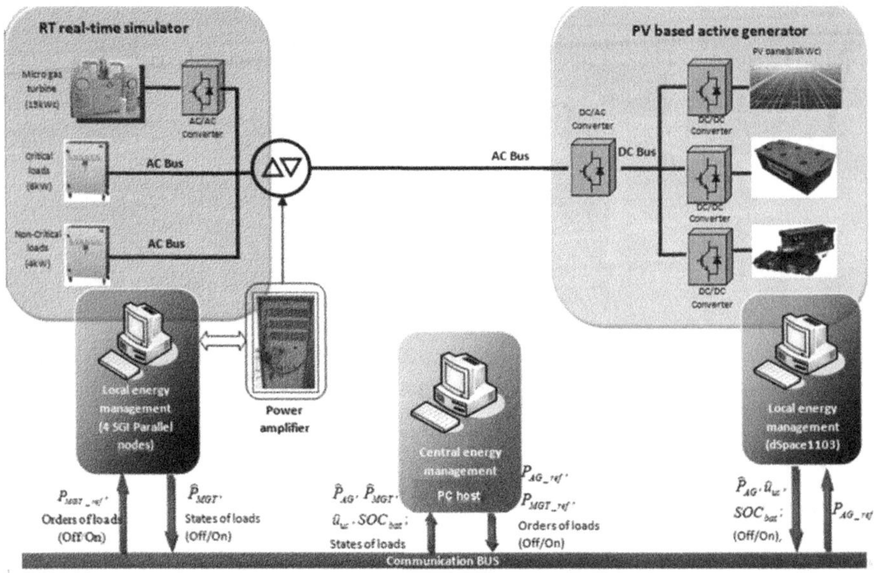

Figure 7.44. *Hardware layout [LU 10b]*

Three types of equipment were used:

– the active PV generator;

– virtual equipment simulated using the OPAL RT-Lab simulator, which includes four parallel processor cores. This system was used to emulate the micro-turbine, 4 kW of non-critical loads and 6 kW of passive loads. These passive loads represent critical equipment and must remain connected to the grid during tests;

– a power amplifier, used to connect the active PV generator and interfaced with the simulator via digital/analog converters.

The control system was made up of four elements:

– a DSpace 1103 board for local control of the active PV generator;

– a centralized management system run on a PC using a dedicated supervision program (SCADA), PCView. This system communicates with local control units;

– the local controller for the micro-turbine and the load controllers, directly integrated into the simulation;

– the communications bus between the centralized management system and the local controllers.

An Ethernet connection using a Modbus-type communications protocol was set up in order to exchange data with various equipment, such as a measurement unit to track currents and voltages, and to share data with a computer. The data exchanged in the course of our application is shown in Figure 7.45. The Modbus protocol enables a supervising computer to communicate with a remote terminal (RTU). This system may be extended to include a larger number of distributed generators and/or to manage reserve power within a local energy community [YAN 18].

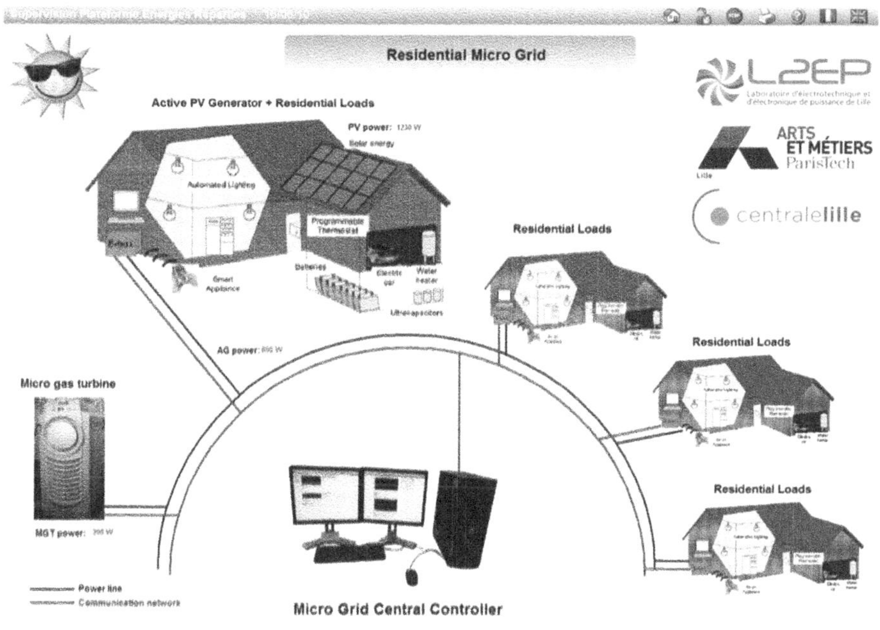

Figure 7.45. *Supervisor screenshot (PCView)*

Several tests were carried out for comparison purposes, using different sizes of PV installation and different loads.

7.10.4. *Analysis of self-consumption in a house*

Our first study concerns a house including a PV power installation with a peak power of 3 kW and local loads of 3 kW; 5 kWh batteries were used, with a maximum charge power of 800 W and discharge power of 500 W. The PV power predictions (Figure 7.46) and load predictions (Figure 7.47) for this test are shown below, with t_0 = 7:00 and $t_0 + \Delta t$ = 18:00.

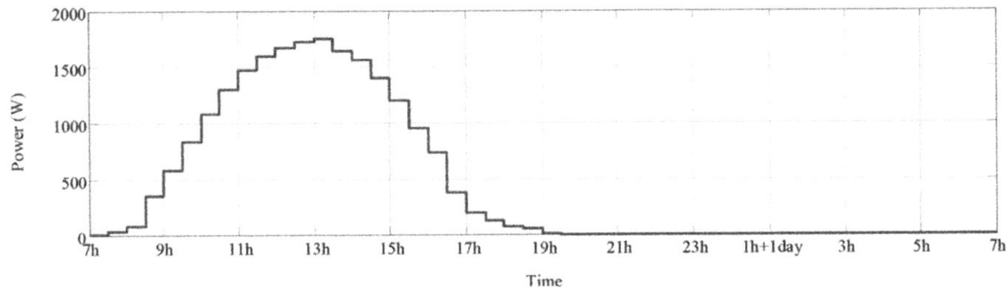

Figure 7.46. *PV predictions for studying self-consumption in a house [LU 10b]*

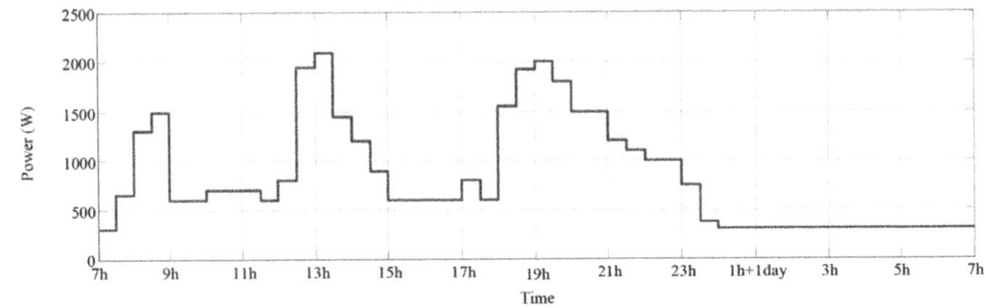

Figure 7.47. *Load predictions for studying self-consumption in a house [LU 10b]*

First, the different energy values were calculated, as described in section 7.6.3, for both daytime and night-time. During the daytime, the energy produced by the PV panels is more than sufficient than that required to power the load (Figure 7.48). The gas turbine remains on at all times, delivering minimum power. The total excess energy produced in the system is automatically stored in the batteries in the active generator and then used to supply power during the night (Figure 7.49).

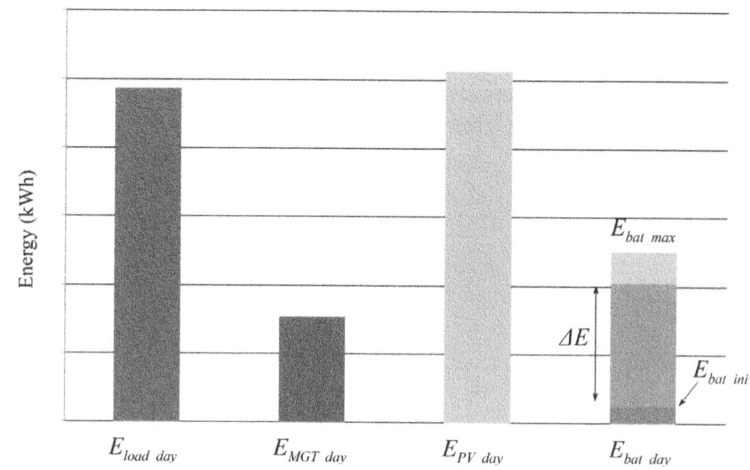

Figure 7.48. *Energy analysis: daytime [LU 10b]*

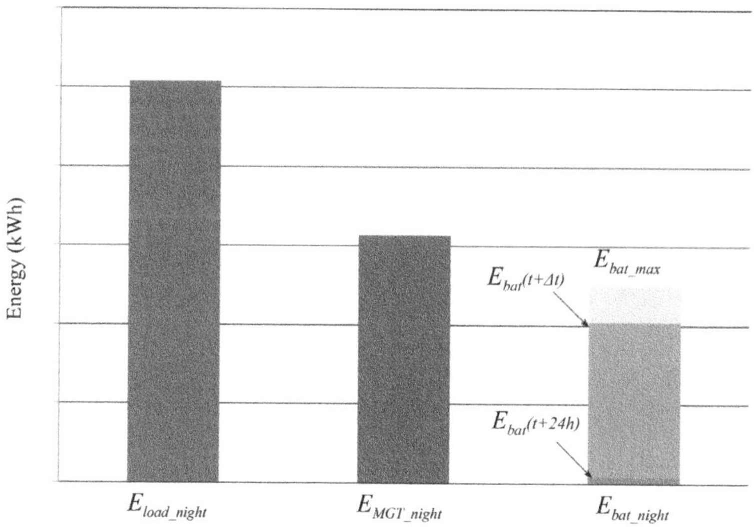

Figure 7.49. *Energy analysis: night-time [LU 10b]*

Power references were calculated by the centralized management system and transmitted to the local controllers. We see that, between 12:30 and 13:30, the PV panels had the capacity to power all loads, but the micro-turbine must remain on at all times, producing minimum power (100 W, case 2 for daytime use, see equations 7.23 and 7.24). The same situation appears in the morning, between 8:00 and 9:00. The calculated power references show that the active PV generator is effectively used over these periods (Figure 7.51) allowing the power reference for the micro-turbine to be kept at a minimum (Figure 7.50).

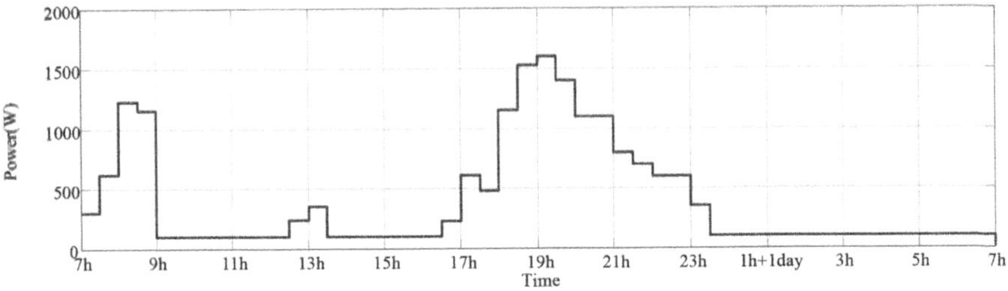

Figure 7.50. *Power reference for the micro gas turbine [LU 10b]*

At 18:00 (nightfall), the batteries were charged and the power reference was calculated (equation 7.26) to discharge them at maximum nominal current until 23:30 (Figure 7.51). Note that the nominal current of the batteries is lower than that of the PV panels. After this time, the load falls and the battery power reference for the remainder of the night is modified accordingly.

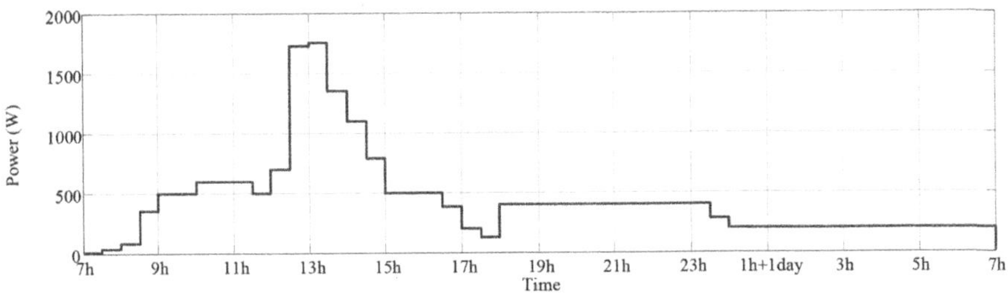

Figure 7.51. *Power reference for the active PV generator [LU 10b]*

The power measurements for the gas turbine (p_{mgt}) and the active generator (p_{ag}) using the experimental platform are shown in Figure 7.52 and are equal to the power references. The measured PV power and that of the battery in the active generator are shown in Figure 7.53.

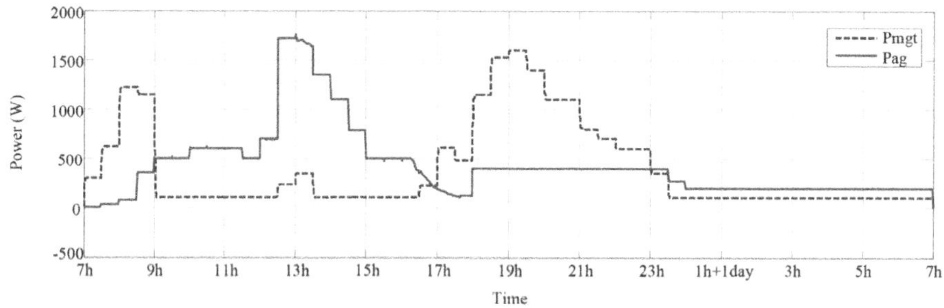

Figure 7.52. *Power generated in the micro-grid [LU 10b]*

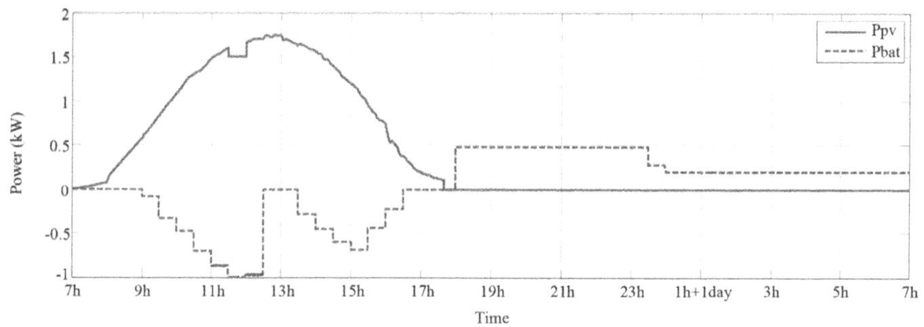

Figure 7.53. *Internal power measurement in the active PV generator for self-consumption in a house [LU 10b]*

We note that part of the PV power is lost at 11:30 and, at the same time, the battery current is at nominal value and in charging mode. Increasing the dimensions of the batteries would prevent this loss. During the night, the batteries supplied almost constant power at nominal value between 18:00 and 23:30, when the demand from the load was high (the nominal discharge value is different from the nominal charge value). Afterwards, as the load demand drops off, the power from the turbine is kept at a minimum after midnight.

Figure 7.54 shows the evolution of the state of charge of the battery in a context of domestic self-consumption. A full charge and discharge cycle takes place over the 24-hour period.

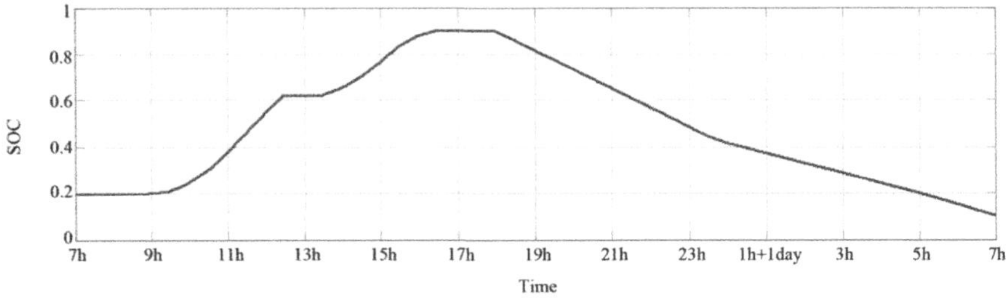

Figure 7.54. *Evolution of the state of charge over time [LU 10b]*

The power plan was established in half-hour steps, and differences were noted. Even in cases where the mean value of power obtained is correct, there can be considerable differences between instantaneous and predicted power references. On close inspection, PV power shown in Figure 7.55 appears to be constant. Batteries are unable to compensate for instantaneous power shortfalls due to their slow dynamics; however, the ultracapacitors used in this context were able to respond to fast power variations.

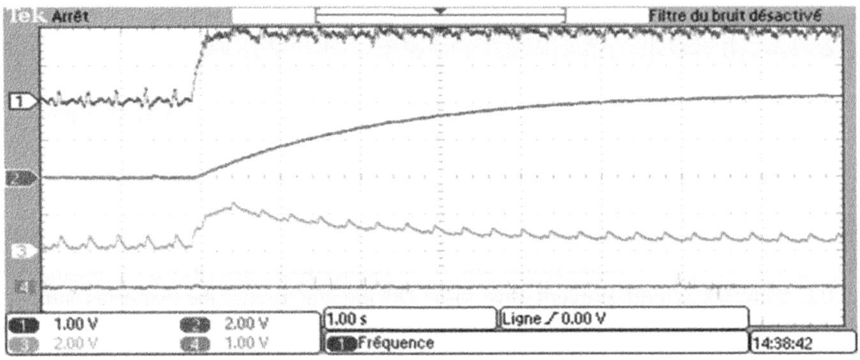

p_{AG_mes}(Line 1): 100 W/div; p_{bat_mes}(Line 2): 100 W/div; p_{uc_mes}(Line 3): 150 W/div; p_{PV_mes} (Line 4) 150 W/div, Timeline: 1 s/div.

Figure 7.55. *Dynamic power compensation by ultracapacitors [LU 10b]*

In Figure 7.56, the load remains constant. When PV power generation increases (CH4, shown in pink), the battery is charged at constant power (CH3, green), but is unable to absorb instantaneous surplus from the PV generator. In this case, the ultracapacitors (CH1, blue) are used for instantaneous balancing.

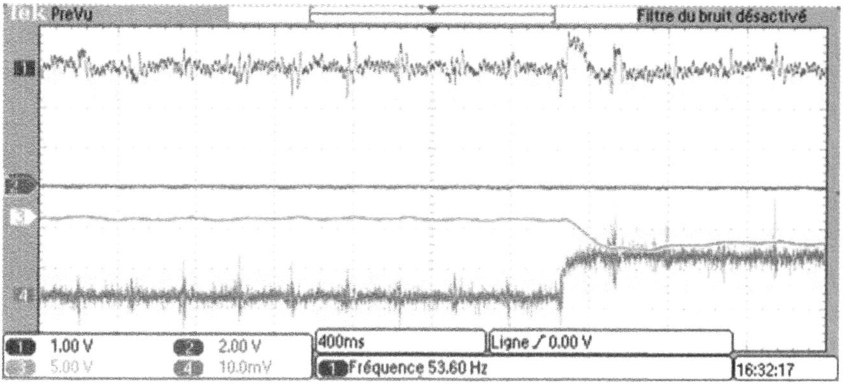

p_{uc_mes}(Line 1): 150 W/div; p_{AG_mes}(Line 2): 100 W/div; p_{bat_mes}(Line 3): 100 W/div; p_{PV_mes} (Line 4) 150 W/div, Timeline: 400 ms/div.

Figure 7.56. *Dynamic instantaneous absorption by ultracapacitors [LU 10b]*

In both cases, the use of ultracapacitors enables instantaneous production of constant power over half-hour steps. In order to show our results clearly, we deactivated primary frequency regulation during these tests.

The electrical energy consumed per generator in the micro-grid can be calculated from our experimental results. In case 1, the PV system provided 53% of the total supply, with the micro-turbine producing the remainder of the required power (Figure 7.57).

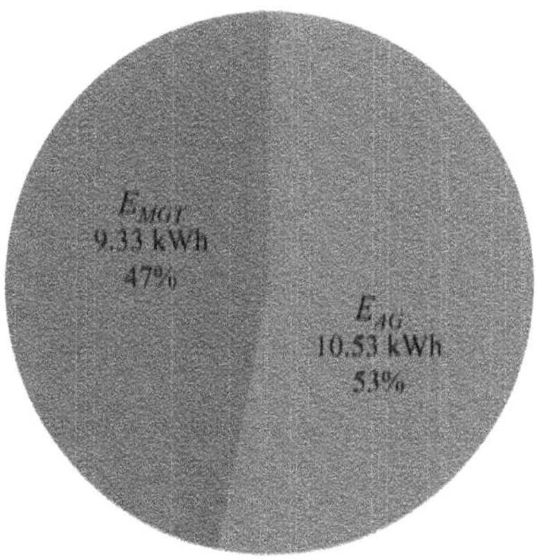

Figure 7.57. *Energy mix for self-consumption in a house*

7.10.5. *Increasing the proportion of PV use in a residential grid*

7.10.5.1. *Impact of a PV producer*

Now, let us consider our 3 kW prosumer (PV generator) in the context of a residential grid involving three other consumers, giving a total local load of 12 kW across the four houses. The PV energy prediction and load prediction for this test are shown in Figure 7.58.

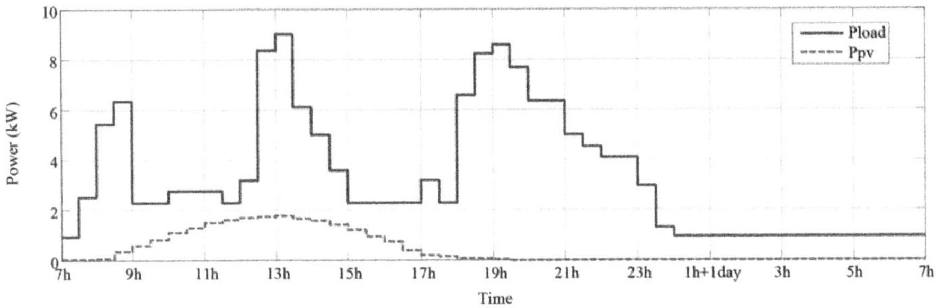

Figure 7.58. *PV power and load predictions [LU 10b]*

According to predictions, PV energy production is much lower than total demand for the load (Figures 7.59 and 7.60). According to this analysis, carried out one day ahead, the batteries will not be used at all.

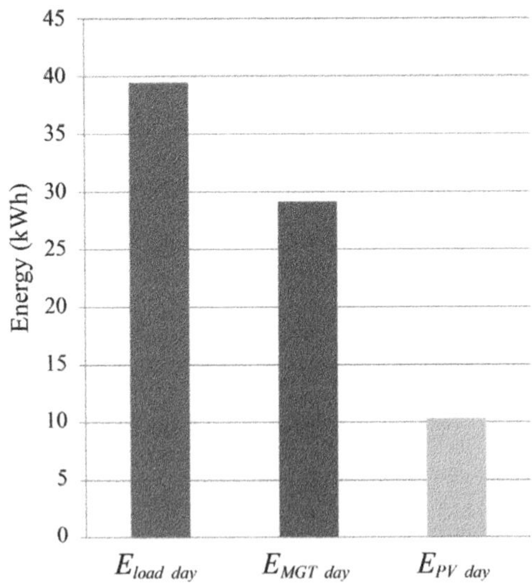

Figure 7.59. *Daytime energy analysis [LU 10b]*

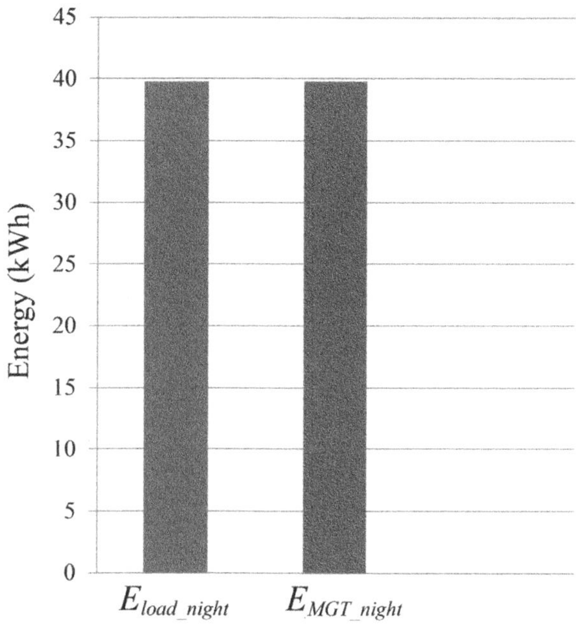

Figure 7.60. *Night-time energy analysis [LU 10b]*

The measured powers are equal to the power references. The power references calculated by the centralized management system show that all generated PV energy is used (with no power limitations), but this energy is less than that produced by the gas turbine (Figure 7.61). In this local scenario, the batteries are not used at all as the required load power is always higher than PV production during the day. This confirms that, in situations with a low proportion of PV power, all production can be injected into the grid and storage systems are useless.

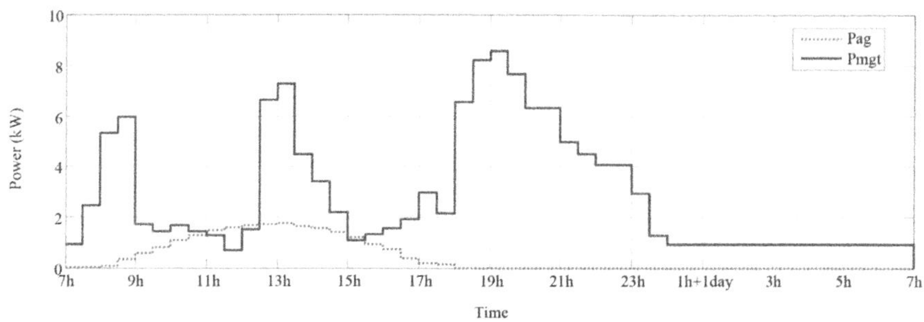

Figure 7.61. *Power references calculated by the centralized management system for a producer [LU 10b]*

The electrical energy produced by each generator may be deduced from these results. The PV generator provides 17% of total power, with the micro gas turbine making up the remaining power (Figure 7.62).

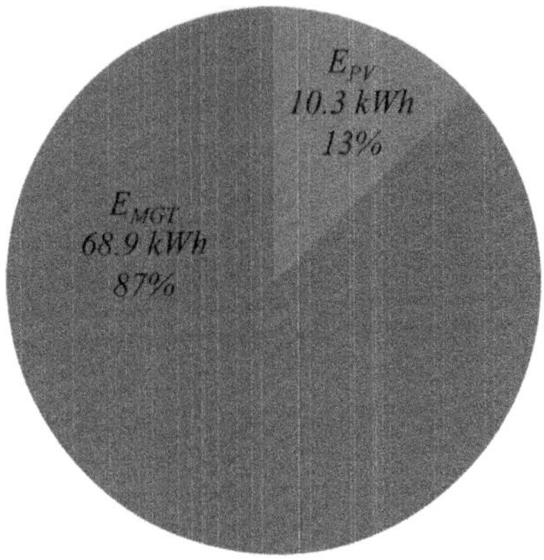

Figure 7.62. *Energy mix with a PV producer*

7.10.5.2. *Impact of a high rate of PV use*

Now, let us consider a case in which the four houses studied above are all equipped with 3 kW PV installations. The load profile is the same as before. In this case, there is no local storage. As the homes are close together, their PV production is almost identical; total PV power is shown in Figure 7.63. Load predictions are the same as before.

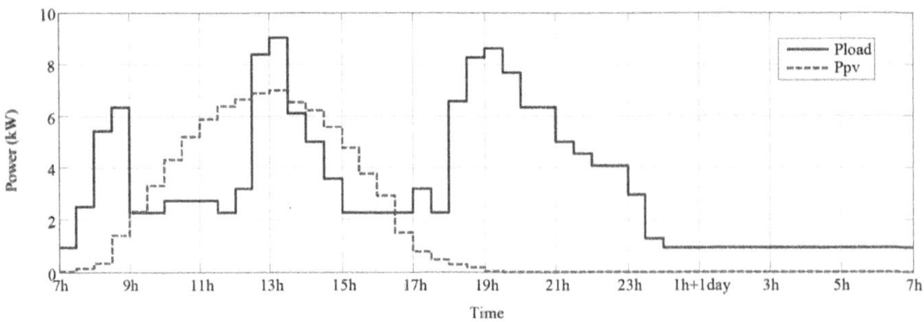

Figure 7.63. *PV power and load predictions [LU 10b]*

For ease of comparison, all four power references for the prosumers in our study are shown in Figure 7.64. Total measured PV power is shown in Figure 7.65.

Between 9:00 and 12:30 and between 13:30 and 16:30, PV power is limited and the gas turbine is run at minimum power as the loads are not sufficient. Part of the available PV power is thus lost. Furthermore, the turbine has to cope with the consumption peak at 19:00 alone. PV power constitutes 36% of the total supply, with the micro-turbine contributing the remaining 64% (Figure 7.66).

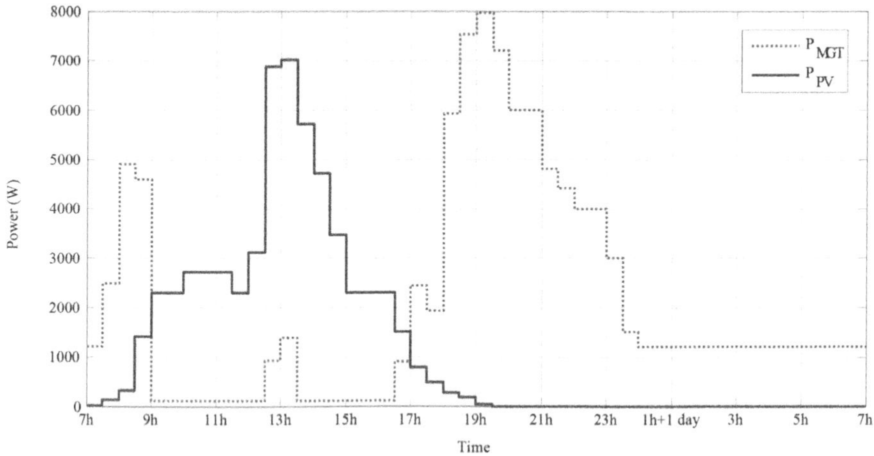

Figure 7.64. *Power references calculated by the central management system [LU 10b]*

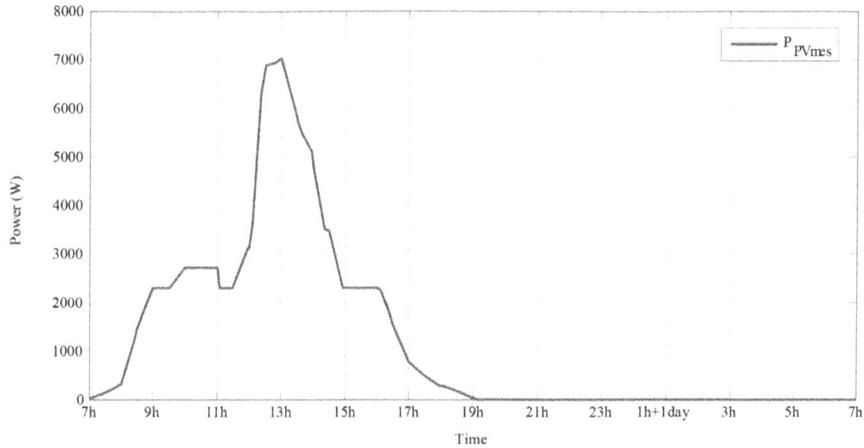

Figure 7.65. *Total measured PV power [LU 10b]*

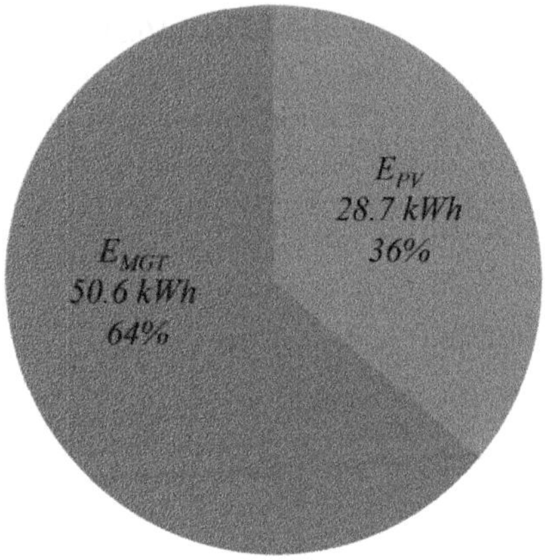

Figure 7.66. *Energy blend obtained for high PV production with no storage*

7.10.5.3. *Impact of prosumers with energy storage*

Four 1272 Wh battery sets are now added to the grid, with a total power of 3,200 W maximum charge and 2 kW maximum discharge. The same PV and load prediction data are used.

Our analysis shows that available PV energy is higher than that needed by the loads during the daytime (Figure 7.67). The energy available to charge the batteries is almost equal to the energy produced by the gas turbine over the course of the day. This energy is then used to supply power during the night, reducing reliance on the gas turbine during this period (Figure 7.68).

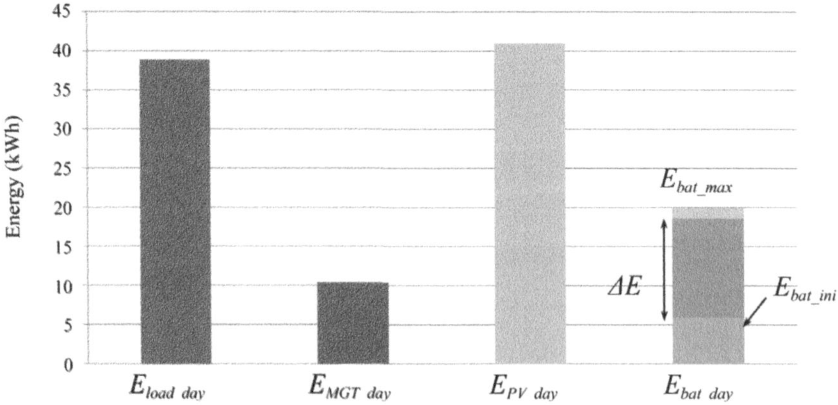

Figure 7.67. *Daytime energy analysis for prosumers [LU 10b]*

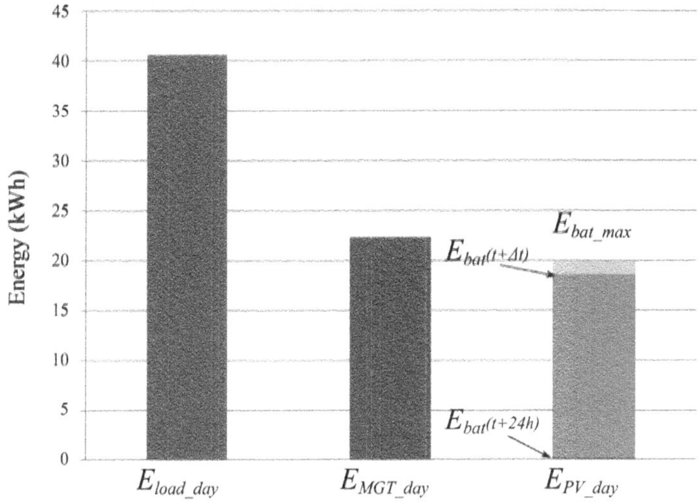

Figure 7.68. *Night-time energy analysis for prosumers [LU 10b]*

PV production during the day is the same as before. At the beginning of the night, power references for the prosumers are used to establish battery discharge rate. The peak power required from the micro-turbine is somewhat lower than before (Figures 7.69 and 7.70).

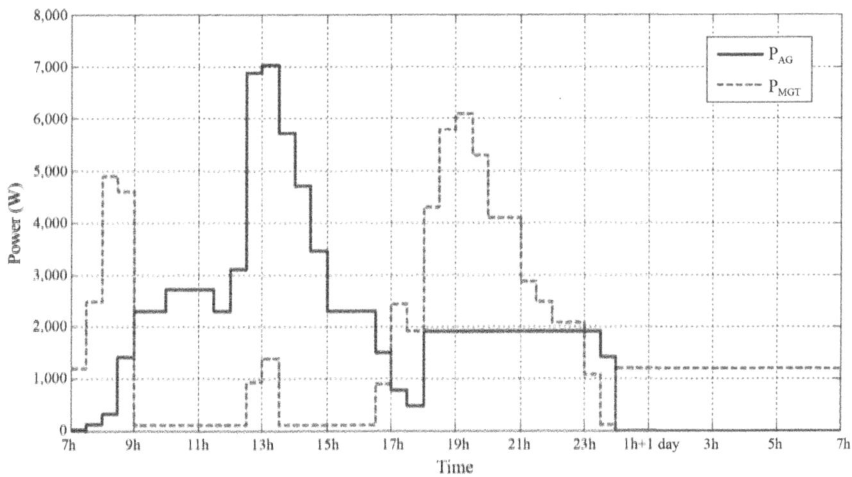

Figure 7.69. *Power produced by the micro-turbine and active PV generators [LU 10b]*

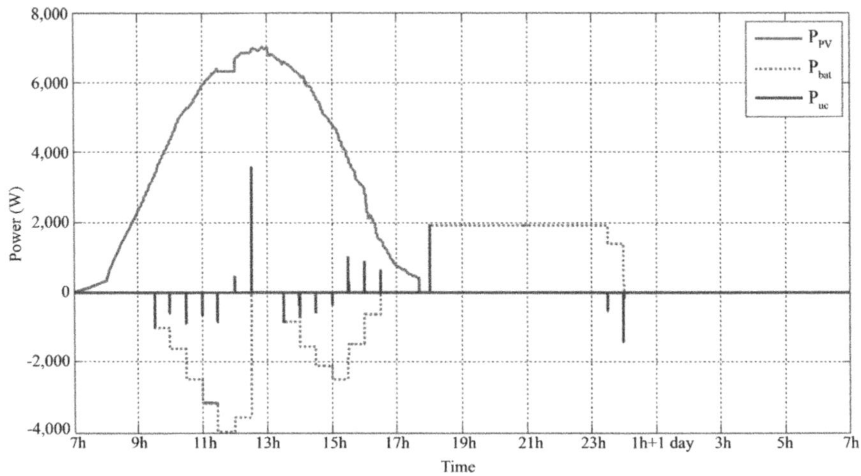

Figure 7.70. *Power from PV panels, batteries and ultracapacitors [LU 10b]*

PV power use is increased thanks to the batteries, and the contribution of PV energy to the blend increases to 41% (Figure 7.71).

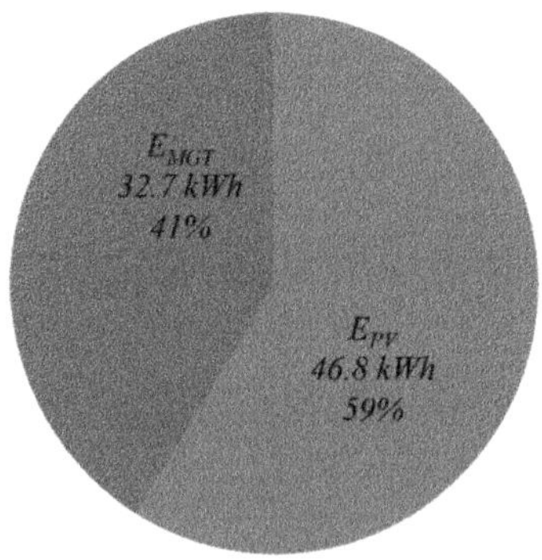

Figure 7.71. *Energy blend obtained for prosumers*

7.11. Review of scientific contributions and methodological summary

The purpose of this chapter was to describe a conceptual technical architecture for a local energy community, designed for local electricity production and consumption within a neighborhood equipped with active PV generators. The system was rendered active, i.e. controllable via the injection of power into the grid, through the use of storage and a local control mechanism. We presented an operational plan and an innovative energy management system for the integration of active generators. A coordinated management system for energy resources was proposed, making use of a communications network. Energy management was broken down into three parts: long term, medium term and short term.

Long-term energy management takes account of photovoltaic energy production and load predictions, planning consumption for the next day. It also determines battery charge and discharge schedules, taking account of available PV power and load demands. A determinist algorithm was proposed, based on predicted load and PV production data.

Medium-term energy management consists of adapting the power references for each generator on a half-hourly basis as a function of actual PV power availability and real loads. An algorithm for battery charging and discharging using constant power was implemented.

Active generators contribute to short-term energy management through primary frequency regulation.

Internal power balancing is carried out in real-time using ultracapacitors.

We tested our generator coordination and power distribution approach using a real-time simulator. One advantage of this technique is that control functions are executed in real-time using real control equipment. This enables energy management algorithms for residential grids to be tested in situations which reflect a variety of real static and dynamic phenomena encountered in the real world. For example, we compared different configurations and system dimensions under identical solar radiation and load profile conditions, showing the impact of battery sizing on the rate of PV energy use in this electrical system.

7.12. Concluding thoughts and research perspectives

The need to integrate renewable energy sources on a large scale while maintaining a stable, high-quality supply represents a significant challenge for electrical grids, requiring radical restructuring of electrical energy supply and usage practices. As we highlighted in this chapter, the success of this transition depends on the development and integration of new equipment at every stage of the electricity production, delivery and usage chain, as well as on new network architectures, new operating approaches, new incentives and business models, new energy security structures and a new form of societal access to this energy.

"Smart" electrical grids assign an active role to all stakeholders, including producers, suppliers, aggregators and consumers, in providing a sustainable, economic and secure electricity supply. ICT is used to improve the reliability, security and efficiency (in both economic and energy terms) of the electrical system. Smart sensors and meters are also being rolled out, collecting data from across the system. The availability of this data may significantly contribute to the integration of distributed energy resources (DER) and storage systems into grid operations, particularly for customers, both consumers and producers of energy, who gain the ability to participate in energy exchange mechanisms. This is key in enabling consumers to play an active role for their own interest (more comfort, cheaper energy, etc.).

In this chapter, we highlighted means of maximizing local PV consumption in a local energy community via the exchange of energy between buildings and mutualized use of domestic storage units. The integration of "citizens' energy" systems may be tested through the creation of smart buildings, smart communities, smart cities, etc., contributing to the development of a zero-carbon electricity system. The European Commission has put forward new rules for the transition to consumer-centered clean energy, in which consumers play an active role at the heart of future energy markets. Customers will enjoy greater choice in terms of supply, alongside the possibility of producing and supplying their own electricity in real-time in response to certain signals (price, etc.).

To improve operations and for optimal energy management, local consumers and producers may be coordinated with decentralized generators (gas turbines, etc.) and/or storage systems within a distribution grid, forming an integrated energy system such as a micro-grid or virtual power plant (VPP). Local consumers and producers may be involved in increasing local energy autonomy or in developing smart communities with the aim of attaining energy-neutral or energy-positive status.

The adoption of complex energy management practices at community level involves multiple energy stakeholders, all contributing to a complex integrated electrical system involving electricity supply, heating, cooling and other energy services.

Energy storage (including the use of electric vehicles) and a certain level of flexibility in demand are essential in order for grid operators to develop effective and efficient electrical distribution systems, improving the performance and profitability of these new energy systems.

In this context, further research is needed in a number of areas, notably:

– innovative design and energy management strategies for micro-grids and community Virtual Power Plant [BEV 17];

– development of flexibility and "system" services;

– evaluation of impacts on society and human behavioral analysis with regard to energy consumption;

– integrated energy systems with local control for energy-neutral or energy-positive zones;

– analysis and use of new energy data in cyber-physical systems;

– mixed energy system design (featuring multiple producers/consumers, DC and hybrid micro-grids);

– development of energy exchange mechanisms and business models for citizen energy autonomy (blockchain, peer-to-peer energy market, etc.);

– evaluation of environmental impacts [KAN 14a];

– studies of best practice and demonstration units, etc.

In any and all cases, it seems likely that solutions will only be reached by collaborative working across scientific disciplines.

8

Reversible Charging from Electric Vehicles to Grids and Buildings

8.1. Introduction

In France in 2010, transport accounted for over 30% of total national energy consumption. Largely based on primary use of fossil fuel, this field is responsible for almost 40% of CO_2 emissions. Responding to environmental concerns and the progressive exhaustion of fossil fuel resources, the automobile industry has begun to propose alternative solutions to traditional vehicles. These include hybrid vehicles which cannot be recharged from the grid (HV), grid-rechargeable hybrid vehicles (RHV) and battery-powered electric vehicles (EV) [ROB 16].

Predictions suggest that there will be 800,000 electric vehicles in France by 2020–2022, with this figure rising to 2.5 million by 2030. This figure would require the creation of four million charging points, most connected to users' homes, i.e. in private spaces. Furthermore, 90% of charging activity would take place at home (mostly at night) or at work (during the day). Figure 8.1 shows the projected evolution of the number of EVs, RHVs and charging points in France, highlighting the predominance of private recharging infrastructures.

Taken together, the batteries of these future EVs and RHVs constitute a sizeable storage capacity which may be exploited by the electric grid when vehicles are connected, or by specific buildings. This approach could reduce the cost of owning an EV and is made possible by the energy and time flexibility of these vehicles. Furthermore, the proportion of fleet vehicles which are parked simultaneously and may potentially be connected to the grid rarely falls below 80% [COD 14].

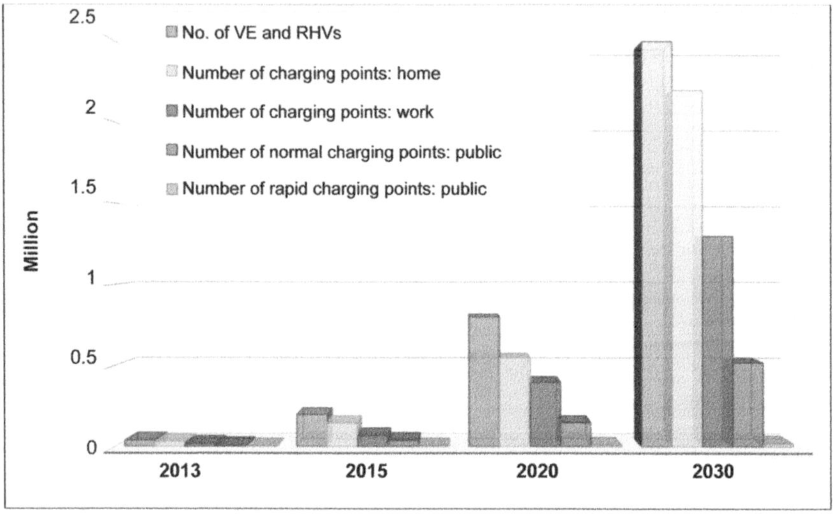

Figure 8.1. *Projected evolution of the numbers of EVs, RHVs and charging points in France (source: Enedis). For a color version of the figures in this chapter see www.iste.co.uk/robyns/buildings.zip*

A variety of different technologies have been, or are soon to be, developed:

– Vehicle to Grid (V2G): the vehicle powers (injects power from its battery into) the electric grid via a charging point on the street or in a carpark (e.g. at a train station, mall, etc.). The acronym G2V is used for cases where the vehicle is recharged from the grid: evidently, this is the most usual situation and should be developed as a priority;

– Vehicle to Home (V2H): the vehicle powers a residential building, generally cut off from the grid;

– Vehicle to Home and Grid (V2HG): the vehicle powers a residential building which is itself connected to an external grid, with the potential to draw power from or inject power into this grid;

– Vehicle to Building (V2B): the vehicle powers a residential or tertiary building. This case evidently concerns multiple vehicles in a parking lot;

– Vehicle to Station (V2S): the vehicle powers a railway station building or railway equipment and vice versa. Again, several vehicles in a parking lot will be involved.

This list of configurations for development is not exhaustive.

In this chapter, we will introduce the different configurations for reversible charging of EVs described above. The potential services and energy management needs of an electric vehicle fleet interacting with an electricity distribution grid will be described in detail. An energy supervisor based on fuzzy logic will be proposed, following the methodology described in Chapter 1 (section 1.8). The contribution of reversible charging of EVs in the context of a railway station will also be discussed.

8.2. Reversible charging of electric vehicles

8.2.1. *Vehicle to Grid*

The traditional approach to connecting and charging an electrical vehicle at a terminal is referred to as Grid to Vehicle (G2V): electricity goes from the grid to the vehicle (unidirectional mode), and the vehicle is considered as a consumer.

The predicted arrival of 2.5 million rechargeable cars on French roads by 2030 will have an impact on the electric grid. Current annual electricity production in France is approximately 500 TWh with a total installed power of approximately 130 GW. Covering an average distance of 15,000 km each year, an electric vehicle consuming 200 Wh/km would require approximately 3 MWh per year. Multiplied by 2.5 million vehicles, this represents somewhere between 6.5 and 7.5 TWh. While current energy production is sufficient, the same cannot be said for the available power. The power required to charge vehicles may create major peaks in the grid. For example, Figure 8.2 shows the power increase which might result from mass recharging of EVs as users return home from work, between 4 pm and 9 pm, with a peak from 6 to 7 pm representing a 3% (2 GW) increase using the predicted figures for 2020 (800,000 vehicles); this peak rises to 10% using the figures for 2030, based on a predicted total of 2.5 million vehicles.

The peak power demand could become a major problem for the electricity grid, requiring reinforcement work; it is also problematic in terms of greenhouse gas emissions, as this peak production is high-carbon. For these reasons, charging at peak times must be avoided; instead, it should take place at times when consumption is lowest, or when excess variable renewable energy production is available, for example. The load thus needs to be controlled. The main purpose of smart load management is to arrange loads in a way which smooths demand, so that charging takes place at off-peak times (at night, when considering nuclear energy production, or in the middle of the day in the case of solar power). This is attained via automatic modulation of load power.

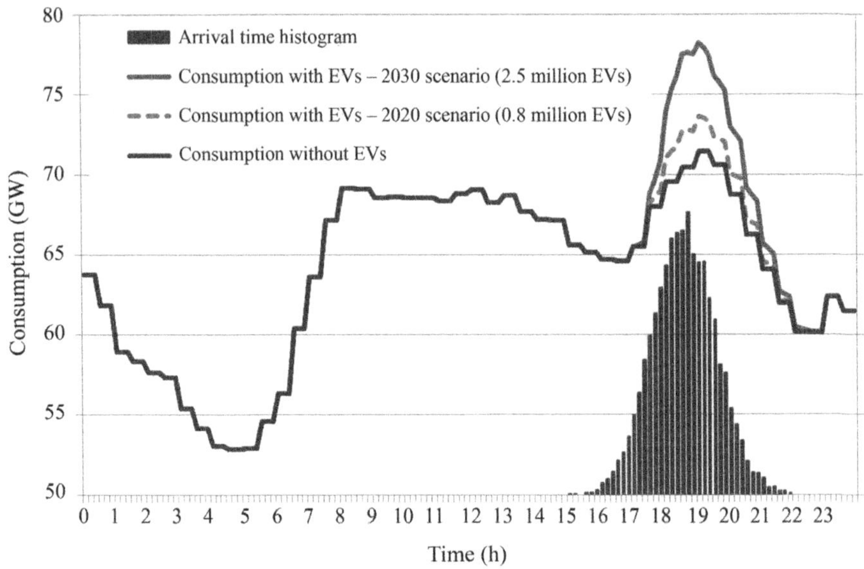

Figure 8.2. *Projected evolution of power requirements from the electric grid in France for mass recharging in the evening, when EV users return home [SAR 16b]*

EV charging modes also influence the impact of recharging on the grid. Different technologies and modes have been developed, including AC and DC options, with different power modes. Currently, there are three levels of charging power, shown in Figure 8.3:

– normal charge (3 kVA, 16 A, terminal connected to a mono-phase grid), with a charge time of approximately 8 hours for 24 kWh;

– accelerated charge (22 kVA, 32 A, terminal connected to a tri-phase grid), with a charge time of approximately 2 hours;

– rapid charging (43 kVA, 63 A, terminal connected to a tri-phase grid), with a charge time of 20 min. There are both AC and DC versions for this mode.

Even faster charging modes, at higher powers – 100 or even 150 kW – are planned for the future.

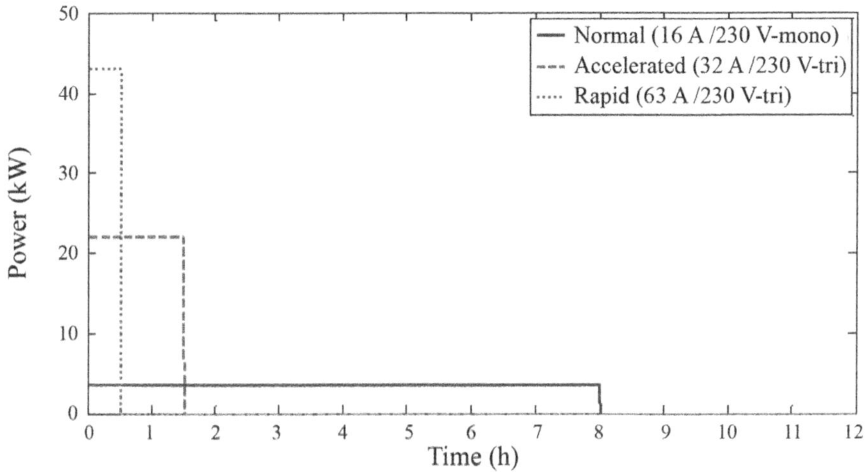

Figure 8.3. *Simplified model of EV charge profiles (24 kWh) for different power values [BOU 15]*

The impact on the electric grid differs according to charge type. For example, charging using a terminal connected to a mono-phase grid may have an effect on current and voltage imbalances in the grid. Accelerated or rapid charging has a greater potential of overloading the grid.

The batteries contained in electric vehicles may be used as a means of offloading energy, given that private vehicles are parked the majority of the time (according to EDF, 50% of vehicles are always parked at home, and 69% of active vehicles are parked in a reserved space for an average of 6 hours per day). It would thus be possible for the grid to draw on the reserves contained in these batteries to respond to peak demands (e.g. in the early evening) or to compensate for a temporary production shortage (e.g. when the weather does not permit the use of renewable sources). This situation is known as Vehicle to Grid (V2G): the EV provides power to the grid in response to system requirements (bi-directional model), offering additional flexibility [CRE].

The main difference between Vehicle to Grid and classic passive (uncontrolled) or dynamic charging (unidirectional charging with all-or-nothing control [ROB 16]) is that the charging process is reversible. As ever, the charger must have the capacity to draw energy from the grid to recharge the battery (Grid to Vehicle, G2V), as well as to inject energy from the battery into the electric grid (Vehicle to Grid, V2G).

In addition to the functions required for dynamic charging, the following elements need to be taken into consideration:

– management of bi-directional energy flows;

– information exchanges between the vehicle, charger and grid;

– additional cost of the charger;

– additional losses;

– issues relating to the injection of power into the grid (including detection of isolated operations);

– complexity of control;

– impact on battery aging, notably due to more frequent charge/discharge cycles.

Normal and accelerated charging power levels appear to be best-suited to V2G applications; rapid charge mode is not suitable, as the terminals used for this purpose are generally designed to increase vehicle autonomy with a short connection time. Vehicles do not stay at the charging terminal for very long (approximately 30 min), meaning that they can provide little or no service to the grid.

Electric vehicles may be called on to provide services to the grid, at least to limit their impact on the grid while charging, and at best, to improve the quality of the electricity supply. There are four different categories of service, in order of influence:

– improving the electricity supply;

– participating in grid operations (ancillary services, such as voltage and frequency control);

– direct benefits for users (economic gain);

– supporting the use of renewable energy sources.

In terms of improving the electricity supply, participating in grid operations and supporting the use of renewable energy sources, it should be noted that the power of each individual vehicle is low in comparison with the grid power, and that these units are distributed across the whole network. For this reason, a new type of actor, the aggregator, comes into play. Aggregators manage a certain number of vehicles in order to provide different services to the grid, and are the point of contact for grid operators. Evidently, this means that a communications connection must be established between each vehicle and the aggregator, enabling charge/discharge control, specification of the next planned usage to ensure that the battery is sufficiently charged, energy metering for billing purposes, etc.

8.2.2. *Vehicle to Home and to Building*

Using V2G, chargers reinject energy into the electric grid. Energy may also be injected into smaller installations, such as a home (V2H, Vehicle to Home), tested in Japan by Nissan, as shown in Figure 8.4, or an apartment or tertiary building (V2B, Vehicle to Building), as shown in Figure 8.5. In this case, one or more vehicles can provide backup power for a given installation. In the case of a home, the EV can provide emergency power to critical equipment over a period from a few hours to four days, depending on the initial state of the battery. Japanese homes consume, on average, 10–12 kWh/day. The battery capacity of a Nissan Leaf is 24 kWh, giving it the potential to power a home for two days, or even longer if the house in question is equipped with PV panels, as shown in Figure 8.4.

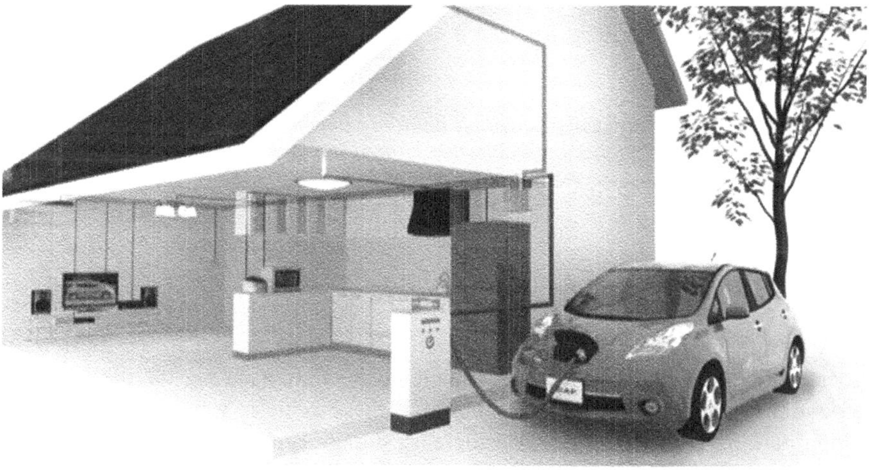

Figure 8.4. *V2H concept developed and implemented by Nissan in Japan (http://www.nissan-global.com/EN/TECHNOLOGY/OVERVIEW/vehicle_to_home.html)*

The charger may only be used to inject energy into an installation, and not into the grid. The V2H interface does not need to have all of the functions required for V2G (e.g. there is no need to communicate with the grid). This technology may be particularly valuable in regions subject to critical situations (earthquakes, weather events, etc.), which may result in homes being cut off from the grid for several days;

in this case, vehicles may provide an emergency power supply. V2H technology may also be used for load reduction purposes at critical peak times, contributing to network reliability.

Figure 8.5. *6 Nissan Leafs powering a tertiary building in Japan, following the V2B concept (http://inhabitat.com/nissan-leaf-vehicle-to-home-technology-powers-office-buildings-in-japan/nissan-leaf-1)*

8.2.3. *Vehicle to Station and energy hubs*

A significant number of commuters drive to a train station in the morning in order to travel to work, leave their vehicle at the station during the day and then retrieve it to complete their journey home. If these vehicles were electric, then the station parking lot would offer considerable storage capacity during the day. This storage potential could be used to provide services to the distribution grid, interacting with installations in or near the station (Vehicle to Station, V2S), such as buildings, renewable (PV or wind) energy sources, or even the rail equipment itself, retrieving the energy emitted by trains when braking. Figure 8.6 provides an illustration of this concept, showing the notion of an energy hub providing an interface between different actors [SAR 16b]. These prospective developments fall under the umbrella of railway smart grids [ROB 16].

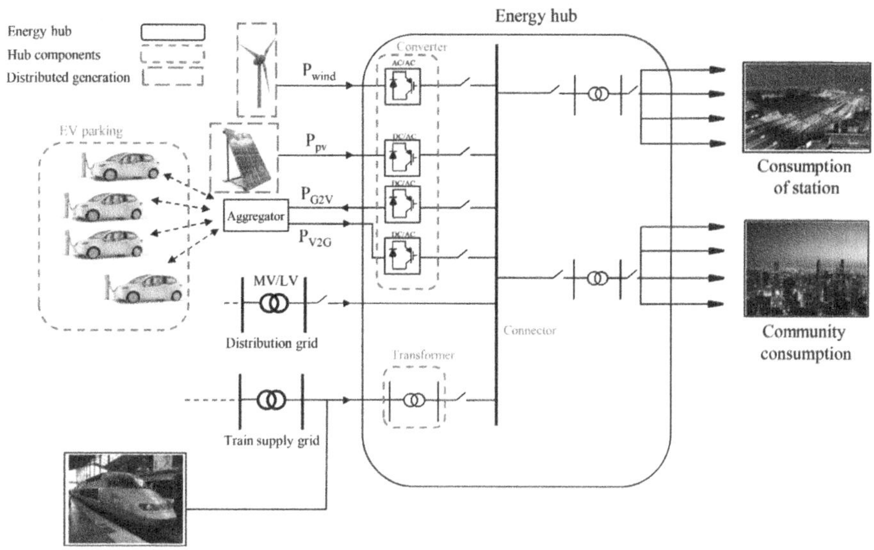

Figure 8.6. *Energy hub concept developed around a train station [SAR 16b]*

8.2.4. *Energy service aggregator*

Given the large numbers of vehicles disseminated across the electric grid, an aggregator is required. This new role provides an interface between EV owners and electric grid operators, and was described in Chapter 6 (section 6.2.2).

8.3. Potential services and energy management of reversible EV fleets

8.3.1. *Services supplied by V2G*

Table 8.1 shows a range of services which a V2G fleet may supply to the electricity system (transmission system operators, TSO, and distribution system operators, DSO), to consumers and to renewable energy producers. This table shows the imposed constraints in terms of available power and service duration, along with an estimation of the number of reversible charge EVs required and maximum geographical distribution. This estimation was made based on a maximum charge/discharge rate of 3.7 kW (under 16A and 230V) and limited to the provision of one service per day in order to reduce the impact on vehicle batteries [SAR 14, SAR 16a, SAR 16b]. This hypothesis would enable services to be provided using any charging mode, with vehicles connected at home or in the workplace, a train station carpark or a supermarket. Increasing the charge/discharge rate would evidently reduce the number of vehicles required to provide each service.

Ancillary services, services for producers and consumers	Number of EVs to Provide Services				
	EVs required	**Power required**	**Discharge time required**	**Geographical area**	**Beneficiaries**
Primary control	5000 EV	1 MW	30 sec – 15 min	3000 km^2	TSO
Secondary control	5000 EV	2 – 3 MW	30 min – 1 h	3000 km^2	TSO
Tertiary control	> 5000 EV	10 MW	30 min – 6 h	3000 km^2	TSO
Peak smoothing	500 EV	500 kW (MT) 100 kW (BT)	2 – 10 h	300 km^2	TSO–DSO consumers
Reactive power	500 EV	100 kvar	As needed	300 km^2	DSO
Support for renewable energy production	≥ 500 EV	100kW – 2MW	20 min – 1 h 30 min	300 km^2	Decentralized producers

Table 8.1. *Services which a V2G fleet may provide to the electric system, consumers and renewable energy producers [SAR 16b]*

8.3.1.1. *Primary frequency control*

The primary power reserve for the synchronous interconnected European grid is provided by all of the European producers connected to transport grids [ROB 15]. This reserve must have the capacity to compensate for the simultaneous loss of two production plants, providing 3000 MW of power. Control is automatic and must take place within a response time of 15–30 s. Any generator providing power in excess of 40 MW is obliged to participate. According to [RTE 13], compensation for primary frequency control is set at €8.04/MW per half-hour step.

To contribute to this service and benefit from financial incentives, offsetting the cost of the energy consumed and the investment in batteries, groups of reversible charge EVs with a collective reserve of 2.4–2.7 MW may be formed. Taking account of vehicle availability statistics, this implies that the fleet must include at least 5000 V2G [PET 13]. Given that these vehicles are generally distributed across the grid, an aggregator is needed to coordinate them for the purposes of service provision.

8.3.1.2. *Secondary frequency control*

The secondary power reserve is activated after primary control, with a time delay of between 100 and 200 s, over the area of a country such as France. All producers with a capacity of 120 MW or more must contribute to this reserve, which has a total capacity of between 500 and 1000 MW in the French case. Remuneration for secondary power control is made up of two parts. The first relates to maintaining a reserve and is set at €8.04/MW per half-hour step. The second corresponds to use of the reserve and is equal to €9.30/MWh [RTE 13]. The financial benefits of participating in this service are thus more attractive than for primary control, on the condition that at least 5000 V2G EVs are involved [PET 13].

8.3.1.3. *Tertiary frequency control*

Contributions to tertiary frequency control are made from 30 min to several hours after an incident. Producers with a capacity of at least 10 MW and consumers with a load reduction capacity of similar size may participate in the power adjustment mechanism. In this situation, a demand and offer process is used, meaning that the management role of the aggregator is more complex; however, it offers potential economic benefits. Once again, a group of 5000 or more EVs is needed in order to provide these services.

8.3.1.4. *Smoothing power peaks*

Power peaks occur at the beginning and end of the day, requiring additional production from costlier and, often, high carbon sources, such as thermal power plants. They may also impose significant constraints on lines and transformers, with a risk of congestion and of line or substation cut-outs. One alternative to investing in reinforcing the lines and substations in question is to act on the load, shifting consumption from peak to off-peak times. As electricity is costlier at peak times, voluntary load reduction or delay creates cost savings; remuneration may also be offered [DEL 09, ROB 15].

The load represented by electric vehicles has the potential to be used in this way. Shifts in charging time can be synchronized with local renewable energy production, improving the environmental integration of EVs [ROB 16]. V2G technology would enable storage capacity to be put to good use at peak times, providing energy at times when it is most expensive, alongside ancillary services. Power peak smoothing offers promising perspectives for grid operators and consumers, particularly as it requires less power (minimum 500 kW) and thus fewer aggregated vehicles (500–1000).

8.3.1.5. *Reactive power compensation*

Reactive power compensation in distribution grids is essentially carried out using equipment fitted with electronic power converters. This service, which contributes to maintaining voltage quality, is becoming increasingly important for renewable sources (wind and PV [ROB 12]). V2G EVs can also provide this service, given a minimum group size of approximately 500 vehicles.

8.3.1.6. *Support for renewable energy production*

The intermittent character of certain renewable sources, such as wind and photovoltaic solar power, makes it difficult to establish precise production forecasts;

for this reason, it is difficult for these installations to contribute to ancillary services. However, the rapid increase in installed power from these sources means that at least partial participation is needed [ROB 12]. One means of responding to these constraints is to associate renewable sources with storage systems. V2G could evidently contribute to this service, on condition that measures are taken to avoid accelerating battery aging through excessive charging and discharging. A group of at least 500 vehicles would be needed to contribute to this service.

8.3.2. *Energy management of a V2G fleet*

8.3.2.1. *Case study*

A section of a real distribution grid, shown in Figure 8.7 [SAR 16b], was modeled as part of a case study, developing an energy management approach for a V2G fleet. This model includes a substation, forming the interface between the transmission grid (90 kV, HV) and the distribution grid (15 kV, MV). Wind and photovoltaic production sources are connected to the medium voltage 15 kV grid, while other photovoltaic sources and V2G EVs, partially associated with buildings, are connected at the LV level with a maximum of 400 V.

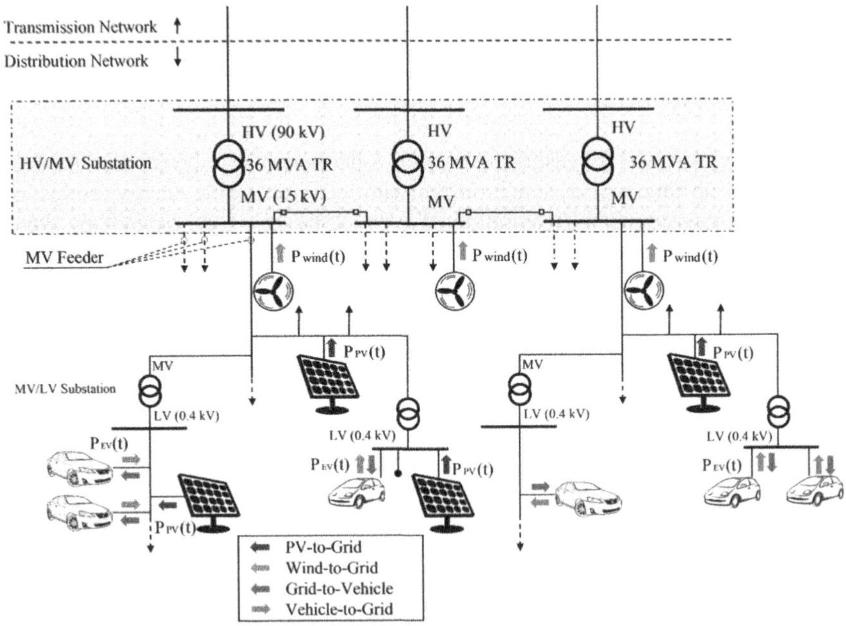

Figure 8.7. *Distribution grid featuring a 90 kV/15 kV substation, wind and PV production and V2G [SAR 16b]*

8.3.2.2. *Specification*

The first stage in designing an energy supervision strategy is to determine the specification, i.e. identify objectives, constraints and available means of action. This makes it possible to identify input and output variables for the supervisor, along with the indicators used to evaluate the degree of success in achieving each objective.

The supervision strategy operates over multiple levels:

– a provisional level, where plans are made on day N-1 for day N: consumption and renewable production forecasts are used to determine an overall estimated power setpoint for the V2G fleet;

– a real-time level, where this overall power setpoint is adapted in response to information received in real-time;

– a level determining a power setpoint for each vehicle, based on the overall real-time power setpoint for the V2G fleet.

The provisional level must respond to two main objectives, each broken down into sub-objectives and with associated constraints:

– minimize the cost of transporting electricity, remaining below the subscribed power limit for the HV/MV substation [ROB 16] and maximizing consumption at times when energy is cheapest. The constraints in this case are subscribed power and variations in consumption;

– minimize environmental impact by minimizing CO_2 emissions, while satisfying user requirements in terms of vehicle autonomy, i.e. battery charge level. The constraints in this case are the availability of renewable energy, the energy requirements of EVs and the need to limit battery aging by a controlled approach to depth of discharge.

As this level of the supervision system works provisionally, creating plans for the next day, for example, the input variables are predicted consumption profiles for the distribution grid in question, potential EV power, and production forecasts for local PV and wind sources. The means of action is a provisional power setpoint for charging or discharging an EV fleet.

Table 8.2 shows global objectives and sub-objectives, constraints, means of action, input and output variables and indicators for the real-time supervision level.

The objectives and constraints encountered at this level are the same as those for the provisional level, but the input variables differ as they are based on information obtained as close as possible to the event. Each sub-objective is associated with an input variable:

– deviation from the subscribed power, ΔP;

– electricity price, which may vary according to the time of day;

– power drawn from or sent back to the source, power balance of the selected area of the distribution grid, P_{PS};

– the state of charge of the battery, SOC.

The means of action is a power setpoint, applied in real-time to the EV fleet. The third level of the supervisor is responsible for breaking this setpoint down by individual vehicle as a function of usage constraints, determined on the basis of arrival and departure times and on the SOC of the vehicle in question.

<table>
<tr><th>Objective</th><th>Sub-objective</th><th>Associated input</th><th>Constraints</th><th>Means of action</th><th>Indicator</th></tr>
<tr><td rowspan="2">Minimize transport cost</td><td>Avoid exceeding subscribed power</td><td>ΔP</td><td>Subscribed power (technical constraint)</td><td rowspan="5">Control battery charging and discharging</td><td rowspan="2">Cost of transport in k€</td></tr>
<tr><td>Maximize consumption at cheapest time</td><td>Price</td><td>Variation in consumption</td></tr>
<tr><td rowspan="3">Environmental efficiency and reduction in CO_2 emissions</td><td>Maximize local consumption of renewable energy</td><td>P_{PS}</td><td>Availability of renewable energy</td><td>Self-consumption of renewable energy</td></tr>
<tr><td rowspan="2">Satisfy EV users</td><td rowspan="2">SOC</td><td>Energy requirement of EVs (battery fully charged before departure)</td><td>Battery charge rate, aging</td></tr>
<tr><td>DoD of the battery (battery aging)</td><td>CO_2 emissions</td></tr>
</table>

Table 8.2. *Global objectives and associated sub-objectives, means of action, input and output variables and indicators for the real-time level of the supervisor [SAR 16b]*

Indicators are identified for each objective in order to evaluate attainment and optimize the supervisor:

– electricity transport cost;

– rate of local self-consumption of renewable energy;

– battery charge rate, used to evaluate aging;

– CO_2 emissions.

8.3.2.3. *Structure of the supervisor*

Using the specification set out above, we designed a three-level supervisor structure, illustrated in Figure 8.8. P_{sc} is the subscibed power at the HV/MV substation.

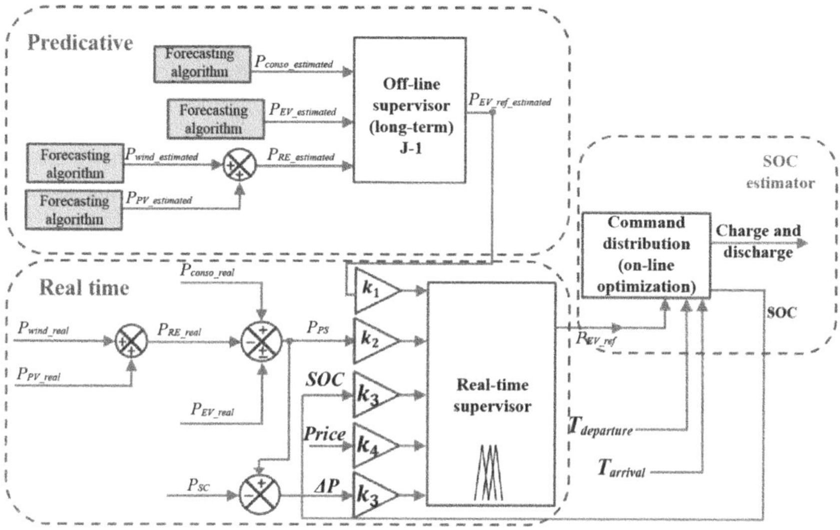

Figure 8.8. *Three-level V2G supervisor structure [SAR 16b]*

The predictive supervisor estimates an overall setpoint for the next day, meaning that the algorithm does not face significant time constraints; explicit optimization methods can be used [SAR 16b].

The real-time supervisor will be developed following the methodology set out in Chapter 1 (section 1.8). This development is described below. When this supervisor is used alongside the provisional supervisor, it receives a fifth input variable in the form of a provisional charge or discharge power setpoint for the EV fleet.

The breakdown of the global setpoint into individual setpoints for each vehicle also needs to be carried out in real-time, following a methodology described in [SAR 16b].

8.3.2.4. *Functional graphs*

The global functional graph for the real-time supervision level is shown in Figure 8.9. Different modes are used to take account of different sub-objectives:

– mode L1: satisfies user needs in terms of vehicle autonomy, i.e. battery SOC. This mode is implemented at the third level of the supervisor, which distributes charge or discharge setpoints for each EV based on SOC, the energy required by the vehicle, the power setpoint and the time remaining before the vehicle is due to depart. This mode does not make use of fuzzy logic;

– mode L2.1.1: limits violations of the subscribed power threshold for the HV/MV substation;

– mode L2.1.2: maximizes consumption during periods when electricity is cheapest, when power at the source station P_{PS} is lower than the subscribed power P_{SC};

– mode L2.2: minimizes environmental impact by minimizing CO_2 emissions, i.e. prioritizing local consumption of renewable energy, particularly when the power at the source station P_{PS} is negative: this corresponds to excess local energy production being injected back into the HV grid.

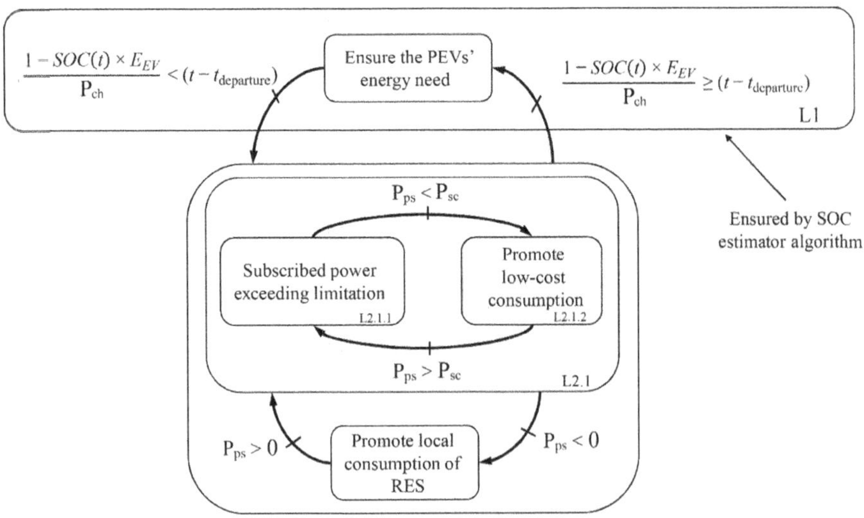

Figure 8.9. *Global functional graph of the real-time supervision level [SAR 16b]*

Detailed functional graphs for the real-time supervision system are provided in Appendix 1.

8.3.2.5. *Membership functions*

Figure 8.10 shows the membership functions for the five input variables and the single output variable for the real-time supervisor.

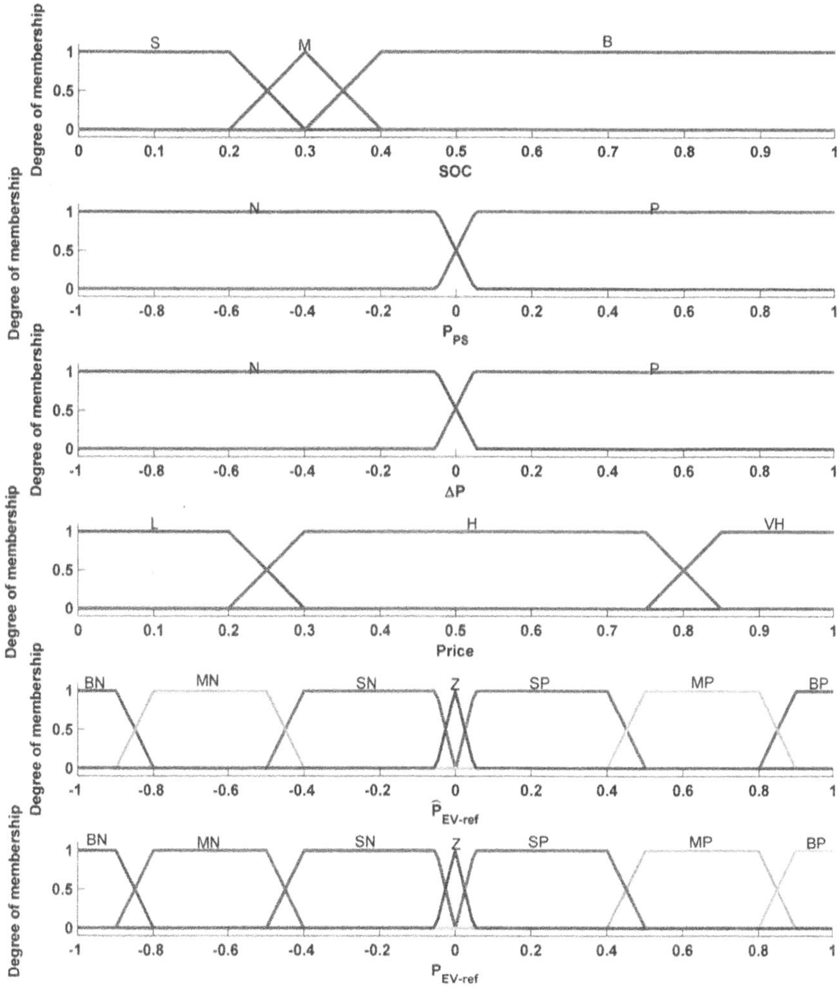

Figure 8.10. *Membership functions of the five inputs and single output for the real-time supervisor [SAR 16b]*

From top to bottom, these represent:

– the SOC, with three fuzzy subsets: Small (S), Medium (M) and Big (B);

– the power at the source station P_{PS}, with two fuzzy subsets: Negative (N) and Positive (P);

– the deviation from the subscribed power ΔP with two fuzzy subsets: Negative (N) and Positive (P);

– electricity price, with three fuzzy subsets: Low (L), High (H) and Very High (VH);

– the provisional reference power $\hat{P}_{EV\text{-}ref}$, output from the provisional supervision level and input into the real-time level, with seven fuzzy subsets: Big Negative (BN), Medium Negative (MN), Small Negative (SN), Zero (Z), Small Positive (SP), Medium Positive (MP) and Big Positive (BP);

– the output from the real-time supervision level is the final reference power $P_{EV\text{-}ref}$ with seven fuzzy subsets: Big Negative (BN), Medium Negative (MN), Small Negative (SN), Zero (Z), Small Positive (SP), Medium Positive (MP) and Big Positive (BP).

The maximum number of fuzzy laws is determined by multiplying the number of fuzzy subsets for each input variable: 3×2×2×3×7 = 252.

Taking the real-time supervision level alone without the provisional level, the maximum number of fuzzy laws is 3×2×2×3 = 36. The two cases are compared below.

8.3.2.6. *Operational graph and fuzzy rules*

The operational graph is obtained from the functional graphs, taking account of the number of fuzzy subsets of each input and output variable. Figure 8.13 shows the operational graph for the real-time supervisor alone, not taking account of the provisional setpoint generated by the provisional level.

Based on this operational graph, the fuzzy laws can be written in classic linguistic form. Table 8.3 shows the fuzzy laws for the real-time supervisor alone, not taking account of the provisional setpoint generated by the previous level, corresponding to the graph in Figure 8.11. We see that a total of 15 fuzzy laws are needed to implement mode L2, i.e. around half of the maximum number of possible fuzzy laws, which was 36 (mode L1 is not treated using fuzzy logic).

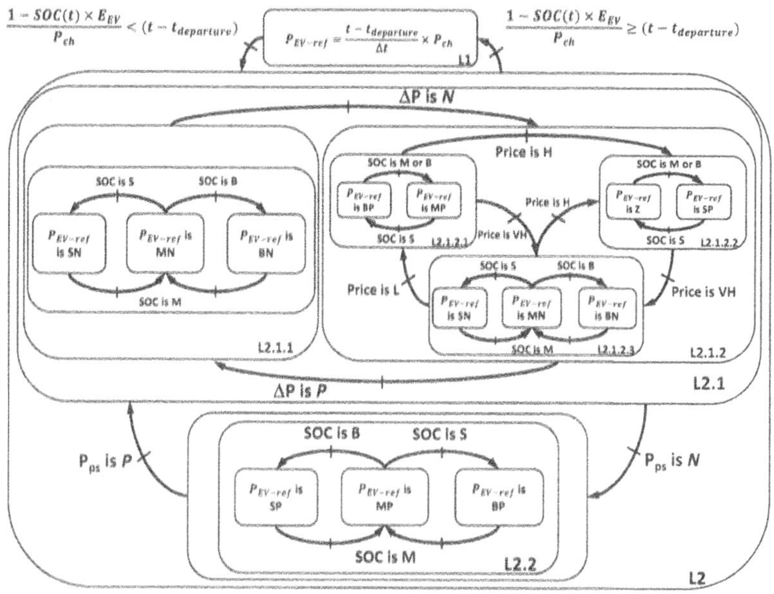

Figure 8.11. *Operational graph for the real-time supervisor alone, without the provisional setpoint generated by the provisional level [SAR 16b]*

Table 8.4 shows the fuzzy laws for the real-time supervisor including the provisional setpoint generated by the provisional level. We see that a total of 29 fuzzy laws are needed to implement mode L2, considerably fewer than the highest possible number, 252. This reduction in the number of fuzzy laws highlights the interest of our supervisor design approach.

L1	Charge PEVs as needed based on their SOC			
L2	L2.1	L2.1.1		If P_{PS} is P and ΔP is P and SOC is B then P_{EV-ref} is BN
				If P_{PS} is P and ΔP is P and SOC is M then P_{EV-ref} is MN
				If P_{PS} is P and ΔP is P and SOC is S then P_{EV-ref} is SN
		L2.1.2	L2.1.2.1	If P_{PS} is P and ΔP is N and Price is L and SOC is S then P_{EV-ref} is BP
				If P_{PS} is P and ΔP is N and Price is L and SOC is M then P_{EV-ref} is MP
				If P_{PS} is P and ΔP is N and Price is L and SOC is B then P_{EV-ref} is MP
			L2.1.2.2	If P_{PS} is P and ΔP is N and Price is H and SOC is S then P_{EV-ref} is Z
				If P_{PS} is P and ΔP is N and Price is H and SOC is M then P_{EV-ref} is SP
				If P_{PS} is P and ΔP is N and Price is H and SOC is B then P_{EV-ref} is SP
			L2.1.2.3	If P_{PS} is P and ΔP is N and Price is VH and SOC is S then P_{EV-ref} is SN
				If P_{PS} is P and ΔP is N and Price is VH and SOC is M then P_{EV-ref} is MN
				If P_{PS} is P and ΔP is N and Price is VH and SOC is B then P_{EV-ref} is BN
	L2.2			If P_{PS} is N and SOC is S then P_{EV-ref} is BP
				If P_{PS} is N and SOC is M then P_{EV-ref} is MP
				If P_{PS} is N and SOC is B then P_{EV-ref} is SP

Table 8.3. *Fuzzy laws for the real-time supervisor alone, without the provisional setpoint generated by the provisional level [SAR 16b]*

L1	Charge PEVs as needed based on their SOC				
L2	L2.1	L2.1.1	If P_{PS} is P and ΔP is P and SOC is B and $\hat{P}_{EV-ref}$ is SN then P_{EV-ref} is SN		
			If P_{PS} is P and ΔP is P and SOC is B and $\hat{P}_{EV-ref}$ is MN then P_{EV-ref} is MN		
			If P_{PS} is P and ΔP is P and SOC is B and $\hat{P}_{EV-ref}$ is BN then P_{EV-ref} is BN		
			If P_{PS} is P and ΔP is P and SOC is M and $\hat{P}_{EV-ref}$ is Z then P_{EV-ref} is Z		
			If P_{PS} is P and ΔP is P and SOC is S and $\hat{P}_{EV-ref}$ is Z then P_{EV-ref} is Z		
		L2.1.2	L2.1.2.1	If P_{PS} is P and ΔP is N and Price is L and $\hat{P}_{EV-ref}$ is SP then P_{EV-ref} is SP	
				If P_{PS} is P and ΔP is N and Price is L and $\hat{P}_{EV-ref}$ is MP then P_{EV-ref} is MP	
				If P_{PS} is P and ΔP is N and Price is L and $\hat{P}_{EV-ref}$ is BP then P_{EV-ref} is BP	
			L2.1.2.2	L2.1.2.2.1	If P_{PS} is P and ΔP is N and Price is H and SOC is S and $\hat{P}_{EV-ref}$ is SP then P_{EV-ref} is SP
					If P_{PS} is P and ΔP is N and Price is H and SOC is S and $\hat{P}_{EV-ref}$ is MP then P_{EV-ref} is MP
					If P_{PS} is P and ΔP is N and Price is H and SOC is S and $\hat{P}_{EV-ref}$ is BP then P_{EV-ref} is BP
				L2.1.2.2.2	If P_{PS} is P and ΔP is N and Price is H and SOC is M and $\hat{P}_{EV-ref}$ is SN then P_{EV-ref} is SN
					If P_{PS} is P and ΔP is N and Price is H and SOC is M and $\hat{P}_{EV-ref}$ is MN then P_{EV-ref} is MN
					If P_{PS} is P and ΔP is N and Price is H and SOC is M and $\hat{P}_{EV-ref}$ is SP then P_{EV-ref} is SP
					If P_{PS} is P and ΔP is N and Price is H and SOC is M and $\hat{P}_{EV-ref}$ is MP then P_{EV-ref} is MP
					If P_{PS} is P and ΔP is N and Price is H and SOC is M and $\hat{P}_{EV-ref}$ is BP then P_{EV-ref} is BP
				L2.1.2.2.3	If P_{PS} is P and ΔP is N and Price is H and SOC is B and $\hat{P}_{EV-ref}$ is BN then P_{EV-ref} is BN
					If P_{PS} is P and ΔP is N and Price is H and SOC is B and $\hat{P}_{EV-ref}$ is SP then P_{EV-ref} is SP
					If P_{PS} is P and ΔP is N and Price is H and SOC is B and $\hat{P}_{EV-ref}$ is MP then P_{EV-ref} is MP
					If P_{PS} is P and ΔP is N and Price is H and SOC is B and $\hat{P}_{EV-ref}$ is BP then P_{EV-ref} is BP
			L2.1.2.3	L2.1.2.3.1	If P_{PS} is P and ΔP is N and Price is VH and SOC is M and $\hat{P}_{EV-ref}$ is SN then P_{EV-ref} is SN
					If P_{PS} is P and ΔP is N and Price is VH and SOC is M and $\hat{P}_{EV-ref}$ is MN then P_{EV-ref} is MN
					If P_{PS} is P and ΔP is N and Price is VH and SOC is M and $\hat{P}_{EV-ref}$ is BN then P_{EV-ref} is BN
				L2.1.2.3.2	If P_{PS} is P and ΔP is N and Price is VH and SOC is B and $\hat{P}_{EV-ref}$ is SN then P_{EV-ref} is SN
					If P_{PS} is P and ΔP is N and Price is VH and SOC is B and $\hat{P}_{EV-ref}$ is MN then P_{EV-ref} is MN
					If P_{PS} is P and ΔP is N and Price is VH and SOC is B and $\hat{P}_{EV-ref}$ is BN then P_{EV-ref} is BN
	L2.2	If P_{PS} is N and $\hat{P}_{EV-ref}$ is SP then P_{EV-ref} is SP			
		If P_{PS} is N and $\hat{P}_{EV-ref}$ is MP then P_{EV-ref} is MP			
		If P_{PS} is N and $\hat{P}_{EV-ref}$ is BP then P_{EV-ref} is BP			

Table 8.4. *Fuzzy laws for the real-time supervisor, taking account of the provisional setpoint generated by the provisional level [SAR 16b]*

8.3.2.7. *Supervisor optimization*

Certain supervisor parameters – gains K1 through K4 in Figure 8.8 and the form of the membership functions in Figure 8.10 – are selected using expertise concerning the system in question. This means that choices are not necessarily optimized in relation to objectives. These parameters may be optimized using explicit optimization methods [ROB 16]. For example, the form of membership functions can be determined using different parameters, as shown in Figure 8.12: x_i for a trapezoid and y_i for a triangle. If we consider our membership functions to be symmetrical, then the number of parameters to optimize is significantly reduced.

In [SAR 16b], the membership functions for the supervisor developed in this section were optimized using a genetic algorithm. Figure 8.13 shows optimized membership functions for the real-time supervisor (broken lines) alongside our empirical membership functions from Figure 8.10 (solid lines).

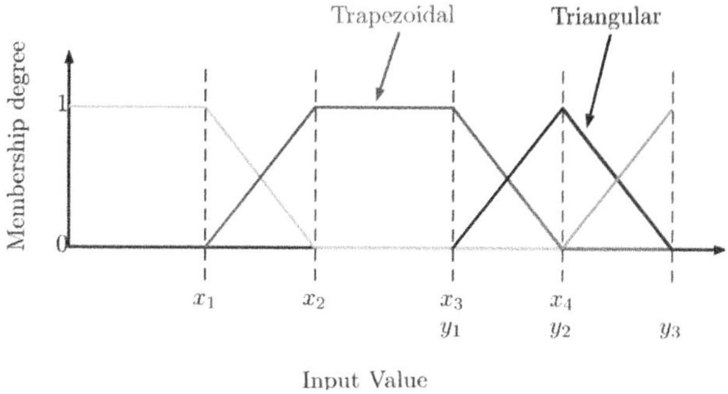

Figure 8.12. *Examples of membership functions with fuzzy subsets in trapezoidal and triangular forms [SAR 16b]*

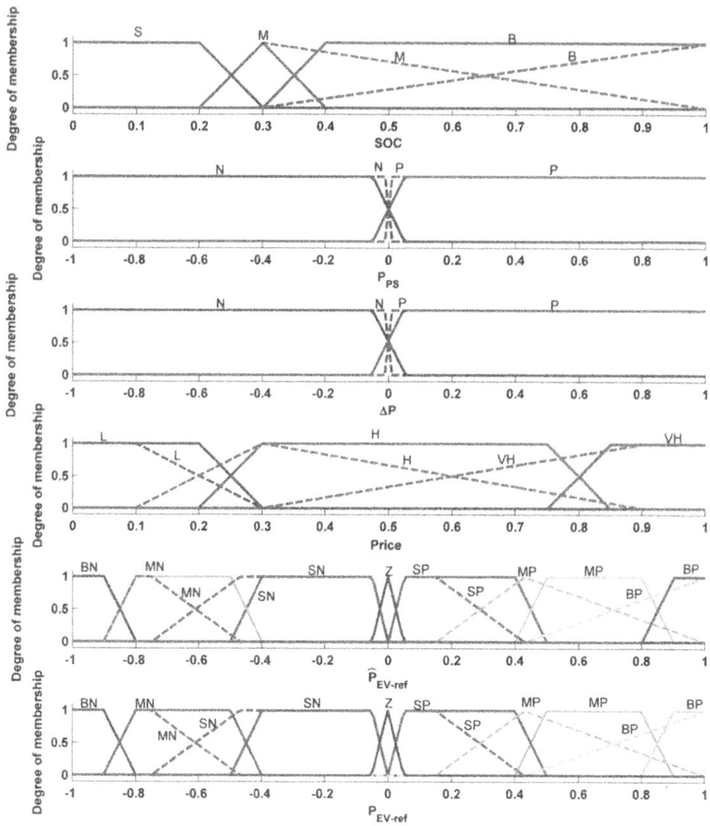

Figure 8.13. *Optimized membership functions for the real-time supervisor, taking account of the provisional setpoint generated by the provisional level [SAR 16b]*

8.3.2.8. *Supervisor performance*

The performance of the supervision system is analyzed by comparing different cases with and without supervision, taking a small area of a distribution grid and a predicted scenario for the year 2030, with 2700 EVs in the area and normal charging at 3.7 kW:

– initial situation without EVs;

– situation with EVs without supervision;

– situation with EVs and predictive supervision alone;

– situation with EVs and empirical real-time supervision alone;

– situation with EVs and optimized real-time supervision alone;

– situation with EVs with combined predictive and optimized real-time supervision.

Table 8.5 shows the results calculated for different indicators: electricity transport cost in the area of the distribution grid considered, optimal subscribed power P_{SC} corresponding to maximum draw on the HV grid, local renewable energy injected into the HV grid (to be minimized), proportion of renewable energy in EV charging in % and in MWh, CO_2 emission minimization rate obtained using the supervisor, compared to the non-supervised EV results and the quantity of CO_2 generated.

The integration of EVs without load supervision results in a 9% increase in electricity transport costs and an 11% increase in maximum power draw, requiring an increase in subscribed power. Supervision creates significant savings in terms of these costs and in the associated CO_2 emissions by coordinating EV charging with local renewable energy production. Predictive supervision provides the best results if we assume that the reality corresponds to predictions: for example, the increase in transport costs is limited to 0.02%. In reality, however, renewable energy production and consumption will inevitably deviate from the forecasts. This explains the need for a real-time supervision level, which adjusts provisional setpoints as a function of real-time measured data. The empirical real-time supervisor, used alone, limits added transport costs to 4.3%; using the optimized version, this figure falls to 3.8% (representing a 4.9% reduction compared to the unsupervised situation). Combining the two levels of supervision, this figure falls still further, to 1.9% (6.3% lower than the unsupervised figures). Reductions in CO_2 emissions follow the same downward trend. Deviations from forecasted renewable energy production may also be taken into account using a stochastic modeling approach [LEG 16].

	2030 scenario – 2700 electric vehicles – normal charge 3.7 kW Maximum available power around 10 MW						
	Supply cost (k€/year)	**Optimal Psc (kW)**	**RE injected into HV grid (MWh)**	**EV coordination (%)**	**EV coordination MWh**	**Minimization of CO_2 emissions (tons)**	**Quantity of CO_2 generated (tons)**
P_PS without EV	2591.5	54980	4136	–	–	–	777
P_PS + EV without supervision	2829	61073	3990	3.5	146	–	851
P_PS + EV forecasts (off-line optimization)	2592	54956	3572	13.6	564	21 (2.4%)	830
P_PS + EV Empirical supervision – real-time	2703	55982	3894	5.8	242	5 (0.5%)	846
P_PS + EV optimized supervision – real-time	2690 (4.9%)	55666	3818	7.6	318	15 (1.7%)	836
P_PS + EV real-time and predictive supervision	2636 (6.8%)	55286	3693	10.7	443	16 (1.9%)	835

Table 8.5. *Performance indicators for different versions of the supervisor [SAR 16b]*

8.3.2.9. *Impact on battery aging*

A battery degradation model, based on electrochemical behaviors and the rainflow model, was developed in [SAR 16b] with the aim of quantifying the impact of V2G on vehicle battery aging. The rainflow algorithm, presented in Chapter 4 (section 4.4.3), allows us to count the number of charge/discharge cycles for a battery based on the SOC profile.

Using the algorithm presented in [SAR 16b], the number of cycles and half-cycles over a given time period is counted using rainflow. A mathematical degradation model is then used to calculate a value representing the reduction in battery life expectancy [XU 13, MIL 10, ZHU 84]. Battery life expectancy is then calculated as a function of the characteristics provided in manufacturer documentation.

To evaluate the impact of V2G charging on battery degradation, we established a number of scenarios comparing situations with and without V2G, with normal charging at home and with mixed workplace/home charging. The results in Table 8.6 show a non-negligible impact on battery life in cases where V2G is used to minimize electricity transport costs. A vehicle used to provide this service over the course of a year will lose 18 months of battery life expectancy compared to a vehicle not used for V2G.

New economic models for EV battery use must be defined in order to take account of the impact of different V2G services on battery aging. The savings made and payment for services should at least compensate for the cost of more intensive battery use.

Scenarios	Nb. cycles	Mean of amplitude (DOD %)	Degradation per year	Lifespan
V2G charging	841	20.31 %	8.85 %	9 years
Normal charging at home/office	732	16.8 %	8.04 %	10 years
Normal charging at home	366	33.6 %	7.66 %	10.5 years

Table 8.6. *Impact of V2G on battery life expectancy when providing transport cost optimization services over the course of 1 year, DOD = depth of discharge [SAR 16b]*

8.4. Vehicle to Station: V2S

In this section, we will discuss several potential contributions of an EV charging management system in a station parking lot. The first of these is the coordination of EV charging in order to optimize subscribed power. The second is to minimize electricity bills in two different situations: regulated and spot-price markets [SAR 14, SAR 16a, SAR 16b].

8.4.1. *Impact and contribution of EVs in a railway station carpark*

Railway stations pay energy suppliers for their local consumption. The presence of EV charging terminals (Figure 8.14) will increase the energy consumed by the station. A considered approach to EV charging and discharging can be used to optimize bills.

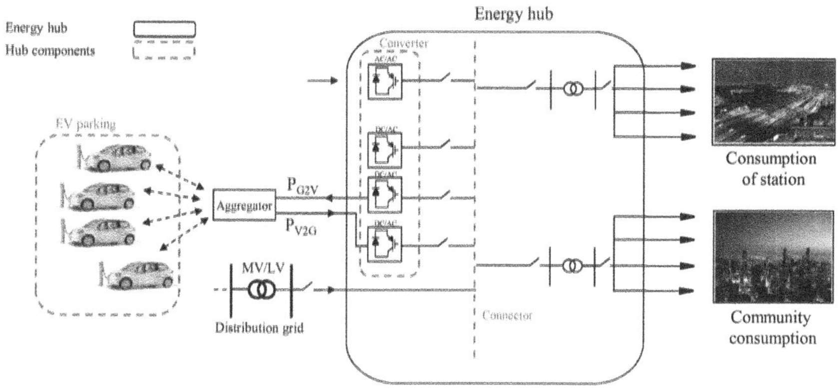

Figure 8.14. *Integration of electric vehicles into an energy hub centered on a railway station*

8.4.1.1. *Modeling and profile scenarios for EVs in a train station parking lot*

Railway stations in France have been the focus of smart grid developments, examining the potential for energy interactions between the electricity grid and its users. They are connected to the distribution grid via an MV/LV substation with a ratio of 20 kV or 15 kV/0.4 kV. These stations may act as energy hubs, with different components such as local renewable energy sources and electric vehicle charging points (Figure 8.6). Several different scenarios are considered in order to analyze the impact of EVs parked in the station parking lot on a daily basis, using vehicle arrival and departure times. Vehicle arrival times follow a normal distribution over the period from 7 am to 9 am, while departure times follow a normal distribution from 4 pm to 8 pm (Figure 8.15). The average duration of availability is approximately 10 hours, and during this time the EV may be used for charge coordination purposes. The capacity of the station is also taken into account, with 1–50 charging points. These terminals themselves are grouped into different types, priced differently:

– normal charge (NC): this is the main charging option for the majority of EVs, with a charge power from 3 to 3.7 kW. For an EV with a capacity of 20 kWh, 6–8 hours are required for a full charge. In this scenario, all vehicles arriving in the parking lot choose NC charging mode;

– accelerated charge (AC): this is carried out using a tri-phase connection with a power of 23 kW. An EV with a battery capacity of 20 kWh can be fully charged in an hour. In this scenario, all EVs arriving in the parking lot choose AC mode;

– rapid charge (RC): this mode is used in highway service stations, where batteries need to be fully charged in a very short period of time. With a charge rate of 43 kW, a 20 kWh battery can be charged within 30 min. In this scenario, we consider that all EVs in the parking lot have selected this option;

– mixed charging (MC): in this scenario, 40% of the vehicles in the station parking lot use RC mode, 20% use AC and 40% use NC.

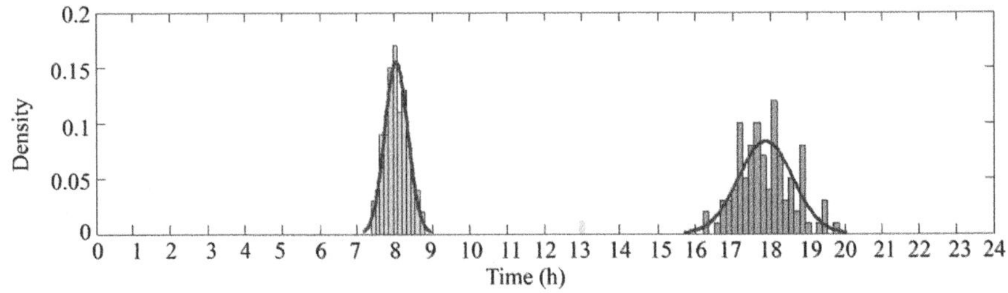

Figure 8.15. *Histogram showing arrival and departure times of EVs for station parking lot charging scenarios [SAR 16b]*

Figure 8.16 shows the charging profile for 20 electric vehicles arriving at a station parking lot in the morning without charge coordination, for a mixed charging scenario.

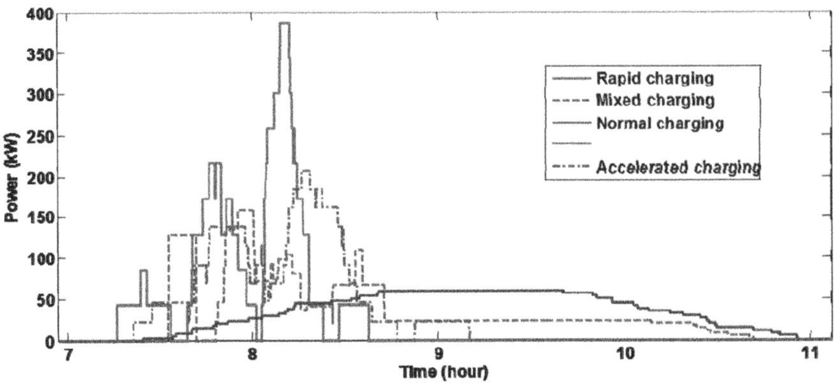

Figure 8.16. *Charging profile for 20 electrical vehicles arriving at a station parking lot in the morning without charge coordination [SAR 16b]*

8.4.1.2. *Optimization of subscribed power (SP)*

Power subscriptions are established on an annual basis and must be chosen with care. The choice of a high subscribed power means a higher subscription cost, but if a lower power is selected, actual consumption will exceed this limit more often, leading to hefty financial penalties. In choosing an appropriate subscribed power, information about the quantity of energy consumed and the mode of consumption must be taken into account.

Figure 8.17 shows a comparison of optimal subscribed power evolution as a function of vehicle numbers, for both uncoordinated and coordinated EV charging scenarios, obtained through explicit optimization. Cases including 10, 20 and 30 EVs are shown. Due to the consumption peak in the morning, those EVs which arrive earliest are called upon to provide V2G service to reduce the station's peak consumption.

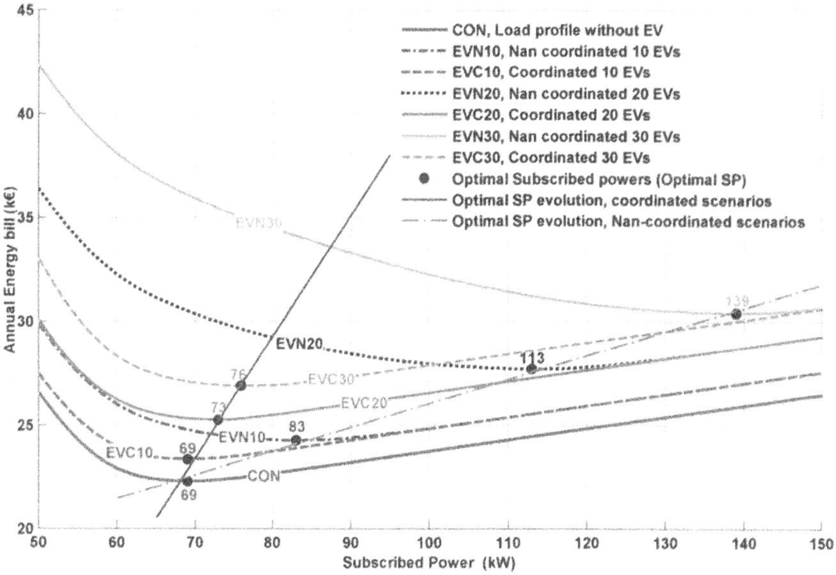

Figure 8.17. *Comparison of optimal subscribed power (SP) evolution as a function of vehicle number for non-coordinated and coordinated scenarios [SAR 16b]*

The SP for the station without EVs (CON) is 69 kW. Adding non-coordinated charging for 10 EVs (EVN10), the SP would need to increase to 83 kW; with coordination (EVC10), an SP of 69 kW remains sufficient. In the scenarios featuring 20 and 30 EVs with non-coordinated charging (EVN20 and EVN30), the SPs would need to increase substantially to 113 kW and 139 kW respectively. With charge coordination (EVC20 and EVC30), the new subscribed power is considerably lower: 73 and 76 kW respectively.

8.4.2. *V2S: contribution of V2G technology in a station parking lot*

8.4.2.1. *V2G for railway stations*

The supervisor presented in section 8.3.2 was applied to the case of V2S. One service which can be provided by V2G is energy bill minimization for the railway station. We considered the contribution made by this service in two different pricing modes: regulatory sales tariff (RST) and spot market pricing (SM). Bills were reduced in both cases, and the pricing mode had a visible effect on bill components. In RST, consumption over the subscribed power limit is the element with the greatest impact, while in the spot pricing market, this does not happen as there is no excess component.

Based on the scenarios laid out in the previous section, we considered a case in which EVs arrive at the station around 8 am and leave around 6 pm. During the intervening period, EV batteries are used for load balancing in the station. The objective of the energy management system in this case is to reduce annual electricity bills.

Our study covers two energy billing contract scenarios:

– regulatory sales tariff (RST);

– spot market pricing (SM).

The first type of contract was available to French industrial consumers (as the *tarif vert* up until January 1, 2016 [CRE 16], when it was replaced by the second type). The advantages of participating in each type of market using V2G technology will be analyzed below in the case of a station with 30 EVs and a subscribed power of 270 kW.

For each energy billing contract scenario, a supervision strategy including both provisional and real-time levels will be considered. The results in terms of energy bill minimization will be presented for each scenario.

8.4.2.2. *Supervision with regulated tariffs*

Within the RST market, different types of contract are available to consumers as a function of their consumption levels. The subscribed power will vary depending on the size of the station, affecting access to different market types. In France, stations with a consumption in excess of 250 kVA are able to access a regulated tariff, the *tarif vert*.

Bills of this type include three components: consumption in excess of the SP, consumed energy and a subscription component. The SP is established once a year, and the subscription component cannot be optimized. Energy bills can be reduced by minimizing consumption in excess of the SP and the cost of energy consumed.

Real-time control of load consumption using the storage capacity offered by EVs takes account of the following variables:

– SOC (state of charge of the set of EVs);

– ΔP (difference between consumption and SP);

– electricity price;

– $\hat{P}_{EV\text{-}ref}$: the power reference of the predictive input.

The supervision system produces a single output variable, the real-time power reference. It is designed using fuzzy logic with 32 rules, three membership functions for the SOC, four for ΔP, three for price and seven for $\hat{P}_{EV\text{-}ref}$. The predictive element is implemented by a binary linear programming algorithm, and the SP is also optimized.

The performance indicators used are shown in Figure 8.18, which include annual bills, the annual cost of excess consumption above the SP (SPE), annual energy consumption and the peak-to-average power ratio (PAR). Three scenarios are considered: without EVs (CON), with EVs and supervision (EVS), and with EVs but without supervision (EVN).

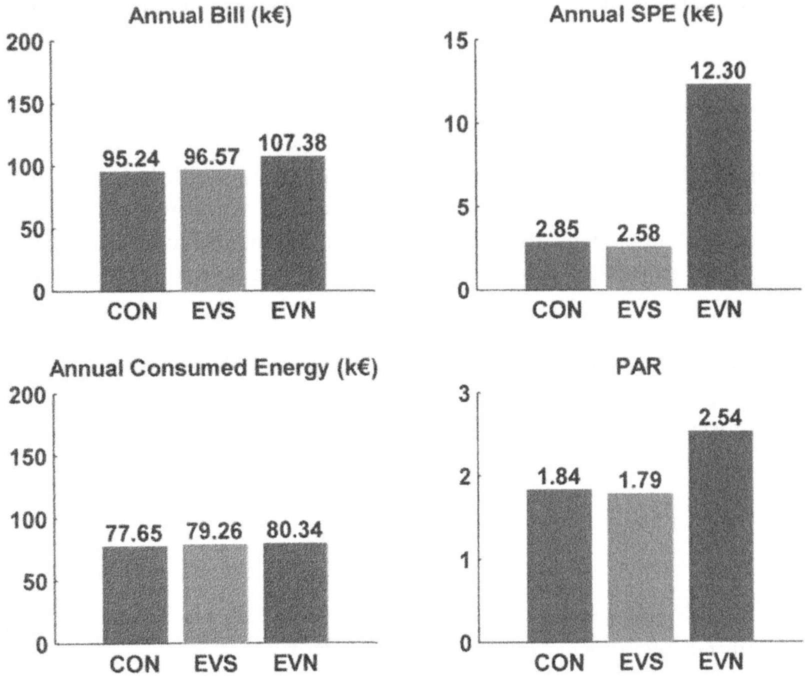

Figure 8.18. *Value of indicators in the regulatory tariff case for three different scenarios: without EVs (CON), with EVs and supervised charging (EVS) and with unsupervised EVs (EVN) [SAR 16b]*

EV charge supervision enables a reduction of up to 10% in annual bills (€10,810) in comparison with the unsupervised approach. Most of this saving is due to an 80% reduction in violations of the SP threshold (saving €9,720). Energy consumption is similar in the EVS and EVN scenarios.

The fourth indicator, the peak-to-average ratio (PAR), highlights the interest of V2G technology in smoothing the station's load profile. This ratio is reduced by 30% by EV charge supervision, even falling below the levels observed for the system without EVs.

8.4.2.3. *Supervision in the context of spot market pricing*

In this case, the station power supply comes from the electricity spot market. Our scenario is based on French spot data. In this case, customers are free to choose their own energy supplier and pay for their consumption as a function of market price.

There are only three input variables for the supervision in this case, as the SP limit violation component is no longer present:

– SOC (state of charge of the set of EVs);

– market price of electricity;

– $\hat{P}_{EV\text{-}ref}$ (reference power of the predictive input).

Figure 8.19 shows the value of two indicators, the annual energy bill and power smoothing (PAR), for the same three scenarios as before: without EVs (CON), with supervised EV charging (EVS) and with unsupervised EV charging (EVN). The objective of the supervision system is to charge EVs when the spot price of electricity is low, and, inversely, to inject power from the EVs into the grid when this price is high. Figure 8.19 highlights the impact of supervision on the annual bill, which is reduced (EV supervision results in a 1.1% gain in comparison with the unsupervised case); however, the impact on power smoothing is more notable (7.2%).

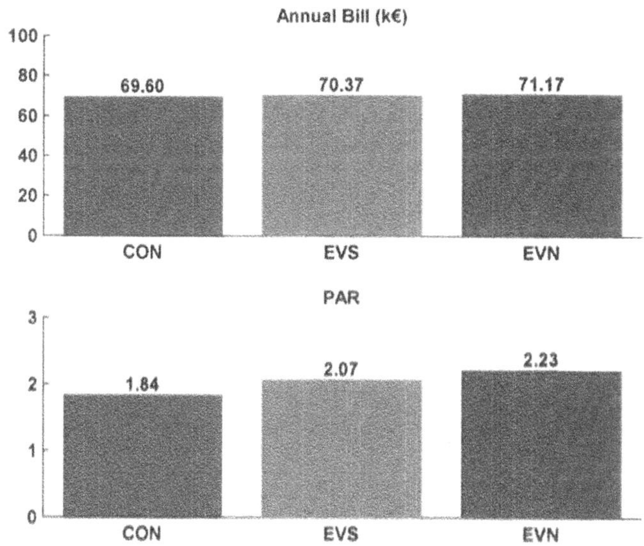

Figure 8.19. *Values of indicators in the spot market pricing case for three different scenarios: without EVs (CON), with EVs and supervised charging (EVS) and with unsupervised EVs (EVN) [SAR 16b]*

Figure 8.20 shows the evolution of the EV charging profile over 48 hours for the three scenarios: without EVs (CON), with supervised EV charging (EVS) and with unsupervised charging (EVN). The evolution of electricity prices in the spot market over the same period is also shown. The V2G case is shown in green, and the G2V case in yellow. Early in the morning, when the price of electricity is high, EVs supply energy; in the early afternoon, when the price is lower, the EVs are recharged. The morning peak is thus smoothed in comparison with the non-coordinated profile. Note that, in the case of a station parking lot, we must consider the price of electricity during the day. Prices are lowest during the night, around 4 am; at this time, the EVs are generally at home and may be recharged at a lower cost.

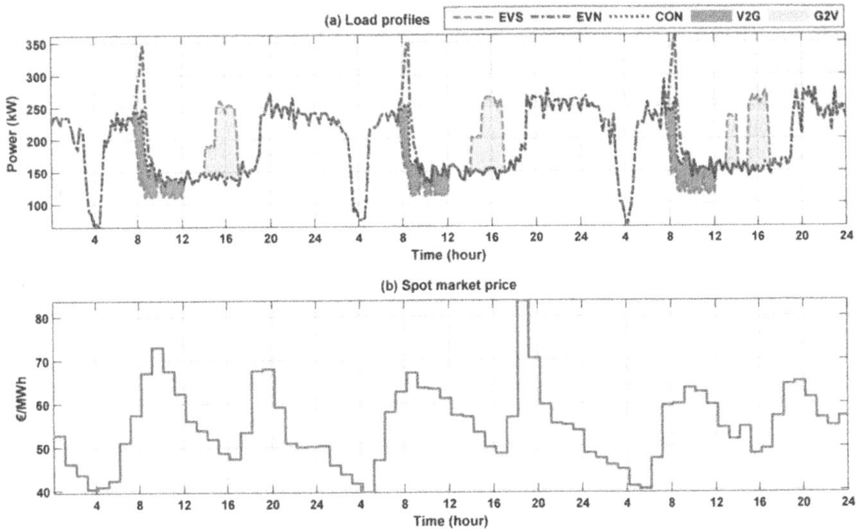

Figure 8.20. *Evolution of EV charging profiles for three scenarios: without EVs (CON), with supervised EV charging (EVS) and with non-supervised charging (EVN). Bottom: evolution of prices in the electricity spot market over 48 hours [SAR 16b]*

8.5. V2H

The Nissan Vehicle-to-Home (V2H) system consists of using electricity stored in the battery of an electric vehicle to power a house or small business when the electrical grid is unable to do so (Figure 8.21). It was developed in the wake of an earthquake and tsunami which paralyzed the electrical grid and led to a halt in nuclear production in Japan in 2011.

Figure 8.21. *Nissan Vehicle-to-Home (V2H) system [http://gas2.org/2014/11/03/nissan-tests-vehicle-home-system/ (2014)]*

Toyota is also working on a V2H system, including communication links between the vehicle, the domestic charging station and the home, which make it possible to control energy flow exchanges (Figure 8.22). With a fully charged battery and a full fuel tank, the rechargeable hybrid Prius can provide the equivalent of four days of consumption for an average Japanese home (although CO_2 emissions will be generated by the thermal motor). The manufacturer plans to synchronize their domestic smart grids with regional electricity distribution management grids, creating a safeguard against mass power cuts – this appears to have been developed in response to the Fukushima disaster.

The concept has also been taken up by European and American manufacturers, and prototypes are beginning to emerge. That said, a typical US household consumes three times more electricity than an average Japanese home.

Another possibility to consider is that a home might be supplied by the grid while retaining the option to draw power from an electric vehicle. This concept is known as V2HG (Vehicle to Home and Grid). In this case, a vehicle powers a residential building which is connected to an external grid, meaning that energy may travel in either direction. The development of renewable production, particularly rooftop PV generation, means that increasing numbers of homes may have an additional power source.

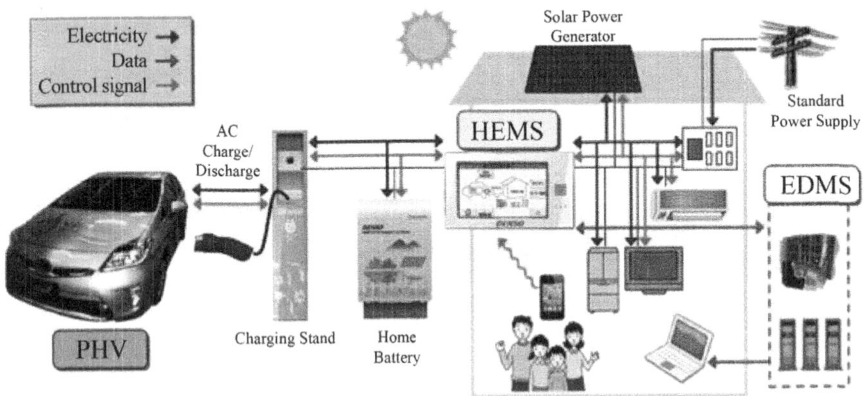

Figure 8.22. *Toyota Vehicle-to-Home (V2H) system [http://www.avem.fr/actualite-toyota-va-tester-son-systeme-d-interface-energetique-vehicle-to-home-au-japon-sur-la-prius-phv-3227.html (2012)]*

Figure 8.23 shows a PV installation connected directly to the grid, in parallel with the electrical network in the home. This system was used in the majority of cases in France up until 2016. In this situation, the EV charging system is non-reversible.

The development of self-consumption in France has led to an increase in direct connections between PV generators and the home electricity network. This configuration is shown in Figure 8.24, which also makes use of the V2HG to optimize autonomous operation of the home; it also serves to increase the quality of energy consumed from the grid, which might otherwise be disrupted by the variability of PV production [ROB 12c]. In cases where the home does not have the capacity to inject electricity back into the grid, this configuration needs to be managed with care and certain safety devices should be installed to protect the home network. This concept is being studied by the Renault automobile company [VEN 16, VEN 17].

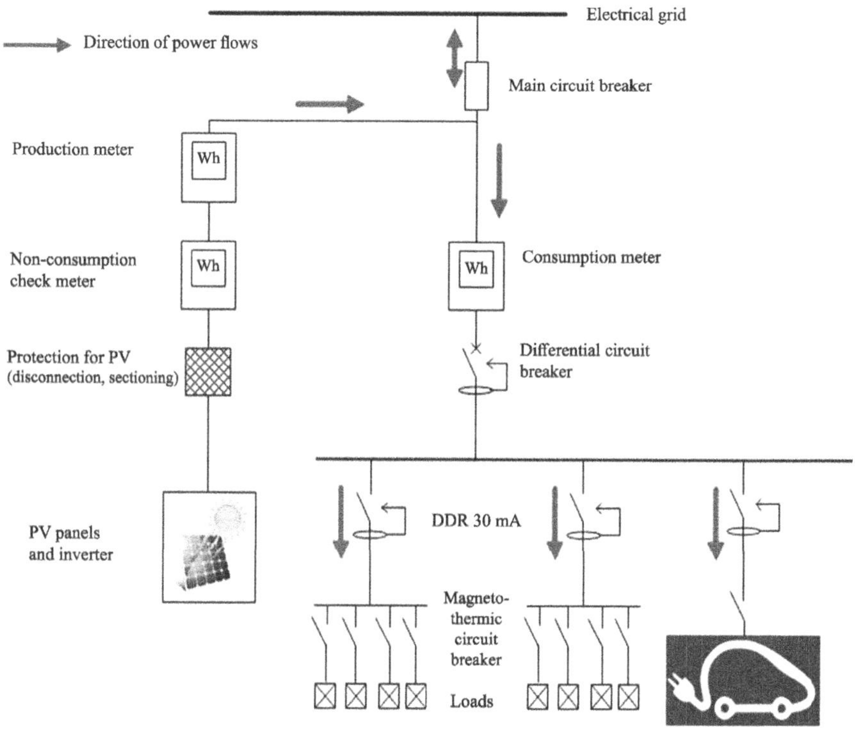

Figure 8.23. *Direct injection of photovoltaic production into the distribution grid*

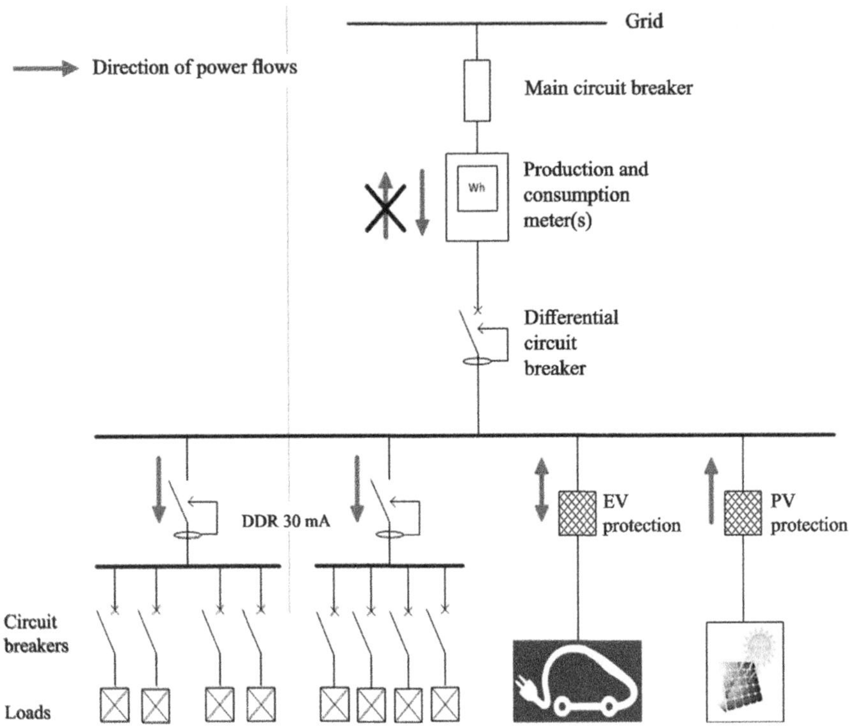

Figure 8.24. *Direct connection of a photovoltaic system to a domestic internal distribution network*

8.6. Conclusion

In a relatively efficient battery system, charge yield is approximately 80%. If energy is injected back into the grid from the battery, it must go through a DC/AC transformer, with a yield of approximately 90%. Hence, approximately 72% of the original energy is returned to the system. In addition to economic considerations, the effects on battery aging need to be taken into account. The potential increase in CO_2 emissions, if the primary energy source is based on fossil fuels (as in the case of rechargeable hybrid vehicles), is even more significant. This level of energy efficiency is to comparable to that provided by large-scale hydroelectric pumped storage systems, with values of approximately 70–80%, but on a smaller scale. Hydroelectric pumped storage is subject to significant geographical constraints; it may thus be more practical to aggregate production from a large number of small installations using RHV and EV batteries, connected to and distributed across the grid. 1000 vehicles providing 1 kW each provide a total of 1 MW of storage power.

However, certain challenges still need to be overcome to enable the widespread adoption of V2G:

– development of a large-scale, reliable communicating infrastructure for real-time management applications;

– vehicle adaptation (reversible chargers) and improved mastery of battery aging;

– increasing the economic value of services and storage;

– implementation of suitable market mechanisms;

– installation of service and actor aggregators;

– social acceptability.

8.7. Acknowledgments

The results presented in this chapter come from studies carried out with support from the ADEME (*Agence française pour le développement de l'environnement et de la maîtrise de l'énergie*: French Agency for Environmental Development and Energy Mastery) and the Hauts de France regional authority, to whom the authors extend their warmest thanks.

8.8. Appendix

8.8.1. *Detailed functional graphs for the V2G application*

Figure 8.25 provides a detailed view of the functional graphs for the real-time supervision layer developed in section 8.3.2.

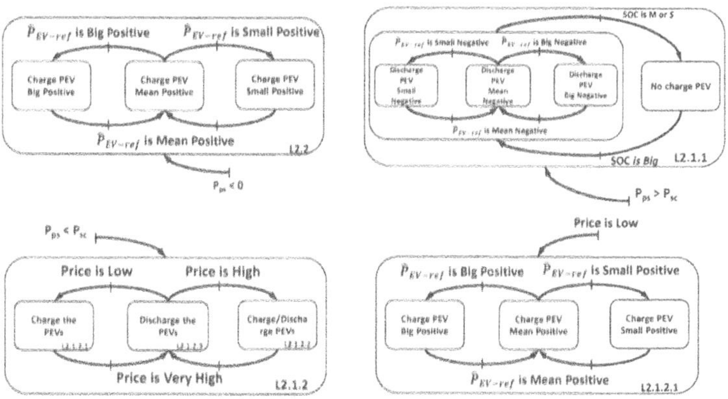

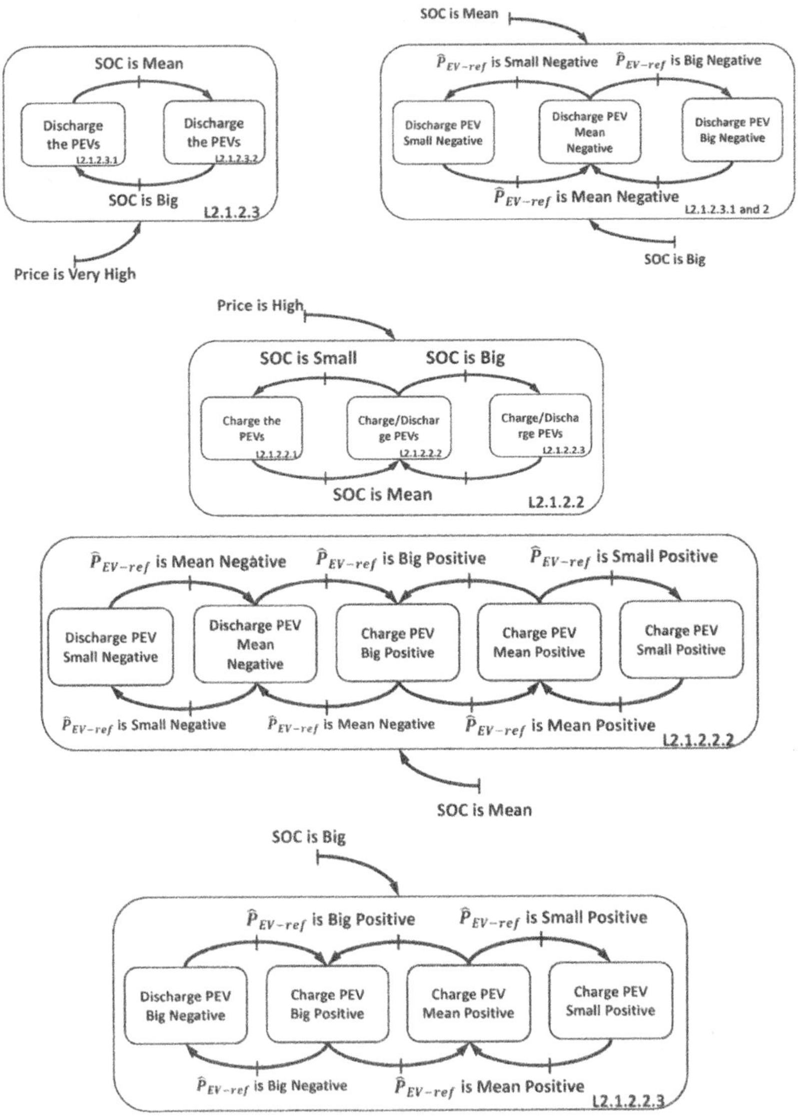

Figure 8.25. *Extract from the functional graphs for the real-time supervision layer [SAR 16b]*

References

[ABB 15] Abbes D., Bensmaine F., Labrunie A. *et al.*, "Energy management and batteries lifespan estimation in a photovoltaïc system with hybrid storage a comparative study", *Power Electronics Conference and 1st Southern Power Electronics Conference (COBEP/SPEC)*, 2015 IEEE 13th Brazilian, pp. 1–6, 2015.

[ACC 10] Acceptability data: Understanding Consumer Preferences in Energy Efficiency, Accenture end–consumer observatory on electricity management, pp. 28–31. Available at: www.accenture.com, 2010.

[ADE 16] ADEME, "Mix électrique 100% renouvelable? Analyses et optimisations", Ref. 8891–8892, June 2016.

[ADE 17] ADEME, L'effacement de consommation électrique en France, Evaluation du potentiel d'effacement par modulation de process dans l'industrie et le tertiaire en France métropolitaine. Available at: www.ademe.fr, September 2017.

[AFE 18] AFE, Association Française de l'Eclairage. Available at: http://www.afe-eclairage.fr/afe/l-eclairage-en-chiffres-26.html, August 2018.

[ALL 10] Allègre A.L., Bouscayrol A., Delarue P. *et al.*, "Energy storage system with supercapacitor for an innovative subway", *IEEE Transaction on Industrial Electronics*, vol. 57, no. 12, pp. 4001–4012, December 2010.

[ARR 15] Decree of January 11 2015 establishing the value of the bonus paid to balancing actors. Available at: http://www.legifrance.gouv.fr/affichTexte.do?cidTexte=JORFTEXT000030132998&dateTexte=&categorieLien=id, 2015.

[ATK 76] Atkinson A.B., Stiglitz J.E., "The design of tax structure: Direct versus indirect taxation", *Journal of Public Economics*, vol. 6, pp. 55–75, 1976.

[AYA 07] Ayad M.Y. *et al.*, "Voltage regulated hybrid DC power source using supercapacitors as energy storage device", *Energy Conversion and Management*, vol. 48, pp. 2196–2202, 2007.

[BAC 08] BACHA S., OCNASU D., "RT-lab user testimony", *1st Annual Opal-RT International Users Conference*, Montreal, Canada, June 12–13, 2008.

[BEA 15] BEAUDE O., Modélisation et optimisation de l'interaction entre véhicules électriques et réseaux d'électricité: apport de la théorie des jeux, Thesis, Paris Saclay, 2015.

[BEN 13] BENSMAINE F., TNANI S., CHAMPENOIS G. *et al.*, "Modeling and parameter identification of supercapacitors using a bidirectional DC-AC converter", In *Industrial Electronics Society, IECON 2013-39th Annual Conference of the IEEE*, pp. 1019–1024, 2013.

[BEN 16] BENSMAINE F., ABBES D., LABRUNIE A. *et al.*, "Sizing and techno-economic analysis of a grid connected photovoltaic system with hybrid storage", *UPEC 2016*, Coimbra, Portugal, 2016.

[BER 10] BERSENEFF B., Réglage de la tension dans les réseaux de distribution du futur, Doctoral thesis, Institut Polytechnique de Grenoble, France, 2010.

[BEV 17] BEVRANI H., FRANCOIS B., ISE T., *Microgrid Dynamics and Control*, Wiley, 2017.

[BLU 15] BLUME S., Global Energy Storage Market Overview & Regional Summary Report, Energy Storage Council, 2015.

[BOI 56] BOITEUX M., "Sur la gestion des Monopoles Publics astreints à l'équilibre budgétaire", *Econometrica*, vol. 24, no. 1, pp. 22–40, 1956.

[BOR 00] BORENSTEIN M. *et al.*, "A basic introduction to fixed-effect and random-effects models for meta-analysis", *Research Synthesis Methods*, pp. 97–111, 2000.

[BOR 98] BORNE P., ROZINOER J., DIEULOT J.-Y. *et al.*, *Introduction à la commande floue*, Technip, 1998.

[BOU 15] BOUALLAGA A., Gestion énergétique d'une infrastructure de charge intelligente de véhicules électriques dans un réseau de distribution intégrant des énergies renouvelables. PhD thesis, Université de Lille 1, June 2015.

[BOU 16] BOUALLAGA A., DAVIGNY A., COURTECUISSE V. *et al.*, "Methodology for technical and economic assessment of electric vehicles integration in distribution grid", *Mathematics and Computers in Simulation*, vol. 131, pp. 172–189, Elsevier, 2016.

[BOU 98] BOUCHON-MEUNIER B., FOULLOY L., RAMDANI M., *Logique floue, Exercices corrigés et exemples d'applications*, Cepaduès, 1998.

[BUH 94] BUHLER H., *Réglage par logique floue*, Presses Polytechniques et Universitaires Romandes, 1994.

[CAR 97] CARDELL E.L. *et al.*, "Market power and strategic interaction in electricity networks", *Resources and Energy Economics*, vol. 19, nos 1–2, pp. 109–138, 1997.

[CEL 17] CELIK B., ROCHE R., BOUQUAIN D. *et al.*, "Decentralized neighborhood energy management with coordinated smart home energy sharing", *IEEE Transactions on Smart Grid*, June 2017.

[CER 10] CERTS website, (URL accessed 15 February, 2010). Available at: http://certs.lbl.gov, 2010.

[CHE 00] CHEDID R.B., KRAKI S.H., EL-CHAMALI C., "Adaptive fuzzy control for wind – diesel weak power systems", *IEEE Transaction on Energy Conversion*, vol. 15, no. 1, pp. 71–78, 2000.

[CLA 11] CLASTRES P., "Smart grids: Another step towards competition, energy security and climate change objectives", *Energy Policy*, vol. 39, no. 9, pp. 5399–5408, September 2011.

[COD 14] CODANI P., PETIT M., PERREZ Y., "Participation d'une flotte de véhicule électrique au réglage de fréquence", *Symposium de Génie Electrique, SGE'14*, Cachan, France, 8–10 July, 2014.

[COM 15] COMMISARIAT GÉNÉRAL AU DÉVELOPPEMENT DURABLE, "Chiffres clés de l'énergie" Édition 2014, Service de l'observation et des statistiques, February 2015.

[COU 07] COURTECUISSE V., EL MOKADEM M., ROBYNS B. *et al.*, "Association of wind turbine based dispersed generators and storage systems to participate in primary frequency control", *12th Conference EPE 2007*, Aalborg, Denmark, 2–5 September, 2007.

[COU 08] COURTECUISSE V., Supervision d'une centrale multisources à base d'éoliennes et de stockage d'énergie connectée au réseau électrique, PhD thesis, L2EP, ENSAM, 20 November, 2008.

[COU 10] COURTECUISSE V., SPROOTEN J., ROBYNS B. *et al.*, "A methodology to design a fuzzy logic based supervision of hybrid renewable energy systems", *Mathematics and Computers in Simulation*, vol. 81, no. 2, pp. 208–224, 2010.

[CRE 16] CRE, La fin des tarifs réglementés de vente pour les consommateurs professionnels: s'informer et anticiper. Available at: http://www.cre.fr/infos-consommateurs/s-informer-sur-la-fin-des-tarifs-reglementes-pour-les-conso.-pro, 2016.

[CRE 18] CRE, Commission de régulation de l'énergie, Délibération de la CRE du 15 February 2018 portant orientations et recommandations sur l'autoconsommation, Délibération No. 2018-027, 2018.

[CRE] CRE, Commission de régulation de l'énergie, www.smartgrid-cre.fr.

[DAN 14] DANG-XUAN L., Contribution à l'étude des systèmes PV/Stockage distribués: impact de leur intégration à un réseau fragile, PhD thesis, École normale supérieure de Cachan-ENS Cachan, 2014.

[DEL 09] DELILLE G., FRANCOIS B., MALARANGE G., "Construction d'une offre de services du stockage pour les réseaux de distribution dans un contexte réglementaire dérégulé", *European Journal of Electrical Engineering*, vol. 12, nos 5–6, pp. 733–762, 2009.

[DEQ 11] DEQUIEDT V., DURIEU J., SOLAL P., *Théorie des jeux et application*, corpus Economie, Economica, 2011.

[DIM 05] DIMEAS A., HATZIARGYRIOU N., "A multiagent system for microgrids", *IEEE Transactions on Power Systems*, vol. 20, no. 3, pp. 1447–1455, 2005.

[DOB 13] DOBROWSKI B., BLANC F. DIEMUNSCH V., Le véhicule électrique: un vrai challenge induisant des réels changements d'habitudes, *REE*, no. 1, pp. 40–48, 2013.

[DUR 18] DURILLON B., DAVIGNY A., KAZMIERZCAK S. *et al.*, "Demand side management considering consumers sensitivities using a game theory approach", *5th IEEE International Energy Conference (Energycon)* , Limassol, Chypre, June 03–07, 2018.

[EC 16] EC ELECTRICITY DIRECTIVE COM, Directive of the European Parliament and of the Council on common rules for the internal market for electricity, 864 final/2, 2016.

[EDF 11] EDF, Electricity price table for the user less or equal than 36 kVA, 1 July, 2011.

[ENT 13] ENTSO-E, ENTSO-E Network Code for Requirements for Grid Connection Applicable to all Generators, Report, European Network of Transmission Operators for Electricity, Brussels, 2013.

[EOS] Available at: http://www.eosolar.fr/solaire-photovoltaique-vendee.php.

[ERD 10] ERDF (Électricité Réseau Distribution France). Available at: www.erdfdistribution.fr, 2010.

[ERN 09] ERNEUERBARE ENERGIEN GESETZ, Gesetz zur Neuregelung des Rechts der Erneuerbaren Energien im Strombereich und zur Änderung damit zusammenhängender Vorschriften, 2009.

[EUR 03] EUROPEAN, "Directive 2006/32/ec of the european parliament and of the council of 5 April 2006 on energy end-use efficiency and energy services and repealing council directive 93/76/eec", *Official Journal of the European Union*. Available at: http://europa.eu/, 27 April 2003.

[EUR 06a] EUROPEAN COMMUNITIES, "Action plan for energy efficiency: Realising the potential", *Commission of the European Communities*, Brussels, COM (2006) 545, 19 October, 2006.

[EUR 06b] EUROPEAN STRATEGY FOR SUSTAINABLE, "Green paper. A European strategy for sustainable, competitive and secure energy", *Commission Green Paper of 8 March 2006: "A European Strategy for Sustainable, Competitive and Secure Energy", COM (2006)*, vol. 105. Available at: http://europa.eu/, 2006.

[FAK 07] FAKHAM H., DEGOBERT P., FRANCOIS B., "Control system and power management for a PV based generation unit including batteries", *International Aegean Conference on Electrical Machines and Power Electronics*, pp. 141–146, Bodrum, Turkey, 10–12 September 2007.

[FAK 11] FAKHAM H., LU D., FRANCOIS B., "Power control design of a battery charger in a hybrid active PV generator for load-following applications", *IEEE Transaction on Industrial Electronics*, vol. 58, no. 1, pp. 85–94, January 2011.

[FAR 10a] FARUQUI A, SERGICI S., "Household Response to Dynamic Pricing of Electricity - A Survey of the Empirical Evidence", *SSRN Electronic Journal*, vol. 38, February, 2010.

[FAR 10b] FARUQUI A. *et al.*, "Fostering economic demand response in the midwest ISO", *Energy*, vol. 35, no. 4, pp. 1544–1552, 2010.

[FEI 08] FEINBERG E.A., GENETHLIOU D., "Applied mathematics for power systems: Load forecasting", *Applied Mathematics*, Chapter 12, Springer, pp. 269–285, 2008.

[FOG 09] FOGGIA G., Pilotage Optimal de Système Multi-sources pour le Bâtiment, Thesis, Institut Polytechnique de Grenoble, 17 July, 2009.

[FRA 96] FRANÇOIS B., HAUTIER J.P., "Hierarchical control design using structural decomposition of a rectifier converter model", *Proceedings of Electrimacs'96*, September 17–18, Saint-Nazaire, France, vol. 1, pp. 255–260, 1996.

[FRA 13] FRACHET L., Tarifs résidentiels pour la réduction de la consommation électrique: une évaluation expérimentale d'acceptation et d'impact, Doctoral thesis, Université de Grenoble, 2013.

[GEO 14] GEOFFRON P., "Quels rôles pour les villes dans la transition énergique? Modèles économiques pour les smart grids", La fabrique de la cité, 2014.

[GIO 18] GIORGINI P., DEPREZ S., WALD LASOWSKI A., *La tentation d'Eugénie, L'humanité face à son destin*, Bayard, 2018.

[GLA 08] GLAVIN M.E., CHAN P.K.W., ARMSTRONG S. *et al.*, "A stand-alone photovoltaic supercapacitor battery hybrid energy storage system", *Power Electronics and Motion Control Conference, 2008. EPE-PEMC 2008. 13th. IEEE*, pp. 1688–1695, 2008.

[GRE 97] GREEN R., "Electricity transmission pricing: an international comparaison", *Utilities Policy*, vol. 6, no. 3, pp. 177–184, 1997.

[GUI 99] GUILLEMAUD L., GUGUEN H., "Extending grafcet for the specification of control of hybrid systems", *IEEE International Conference on Systems, Man, and Cybernetics*, Tokyo, pp. 171–175, 1999.

[HAD 09] HADJIPASCHALIS I., POULLIKKAS A., EFTHIMIOU V., "Overview of current and future energy storage technologies for electric power applications", *Renewable and Sustainable Energy Reviews*, vol. 13, nos 6–7, pp. 1513–1522, 2009.

[HAJ 07] HAJIZADEH A., GOLKAR M.A., "Intelligent power management strategy of hybrid distributed generation system", *Electrical Power and Energy Systems*, vol. 29, pp. 783–795, 2007.

[HAS 14] HASELHUHN R., HARTMANN U. (eds), *Le guide du photovoltaïque*, Edisun Power France, 2014

[HAT 07] HATZIARGYRIOU N., ASANO H., IRAVANI R. *et al.*, "Microgrids-an overview of ongoing research, development and demonstration projects", *IEEE Power & Energy Magazine*, pp. 78–94, July/August 2007.

[HER 10] HERTER K., WAYLAND S., "Residential response to critical-peak pricing of electricity: California evidence", *Energy*, vol. 35, no. 4, pp. 1561–1567, 2010.

[HES 18] HESPUL. Available at: http://www.photovoltaique.info/, 2018.

[HOG 10] HOGAN R. *et al.*, "Towards a combined merchant-regulatory mechanism for electricity transmission expansion," in ROSELLÓN J., VOGELSANG I. (eds), *Journal of Regulatory Economics*, vol. 38, no. 2, October 2010.

[HOU 15] HOUARI A., ABBES D., LABRUNIE A. *et al.*, "Hybridization of electrical energy storage for intelligent integration of photovoltaics in electric networks", *2015 17th European Conference on Power Electronics and Applications (EPE'15 ECCE-Europe)*, IEEE, pp. 1–10, 2015.

[JOS 11] JOSEPH H., ETO, Use of frequency response metrics to assess the planning and operating requirements for reliable integration of variable renewable generation, Report 2011.

[JOS 05] JOSKOW P., TIROLE J., "Merchant transmission investment", *Journal of Industrial Economics*, vol. 53, no. 2, pp. 233–264, 2005.

[JOS 07] JOSKOW P., TIROLE J., "Reliability and competitive electricity markets", *Rand Journal of Economics*, vol. 38, no. 1, pp. 60–84, 2007.

[JOS 97] JOSKOW P., "Restructuring, competition and regulatory reform in the US electricity sector", *The Journal of Economic Perspectives*, vol. 11, no. 3, pp. 119–138, 1997.

[KAD 13] KADRI R., SAUDEMONT C., ROBYNS B. *et al.*, "A power flow control strategy for high energy efficient smart LED lighting system powered by PV and MPPT controlled DC grid", *IEEE PES International Conference POWERTECH 2013*, Grenoble, 16–20 June, 2013.

[KAN 14a] KANCHEV H., COLAS F., LAZAROV V. *et al.*, "Emission reduction and economical optimization of an Urban Microgrid operation including dispatched PV-based active generators", *IEEE Transaction On Sustainable Energy*, vol. 5, no. 4, pp. 1755–1796, October 2014.

[KAN 14b] KANCHEV H., Gestion des flux énergétiques dans un système hybride de sources d'énergie renouvelable: Optimisation de la planification opérationnelle de l'ajustement d'un micro réseau électrique urbain, PhD thesis, Ecole Centrale de Lille, 24 January, 2014.

[KAR 04] KARKLINS-MARCHAY A., *Joseph Schumpeter: vie, oeuvres, concepts*, Ellipses, 2004.

[KAT 08] KATIAREI F. *et al.*, "Microgrids management: Control and operation aspects of microgrids", *IEEE Power & Energy Magazine*, vol. 6, no. 3, pp. 54–65, 2008.

[KIR 14] KIRAN C., KRISTRÖM B., "Greening household behaviour and energy", OECD Environment Working Papers, no. 78, OECD Publishing, 2014.

[KLE 08] KLEISSL J., Solar power forecasting at UC San Diego, Internet Technical Report, Department of Mechanical & Aerospace Engineering, UCSD. Available at: http://maeresearch.ucsd.edu/kleissl/SolarPowerForecasting.pdf, 2008.

[KRO 08] KROPOSKI B., LASSETER R., ISE T. *et al.*, "Making microgrids work", *IEEE Power & Energy Magazine*, May/June 2008.

[LAF 94] LAFONT J.J., TIROLE J., "Access pricing and competition", Working Papers, nos 94–31, Massachusetts Institute of Technology (MIT), Department of Economics, 1994.

[LAG 09] LAGORSE J., SIMOES G.M., MIRAOUI A., "A multiagent fuzzy-logic-based energy management of hybrid systems", *IEEE Transactions on Industry Applications*, vol. 45, no. 6, pp. 2123–2129, November/December 2009.

[LAN 01] LANGEVIN P., "Les notes de Pôle", Université Aix-Marseille. Available at: https://pddtm.hypotheses.org/176, 2001.

[LAY 15] LAYADI T.M., CHAMPENOIS G., MOSTEFAI M. *et al.*, "Lifetime estimation tool of lead–acid batteries for hybrid power sources design", *Simulation Modelling Practice and Theory*, vol. 54, pp. 36–48, 2015.

[LEA 00] LÉAUTIER T.O., "Regulation of an electric power transmission company", *The Energy Journal, International Association for Energy Economics*, vol. 21, no. 4, pp. 61–92, 2000.

[LEA 01] LÉAUTIER T.O., "Transmission constraints and imperfect markets for power", *Journal of Regulatory Economics*, vol. 19, no. 1, pp. 27–54, January 2001.

[LEA 09] LÉAUTIER T.O., THELEN V., "Optimal expansion of the power transmission grid: Why not?", *Journal of Regulatory Economics*, Springer, vol. 36, no. 2, pp. 127–153, October 2009.

[LEA 14a] LÉAUTIER T.O., "The "demand side" effect of price caps: Uncertainty, imperfect competition, and rationing", IDEI Working Paper, no. 815, 27 January, 2014.

[LEA 14b] LÉAUTIER T.O., "Transmission constraints and strategic underinvestment in electric power generation", IDEI Working Paper, no. 816, 6 January, 2014.

[LEC 03] LECLERCQ L., ROBYNS B., GRAVE J.M., "Control based on fuzzy logic of a flywheel energy storage system associated with wind and diesel generators", *Mathematics and Computers in Simulation*, vol. 63, pp. 271–280, 2003.

[LEE 08] LEE D.-J., WANG L., "Small-signal stability analysis of an autonomous hybrid renewable energy power generation/energy storage system part I: Time-domain simulations", *IEEE Transactions on Energy Conversion*, vol. 23, no. 1, pp. 311–320, 2008.

[LEG 95] LE GALESP, "Du gouvernement des villes à la gouvernance urbaine", *Revue française de science politique*, vol. 45, pp. 57–95, 1995.

[LEG 16] LE GOFF LATIMIER R., Gestion et dimensionnement d'une flotte de véhicules électriques associée à une centrale photovoltaïque: co-optimisation stochastique et distribuée, PhD thesis, Université Paris-Saclay, 2016.

[LEN] Available at: http://lenergeek.com/2017/06/02/stockage-stationnaire-electricite-drom-territoires-insulaires/.

[LI 09] LI P., Formalisme pour la Supervision des Systèmes Hybrides Multi-Sources de Générateurs d'Energie Répartie: Application à la Gestion d'un Micro Réseau, PhD thesis report, Ecole Centrale de Lille, June 2009.

[LIV] Live TREE, www.livetree.fr.

[LOC 09] LOCMENT F., SECHILARIU M., FORGEZ C. *et al.*, "Energetic macroscopic representation and maximum control structure of a semi-isolated network. Photovoltaic application for a low energy building", *EJEE (European Journal of Electrical Engineering)*, vol. 12, nos 5–6, pp. 609–637, 2009.

[LU 08] LU D., ZHOU T., FAKHAM H. *et al.*, "Design of a power management system for a PV station including various storage technologies", *13th International Power Electronics and Motion Control Conference, EPE-PEMC 2008*, Poznan, 9-2008, CD-ROM, 1–3 September, 2008.

[LU 10a] LU D., ZHOU T., FAKHAM H. *et al.*, "Application of Petri Nets for the energy management of a photovoltaic based power station including storage units", *Renewable Energy*, vol. 35, no. 6, pp. 1117–1124, 2010.

[LU 10b] LU D., Conception et contrôle d'un générateur PV actif à stockage intégré - Application à l'agrégation de producteurs-consommateurs dans le cadre d'un micro réseau intelligent urbain, Thesis, Ecole Centrale de Lille 2010.

[LUX 15] LUX RESEARCH: China's Lithium-Ion Battery Market To Quadruple By, Lux Research. Available at: http://insideevs.com/lux-research-chinas-lithium-ion-battery-market-quadruple-2025, 2015.

[MAC 07] MACLAY J.D., BROUWER J., SAMUELSEN G.S., "Dynamic modeling of hybrid energy storage systems coupled to photovoltaic generation in residential applications", *Journal of Power Sources*, vol. 163, no. 2, pp. 916–925, 2007.

[MAG 11] MAGRETTA J., *Understanding Michael Porter: The Essential Guide to Competition and Strategy*, Harvard Business Review Press, 2011.

[MAR 09] MARANO V., ONORI S., GUEZENNEC Y. *et al.*, "Lithium-ion batteries life estimation for plug-in hybrid electric vehicles", *Vehicle Power and Propulsion Conference, 2009. VPPC'09.* IEEE, pp. 536–543, 2009.

[MAR 05] MARIE-FRANÇOISE J.N., GUALOUS H., OUTBIB R. *et al.*, "42V power net with supercapacitor and battery for automotive application", *Journal of Power Sources*, vol. 143, pp. 275–283, 2005.

[MAR 07] MARNAY C., FIRESTONE R., "Microgrids: An emerging paradigm for meeting building electricity and heat requirements efficiently and with appropriate energy quality", *The European Council for an Energy Efficient Economy 2007 Summer Study*, La Colle sur Loup, France, 4–9 June, 2007.

[MAR 12] MARTINEZ J.S., JOHN R.I., HISSEL D. *et al.*, "A survey-based type-2 fuzzy logic system for energy management in hybrid electrical vehicles", *Information Sciences*, no. 190, pp. 192–207, 2012.

[MAX 13] Technical documentation for the MAXWELL BMOD0165 P048 BXX supercapacitor. Available at: https://www.mouser.fr/new/maxwell/Maxwell-48V-ultracapacitor/.

[MCM 98] MCMENAMIN J. S., MONFORTE F.A., "Short term energy forecasting with neural networks", *Energy Journal*, vol. 19, no. 4, pp. 43–61, 1998.

[MET 16] METEOROLOGICAL DATA. Available at: http://www.nrel.gov/midc/nwtc2, Accessed February 2016.

[MIL 10] MILLNER A., "Modeling lithium ion battery degradation in electric vehicles", *2010 IEEE Conference on Innovative Technologies for an Efficient and Reliable Electricity Supply (CITRES)*, pp. 349–356. IEEE, 2010.

[MIN 16] MINISTÈRE DE LA TRANSITION ÉCOLOGIQUE ET SOLIDAIRE, Commissariat général au Développement durable, Énergies et climat, "Consommation finale d'énergie liée à l'usage des bâtiments", [ref., August 2018]. Available at: http://www.statistiques.developpement-durable.gouv.fr/energie-climat/s/consommations-secteur-residentiel-tertiaire.html, 2016.

[MIN 17] MINISTÈRE DE L'ENVIRONNEMENT, DE L'ÉNERGIE ET DE LA MER EN CHARGE DES RELATIONS INTERNATIONALES SUR LE CLIMAT, "Chiffres clés de l'énergie, édition 2016", February 2017.

[MIN 18] MINISTÈRE DE LA TRANSITION ÉCOLOGIQUE ET SOLIDAIRE, "Mobilisation pour accélérer le déploiement de l'énergie solaire", Press release, 28 June, 2018.

[MIS 08] MISHRA S., Short term load forecasting using computational intelligence methods, Masters thesis, National Institute of Technology Rourkela, 2008.

[MOR 06] MORENO J., ORTUZAR M.E., DIXON J.W., "Energy management system for a hybrid electric vehicle, using ultracapacitors and neural networks", *IEEE Transactions on Industrial Electronics*, vol. 53, no. 2, pp. 614–623, 2006.

[MOR 07] MOROZUMI S., ARASHIRO Y., INOUE N., "Large-scale PV demonstration projects in Japan", *2nd International Symposium for Distributed Electricity Generation and Smart Grids*, pp. 17–18, 2007.

[MUL 07] MULTON B., AHMED H.B., "Le stockage stationnaire d'énergie électrique: pourquoi et comment ?", *3EI Journal*, pp. 6–16, 2007.

[MUL 13] MULTON B., AUBRY J., HAESSIG P. *et al.*, Système de stockage de l'énergie électrique, *Techniques de l'Ingénieur*, BE 8 100, April 2013.

[NER 11] NERHIR M.H., WANG C., STRUNZ K. *et al.*, "A review of hybrid renewable/alternative energy systems for electric power generation: Configurations, control, and applications", *IEEE Transaction On Sustainable Energy*, vol. 2, no. 4, October 2011.

[NRE 16] NREL, Meteorological data. Available at: http://www.nrel.gov/midc/nwtc2, Accessed February 2016.

[OCO 16] O'CONNOR JOSEPH P., *Off Grid Solar: A Handbook for Photovoltaics with Lead Acid or Lithium-ion Batteries*, CreateSpace Independent Publishing Platform, 2016.

[PER 12] PEREZ Y., RIOUS V., Valeur économique des smart grids. Cas des effacements, ANCRE Smarts-Grid et stockage, opportunités et risques. Quels modèles économiques et sociétaux?, Seminar, 2012.

[PHO] Available at: http://www.photovoltaique.info/Site-isole.html.

[PHO 17] PHOTOVOLTAIC GEOGRAPHICAL INFORMATION SYSTEM (PVGIS). Available at: http://re.jrc.ec.europa.eu/pvgis/, 2017.

[PIC 18] PICBLEU, Les batteries pour panneaux solaires photovoltaïques. Available at: https://www.picbleu.fr/page/les-batteries-pour-panneaux-solaires-photovoltaiques, 2018.

[PIL 02] PILAVACHI P.A., "Mini- and micro-gas turbines for combined heat and power", *Applied Thermal Engineering*, vol. 22, no. 18, pp. 2003–2014, December 2002.

[PIN 12] PINDYNCK R., RUBINFELD D., *Microéconomie*, Pearson, 2012.

[POR 82] PORTER M., *Choix stratégiques et concurrence*, Economica, Paris, p. 426, 1982.

[PVG 17] PHOTOVOLTAIC GEOGRAPHICAL INFORMATION SYSTEM (PVGIS). Available at: http://re.jrc.ec.europa.eu/pvgis/, 2017.

[RAM 27] RAMSEY F., "A contribution to the theory of taxation", *Economic Journal*, vol. 37, no. 145, pp. 47–61, 1927.

[RIC 11] RICALDE L.J., GAMEZ M., SANCHEZ E.N., "Design of a smart grid management system with renewable energy generation", *IEEE Symposium on Computational Intelligence Applications In Smart Grid (CIASG)*, pp. 1–4, 2011.

[RIF 09] RIFFONEAU Y., Gestion des flux énergétiques dans un système photovoltaique avec stockage connecté au réseau, Thesis, Institut Polytechnique de Grenoble, 23 October, 2009.

[RIO 12a] RIOUS V., "Un mécanisme de capacité pour le système électrique français", *Energy Focus*, Microeconomix, 2012.

[RIO 12b] RIOUS V., SAGUAN M., "Rémunération des effacements diffus: doit-on distordre le marché pour en corriger les imperfections ?", *Energy Focus*, Microeconomix, 2012.

[ROB 12a] ROBOAM X. (ed.), *Integrated Design by Optimization of Electrical Energy Systems*, ISTE Ltd, London and John Wiley & Sons, New York, 2012.

[ROB 12b] ROBOAM X. (ed.), *Systemic Design Methodologies for Electrical Energy Systems*, ISTE Ltd, London and John Wiley & Sons, New York, 2012.

[ROB 12c] ROBOAM B., DAVIGNY A., BRUNO F. *et al.*, *Electricity Production from Renewable Energies*, ISTE Ltd, London and John Wiley & Sons, New York, 2012.

[ROB 13a] ROBYNS B., DAVIGNY A., SAUDEMONT C., "Methodologies for supervision of hybrid energy sources based on storage systems – A survey", *Mathematics and Computers in Simulation*, vol. 91, pp. 52–71, May 2013.

[ROB 13b] ROBYNS B., DAVIGNY A., SAUDEMONT C., "Energy management of storage systems based power sources and loads", *Electromotion Journal*, vol. 20, nos 1–4, pp. 25–35, 2013.

[ROB 15] ROBYNS B., BRUNO F., DELILLE G. *et al.*, *Energy Storage in Electric Power Grids*, ISTE Ltd, London and John Wiley & Sons, New York, 2015.

[ROB 16] ROBYNS B., HISSEL D., SAUDEMONT C. *et al.*, *Electrical Energy Storage in Transportation Systems*, ISTE Ltd, London and John Wiley & Sons, New York, 2016.

[RTE 13] RTE, Référentiel Technique de RTE. Available at: http://www.rte-france.com, 2013.

[RTE 17a] RTE, SER, ENEDIS, ADEeF, "Panorama de l'électricité renouvelable au 30 Septembre 2017", November 2017.

[RTE 17b] RTE, Distribution of the electricity production of French Transmission System Operator. Available at: http://clients.rte-france.com/lang/an/visiteurs/vie/prod/realisation_production.jsp, 2017.

[RTE 17c] RTE, consumption, production and CO2 content of the French electricity, RTE. Available at: http://www.rte-france.com/fr/developpement-durable/maitriser-sa-consommation-electrique/eco2mix-consommation-production-et-contenu-co2-de-l-electricite-francaise#emissionCO2, 2017.

[RTE 18] RTE, https://www.rte-france.com/fr/eco2mix/eco2mix, (accessed August 2018).

[SAF 12] SAFT, Technical documentation for SAFT Evolion 74Ah battery: Evolion Li-ion battery. Saft's proven ultra-compact solution for telecom applications, Document No. 21811-2-0212, February 2012.

[SAL 12] SALANIE B., "Théorie des contrats", *Economica*, Janvier, 2012.

[SAR 13] SARABI S., KEFSI L., MERDASSI A. *et al.*, "Supervision of plug-in electric vehicles connected to the electric distribution grids", *International Journal of Electrical Energy*, vol. 1, no. 4, pp. 256–263, December 2013.

[SAR 14] SARABI S., BOUALLAGA A., DAVIGNY A. *et al.*, "The feasibility of the ancillary services for vehicle-to- grid technology", *11th International Conference on the European Energy Market (EEM2014)*, IEEE, Krakow, Poland, 28–30 May, 2014.

[SAR 16a] SARABI S., DAVIGNY A., COURTECUISSE A. *et al.*, "Potential of vehicle-to-grid ancillary services considering the uncertainties in plug-in electric vehicle availability and service/localization limitations in distribution grids", *Applied Energy*, vol. 171, pp. 523–540, Elsevier, June 2016.

[SAR 16b] SARABI S., Contribution of Vehicle-to-Grid (V2G) to the energy management of Plug-in Electric Vehicles' fleet on the distribution network, PhD thesis, Ecole Nationale des Arts et Métiers, November 2016.

[SCH 88] SCHWEPPE F. *et al.*, *Spot Pricing of Electricity*, Kluwer Academic Publishers, Boston, MA, p. 355, 1988.

[SEC 10] SECHILARIU M., LOCMENT F., HOUSSAMO I., "Multi-source power generation system in semi-isolated and safety grid configuration for buildings", *15th IEEE Mediterranean Electrotechnical Conference (MELECON)*, pp. 967–972, 2010.

[SIE 07] SIERRA E., HOSSIAN A., BRITOS P. *et al.*, "Fuzzy control for improving energy management within indoor building environments", *Electronics, Robotics and Automotive Mechanics Conference CERMA 2007*, pp. 412–416, 2007.

[SIN 10] SINGO A.T., Système d'alimentation photovoltaïque avec stockage hybride pour l'habitat énergétiquement autonome, PhD thesis, Groupe de Recherche en Electrotechnique et Electronique de Nancy, Faculté des Sciences et Techniques, 54500 Vandoeuvre-lès-Nancy, 2010.

[SMA 11] SMARTGRIDS – CRE, "Smart Grid City: une gestion locale des sources d'approvisionnement et de consommation", [ref., August 2018]. Available at: http://www.smartgrids-cre.fr/index.php?p=smartcities-smart-grid-city, 2011.

[SOM] So MEL, So Connected project available at: http://www.lillemetropole.fr/mel/institution/competences/energie/so-mel-so-conected.html.

[SPR 09] SPROOTEN J., COURTECUISSE V., ROBYNS B. *et al.*, "Méthodologie de développement de superviseurs à logique floue de centrales multisources à base d'énergie renouvelable", *European Journal of Electrical Engineering (EJEE)*, pp. 553–583, 2009.

[SWI 11] SWIERCZYNSKI M., TEODORESCU R., CORTES R.P., "Lifetime investigations of a lithium iron phosphate (LFP) battery system connected to a wind turbine for forecast improvement and output power gradient reduction", *Battcon Archive Papers*, pp. 20-1–20-8, 2011.

[SYM 03] SYMONS P., "Batteries for stationary standby and for stationary cycling applications, part 4: Charge management", *IEEE Power Engineering Society General Meeting*, vol. 1, p. 157, July 2003.

[TES 18] TESLA, "Powerwall, votre batterie domestique", Available at: https://www.tesla.com/fr_FR/powerwall, August 2018.

[THI 10] THIAUX Y., Optimisation des profils de consommation pour minimiser les coûts économique et énergétique sur cycle de vie des systèmes photovoltaïques autonomes et hybrides-Evaluation de la technologie Li-ion, PhD thesis, École normale supérieure de Cachan-ENS Cachan, 2010.

[THO 09] THOUNTHONG S., RAEL B., DAVAT B., "Energy management of fuel cell/battery/supercapacitor hybrid power source for vehicle applications", *Journal of Power Source*, Elsevier, pp. 376–381, 2009.

[USB 14] U.S. BATTERY TECHNICAL DOCUMENTATION FOR A NIMH (NICKEL-METAL HYDRIDE) BATTERY. Available at: http://usbattery.com/wp-content/uploads/2014/05/usb-expected-life-cycle.pdf.

[VAS 99] VAS P., *Artifical-intelligence-based Electrical Machine and Drive*, Oxford Science, Oxford, U.K., 1999.

[VEN 16] VENTORUZZO G., DAVIGNY A., HENNETON A. *et al.*, "Integration and safety of electric vehicles in a residential electrical installation for V2H services", *18th Conference EPE 2016 ECCE Europe*, Karlsruhe, 5–9 September, 2016.

[VEN 17] VENTORUZZO G., DAVIGNY A., HENNETON A. *et al.*, "Safety analysis of a residential electrical installation for different V2H configurations in case of short-circuit", *12th Conference ELECTRIMACS 2017*, Toulouse, 4–6 July, 2017.

[VOG 01] VOGELSANG I., "Price regulation for independent transmission companies", *Journal of Regulatory Economics*, vol. 20, no. 2, pp. 141–65, Springer, September 2001.

[VOG 06] VOGELSANG I., "Electricity transmission pricing and performance-based regulation," *The Energy Journal, International Association for Energy Economics*, pp. 97–126, 2006.

[XU 13] XU B., ULBIG A., Degradation-limiting optimization of battery energy storage systems operation, Master's thesis, ETH Zurich, 2013.

[YAN 17] YAN X., ABBES D., FRANCOIS B., "Uncertainty analysis for day ahead power reserve quantification in an urban microgrid including PV generators", *Renewable Energy*, vol. 106, pp. 288–297, June 2017.

[YAN 18] YAN X., ABBES D., FRANCOIS B., "Development of a tool for urban microgrid optimal energy planning and management", *Simulation Modelling Practice and Theory*, vol. 89, pp. 64–81, 2018.

[YEL 13] YELLE V., Social Perception of Ecosystem Management in Québec's Black Spruce Forest: Can large harvests emulating fire be acceptable to forest users, stakeholders and the uninformed public?, PhD thesis, Université Laval Québec, 2013.

[ZHA 10a] ZHANG H., MOLLET F., SAUDEMONT C. *et al.*, "Experimental validation of energy management strategies in a local DC power distribution system of more electric aircraft by using hybrid storage and dissipation systems", *IEEE Transactions on Industrial Electronics*, vol. 57, no. 12, pp. 3905–3916, 2010.

[ZHA 10b] ZHANG H., SAUDEMONT C., ROBYNS B. *et al.*, "Comparison of different DC voltage supervision strategies in a local power distribution system of more electric aircraft", *Elsevier Editorial System TM for Mathematics and Computers in Simulation*, vol. 81, no. 2, pp. 263–276, 2010.

[ZHA 12] ZHANG H., DAVIGNY A., COLAS F. *et al.*, "Fuzzy logic based energy management strategy for commercial buildings integrating photovoltaic and storage systems", *Energy and Buildings*, vol. 54, pp. 196–206, Elsevier, 2012.

[ZHO 11] ZHOU T., FRANCOIS B., "Energy management and power control of an hybrid active wind generator for distributed power generation and grid integration", *IEEE Transactions on Industrial Electronics*, vol. 58, no. 1, pp. 85–94, January 2011.

[ZHO 15] ZHONG Q.C., MALINOWSKI M., FRANCOIS B. *et al.*, "Grid-connected photovoltaic generation plants as alternative energy sources", *IEEE Industrial Electronics Magazine*, vol. 9, no. 1, pp. 18–32, March 2015.

[ZHU 84] ZHURKOV S.N., "Kinetic concept of the strength of solids", *International Journal of Fracture*, vol. 26, no. 4, pp. 295–307, 1984.

[ZNI 15] Specification for the call to tender for the production and exploitation of electrical plants using techniques for converting solar radiation at a power above 100 kWc located in zones not connected to the grid. Document released by the ministry responsible for energy applying article 2 of decree no. 2002-1434 of December 4 2002. Document published on the Energy Regulation Commission website, May 20 2015.

[ZUR 94] ZURAWSKI R., ZHOU M., "Petri net and industrial application: A tutorial", *IEEE Transaction on Industrial Electronics*, vol. 41, no. 6, pp. 567–583, 1994.

Index

G, H, I

L, M, N

O, P, R, S

T, U, V

Other titles from

iSTE

in

Energy

2019

BENALLOU Abdelhanine
Energy Transfers by Convection
(Energy Engineering Set – Volume 3)
Energy Transfers by Radiation
(Energy Engineering Set – Volume 4)

LACHAL Bernard
Energy Transition

2018

BENALLOU Abdelhanine
Energy and Mass Transfers: Balance Sheet Approach and Basic Concepts
(Energy Engineering Set – Volume 1)
Energy Transfers by Conduction
(Energy Engineering Set – Volume 2)

JEMEÏ Samir
Hybridization, Diagnostic and Prognostic of Proton Exchange Membrane Fuel Cells: Durability and Reliability

RUFFINE Livio, BROSETA Daniel, DESMEDT Arnaud
Gas Hydrates 2: Geoscience Issues and Potential Industrial Applications

VIALLET Virginie, FLEUTOT Benoit
Inorganic Massive Batteries
(Energy Storage – Batteries, Supercapacitors Set – Volume 4)

2017

BROSETA Daniel, RUFFINE Livio, DESMEDT Arnaud
Gas Hydrates 1: Fundamentals, Characterization and Modeling

LA SCALA Massimo
From Smart Grids to Smart Cities: New Challenges in Optimizing Energy Grids

MOLINA Géraldine, MUSY Marjorie, LEFRANC Margot
Building Professionals Facing the Energy Efficiency Challenge

SIMON Patrice, BROUSSE Thierry, FAVIER Frédéric
Supercapacitors Based on Carbon or Pseudocapacitive Materials
(Energy Storage – Batteries, Supercapacitors Set – Volume 3)

2016

ALLARD Bruno
Power Systems-on-Chip: Practical Aspects of Design

ANDRE Michel, SAMARAS Zissis
Energy and Environment

DUFOUR Anthony
Thermochemical Conversion of Biomass for the Production of Energy and Chemicals

2015

CROGUENNEC Laurence, MONCONDUIT Laure, DEDRYVÈRE Rémi
Electrodes for Li-ion Batteries
(Energy Storage – Batteries, Supercapacitors Set – Volume 2)

LEPRINCE-WANG Yamin
Piezoelectric ZnO Nanostructure for Energy Harvesting
(Nanotechnologies for Energy Recovery Set – Volume 1)

ROBYNS Benoît, FRANÇOIS Bruno, DELILLE Gauthier, SAUDEMONT Christophe
Energy Storage in Electric Power Grids

ROSSI Carole
Al-based Energetic Nanomaterials
(Nanotechnologies for Energy Recovery Set – Volume 2)

TARASCON Jean-Marie, SIMON Patrice
Electrochemical Energy Storage
(Energy Storage – Batteries, Supercapacitors Set – Volume 1)

2013

LALOUI Lyesse and DI DONNA Alice
Energy Geostructures: Innovation in Underground Engineering

2012

BECKERS Benoit
Solar Energy at Urban Scale

ROBYNS Benoît, DAVIGNY Arnaud, FRANÇOIS Bruno, HENNETON Antoine, SPROOTEN Jonathan
Electricity Production from Renewable Energies

2011

GAO Fei, BLUNIER Benjamin, MIRAOUI Abdellatif
Proton Exchange Membrane Fuel Cell Modeling

MULTON Bernard
Marine Renewable Energy Handbook

2010

BRUNET Yves
Energy Storage

2009

SABONNADIÈRE Jean-Claude
Low Emission Power Generation Technologies and Energy Management

SABONNADIÈRE Jean-Claude
Renewable Energy Technologies

Printed and bound by CPI Group (UK) Ltd, Croydon, CR0 4YY

13/09/2026

14961078-0001